Getting the most from this book

Mathematics is not only a beautiful and exciting subject in its own right, but also one that underpins many other branches of learning. It is consequently fundamental to our national well-being.

This book covers the compulsory core content of Advanced Level Further Mathematics study, following on from the Year 1/AS Further Mathematics book. The optional applied content is covered in the Mechanics and Statistics books, and the remaining options in the Modelling with Algorithms, Numerical Methods, Further Pure Maths with Technology and Extra Pure Maths books.

Between 2014 and 2016 A Level Mathematics and Further Mathematics were very substantially revised, for first teaching in 2017. Major changes included increased emphasis on:

- Problem solving
- Mathematical proof
- Use of ICT
- Modelling
- Working with large data sets in statistics.

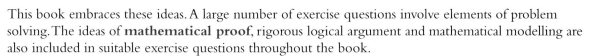

This book embraces these ideas. A large number of exercise questions involve elements of problem solving. The ideas of **mathematical proof**, rigorous logical argument and mathematical modelling are also included in suitable exercise questions throughout the book.

The use of technology, including graphing software, spreadsheets and high specification calculators, is encouraged wherever possible, for example in the Activities used to introduce some of the topics. In particular, readers are expected to have access to a calculator which handles matrices up to order 3×3. Places where ICT can be used are highlighted by a (T) icon. Margin boxes highlight situations where the use of technology – such as graphical calculators or graphing software – can be used to further explore a particular topic.

Throughout the book the emphasis is on understanding and interpretation rather than mere routine calculations, but the various exercises do nonetheless provide plenty of scope for practising basic techniques. The exercise questions are split into three bands. Band 1 questions are designed to reinforce basic understanding; Band 2 questions are broadly typical of what might be expected in an examination, but you should refer to the OCR SAMs and practice papers; Band 3 questions explore around the topic and some of them are rather more demanding. In addition, extensive online support, including further questions, is available by subscription to MEI's Integral website, integralmaths.org.

In addition to the exercise questions, there are three sets of Practice questions, covering groups of chapters. These include identified questions requiring problem solving (PS), mathematical proof (MP), use of ICT (T) and modelling (M).

This book is written on the assumption that readers are studying or have studied A Level Mathematics. It can be studied alongside the Year 2/A Level Mathematics book, or after studying A Level Mathematics. There are also places where an understanding of the topics depends on knowledge from earlier in the book or in the Year 1/AS Mathematics book, the Year 2 Mathematics book or the Year 1/AS Further Mathematics book, and this is flagged up in the Prior knowledge boxes (as well as in specific review sections – as explained in the paragraph below). This should be seen as an invitation to those who have problems with the particular topic to revisit it. At the end of each chapter there is a list of key points covered as well as a summary of the new knowledge (learning outcomes) that readers should have gained.

This book follows on from *MEI A Level Further Mathematics for Year 1 (AS)* and most readers will be familiar with the material covered in it. However, there may be occasions when they want to check on topics in the earlier book: we have therefore included three short Review chapters to provide a condensed summary of the work that was covered in the earlier book, including one or more exercises. In addition there are three chapters that begin with a Review section and exercise, and then go on to new work based on it. Confident readers may choose to miss out the Review material, and only refer to these parts of the book when they are uncertain about particular topics. Others, however, will find it helpful to work through some or all of the Review material to consolidate their understanding of the first year work.

Two common features of the book are Activities and Discussion points. These serve rather different purposes. The Activities are designed to help readers get into the thought processes of the new work that they are about to meet; having done an Activity, what follows will seem much easier. The Discussion points invite readers to talk about particular points with their fellow students and their teacher and so enhance their understanding. Another feature is a Caution icon ❶, highlighting points where it is easy to go wrong.

Answers to all exercise questions and practice questions are provided at the back of the book, and also online at www.hoddereducation.co.uk/MEIFurtherMathsYear2

This is a 4th edition MEI textbook so much of the material is well tried and tested. However, as a consequence of the changes to A Level requirements in further mathematics, large parts of the book are either new material or have been very substantially rewritten.

Catherine Berry

Roger Porkess

Prior knowledge

This book follows on from MEI A Level Further Mathematics Year 1 (AS). It is designed so that it can be studied alongside MEI A level Mathematics Year 2. Knowledge of the work in MEI A Level Mathematics Year 1 (AS) is assumed.

- **Chapter 1: Vectors 1** reviews and develops the work in chapter 7 of MEI A Level Further Mathematics Year 1.

- **Review: Matrices and transformations** reviews the work covered in chapter 1 of MEI A Level Further Mathematics Year 1.

- **Chapter 2: Matrices** reviews and builds on the work in chapter 6 of MEI A Level Further Mathematics Year 1.

- **Chapter 3: Series and induction** reviews and develops the work introduced in chapter 4 of MEI A Level Further Mathematics Year 1. It requires knowledge of partial fractions which is covered in chapter 7 of A Level Mathematics Year 2

- **Chapter 4: Further calculus assumes** knowledge of the calculus from MEI A Level Mathematics Year 1 (chapters 10 and 11). You also need to be able to differentiate and integrate exponential functions, the function $\frac{1}{x}$ and related functions, and trigonometric functions (covered in chapters 9 and 10 of MEI A Level Mathematics Year 2). You need to be able to differentiate a function implicitly (chapter 9 of MEI A Level Mathematics Year 2). You should also be familiar with the inverse trigonometric functions (covered in chapter 6 of MEI A Level Mathematics Year 2). You also need to have covered the work on partial fractions in chapter 7 of A Level Mathematics Year 2.

- **Chapter 5: Polar coordinates** assume knowledge of radians (covered in chapter 2 of MEI A Level Mathematics Year 2). You will need to be familiar with the reciprocal trigonometric functions sec and cosec (introduced in chapter 6 of MEI A Level Mathematics Year 2), and the double angle formulae (introduced in chapter 8 of MEI A Level Mathematics Year 2). You also need to be confident in integration (covered in chapter 11 of MEI A Level Mathematics Year 1) and know how to integrate simple trigonometric functions (chapter 10 of MEI A Level Mathematics Year 2).

- **Chapter 6: Maclaurin series** uses differentiation of simple exponential, logarithmic and trigonometric functions, covered in chapter 9 of MEI A Level Mathematics Year 2.

- **Review: Complex numbers** reviews the work in chapters 1 and 5 of MEI A Level Further Mathematics Year 1.

- **Chapter 7: Hyperbolic functions** uses the ideas of the domain and range of a function, and an inverse function, covered in chapter 4 of MEI A Level Mathematics Year 2. It uses similar techniques to those covered in chapter 4 of this book.

- **Chapter 8: Applications of integration** uses all the calculus techniques covered in MEI A Level Mathematics Year 1 and MEI A Level Mathematics Year 2, and in chapters 4 and 7 of this book.

- **Review: Roots of polynomials** reviews the work covered in chapter 3 of MEI A Level Further Mathematics Year 1.

- **Chapter 9: First order differential equations** uses all the calculus techniques covered in MEI A Level Mathematics Year 1 and MEI A Level Mathematics Year 2, and in chapters 4 and 7 of this book.

- **Chapter 10: Complex numbers** builds on the work reviewed in Review: Complex numbers. You need to be familiar with trigonometric identities such as the double angle formulae (covered in chapter 8 of MEI A Level Mathematics Year 2) and you need to know about geometric series (covered in chapter 3 of MEI A Level Mathematics Year 2).

- **Chapter 11: Vectors 2** builds on the work on lines and planes in chapter 1 of this book and in chapter 7 of MEI A Level Further Mathematics Year 1.

- **Chapter 12: Second order differential equations** uses all the calculus techniques covered in MEI A Level Mathematics Year 1 and MEI A Level Mathematics Year 2, and in chapters 4 and 7 of this book. It follows on from chapter 9 of this book. You also need to know how to write an expression of the form $a\cos\theta + b\sin\theta$ in the form $r\sin(\theta + \alpha)$. This is covered in chapter 8 of MEI A Level Mathematics Year 2.

1 Vectors 1

I pulled out, on the spot, a pocket book, which still exists, and made an entry, on which, at the very moment, I felt it might be worth my while to expend the labour of at least ten (or it might be fifteen) years to come. But then it is fair to say that this was because I felt a problem to have been at that moment solved, an intellectual want relieved, which had haunted me for at least fifteen years before.

William R. Hamilton, writing on 16th October 1858, about his invention of quarternions on 16th October 1843

Discussion points

→ A zip wire can be modelled as a straight line. How can you find the equation of a line in three dimensions?

→ How could you work out the distance between the two zip wires?

Review: Working with vectors

Prior knowledge

You need to be able to use the language of vectors, including the terms magnitude, direction and position vector. You should also be able to find the distance between two points represented by position vectors and be able to add and subtract vectors and multiply a vector by a scalar.

- A vector quantity has magnitude and direction.
- A scalar quantity has magnitude only.
- Vectors are typeset in bold, $\mathbf{a}$ or $\mathbf{OA}$, or in the form $\overrightarrow{OA}$.
  They are handwritten either in the underlined form $\underline{a}$, or as $\overrightarrow{OA}$.

1

- Unit vectors in the x, y and z directions are denoted by $\mathbf{i}$, $\mathbf{j}$ and $\mathbf{k}$ respectively.
- The resultant of two (or more) vectors is found by the sum of the vectors. A resultant vector is usually denoted by a double-headed arrow.
- The position vector $\overrightarrow{OP}$ of a point P is the vector joining the origin to P.
- The vector $\overrightarrow{AB} = \mathbf{b} - \mathbf{a}$, where $\mathbf{a}$ and $\mathbf{b}$ are the position vectors of A and B.
- The length (or modulus or magnitude) of the vector $\mathbf{r}$ is written as r or as $|\mathbf{r}|$.

$$\mathbf{r} = a\mathbf{i} + b\mathbf{j} + c\mathbf{k} \Rightarrow |\mathbf{r}| = \sqrt{a^2 + b^2 + c^2}$$

The scalar product

Figure 1.1 shows two vectors $\begin{pmatrix} a_1 \\ a_2 \end{pmatrix}$ and $\begin{pmatrix} b_1 \\ b_2 \end{pmatrix}$. The angle between these two vectors can be found using the **scalar product**:

$$\mathbf{a}.\mathbf{b} = |\mathbf{a}||\mathbf{b}|\cos\theta$$

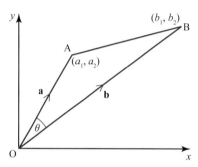

Figure 1.1

Using the column format, the scalar product can be written as:

$$\mathbf{a}.\mathbf{b} = \begin{pmatrix} a_1 \\ a_2 \end{pmatrix} \cdot \begin{pmatrix} b_1 \\ b_2 \end{pmatrix} = a_1 b_1 + a_2 b_2$$

In three dimensions this is extended to:

$$\mathbf{a}.\mathbf{b} = \begin{pmatrix} a_1 \\ a_2 \\ a_3 \end{pmatrix} \cdot \begin{pmatrix} b_1 \\ b_2 \\ b_3 \end{pmatrix} = a_1 b_1 + a_2 b_2 + a_3 b_3$$

Example 1.1

Find the acute angle between the vectors:

(i) $\mathbf{a} = \begin{pmatrix} 2 \\ -3 \end{pmatrix}$ and $\mathbf{b} = \begin{pmatrix} 4 \\ 5 \end{pmatrix}$

(ii) $\mathbf{c} = \begin{pmatrix} 1 \\ 2 \\ 1 \end{pmatrix}$ and $\mathbf{d} = \begin{pmatrix} 0 \\ 4 \\ -3 \end{pmatrix}$

Solution

(i)

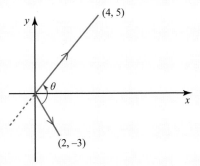

Figure 1.2

$$|\mathbf{a}| = \sqrt{2^2 + (-3)^2} = \sqrt{13} \quad \text{and} \quad |\mathbf{b}| = \sqrt{4^2 + 5^2} = \sqrt{41}$$

Then

$$\mathbf{a}.\mathbf{b} = |\mathbf{a}||\mathbf{b}|\cos\theta$$

$$\Rightarrow (2 \times 4) + (-3 \times 5) = \sqrt{13}\sqrt{41}\cos\theta$$

$$\Rightarrow \frac{-7}{\sqrt{13}\sqrt{41}} = \cos\theta$$

$$\Rightarrow \theta = 107.7°$$

So the acute angle between the vectors is $180° - 107.7° = 72.3°$.

(ii) $$|\mathbf{c}| = \sqrt{1^2 + 2^2 + 1^2} = \sqrt{6} \quad \text{and} \quad |\mathbf{d}| = \sqrt{0^2 + 4^2 + (-3)^2} = 5$$

Then

$$\mathbf{c}.\mathbf{d} = |\mathbf{c}||\mathbf{d}|\cos\theta$$

$$\Rightarrow (1 \times 0) + (2 \times 4) + (1 \times -3) = \sqrt{6} \times 5 \times \cos\theta$$

$$\Rightarrow 5 = 5\sqrt{6}\cos\theta$$

$$\Rightarrow \frac{1}{\sqrt{6}} = \cos\theta$$

$$\Rightarrow \theta = 65.9°$$

Notes

1 The scalar product, unlike a vector, has size but no direction.

2 The scalar product of two vectors is **commutative**. This is because multiplication of numbers is commutative. For example:

$$\begin{pmatrix} 3 \\ -4 \end{pmatrix} . \begin{pmatrix} 1 \\ 5 \end{pmatrix} = (3 \times 1) + (-4 \times 5) = (1 \times 3) + (5 \times -4) = \begin{pmatrix} 1 \\ 5 \end{pmatrix} . \begin{pmatrix} 3 \\ -4 \end{pmatrix}$$

3 When finding the scalar product, using the vectors $\overrightarrow{AB}$ and $\overrightarrow{CB}$ (both directed towards the point B), or the vectors $\overrightarrow{BA}$ and $\overrightarrow{BC}$ (both directed away from the point B) gives the angle θ. This angle could be acute or obtuse.

However, if you use vectors $\overrightarrow{AB}$ (directed towards B) and $\overrightarrow{BC}$ (directed away from B), then you will obtain the angle $180° - \theta$ instead, as shown in Figure 1.3.

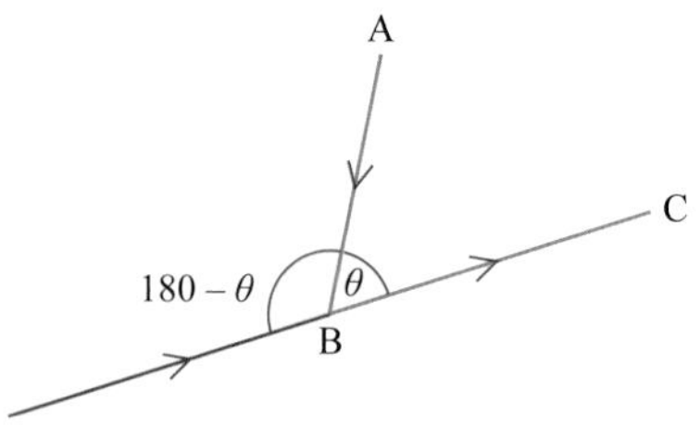

Figure 1.3

4 If two vectors are **perpendicular**, then the angle between them is $90°$.

Since $\cos 90° = 0$, it follows that if vectors **a** and **b** are perpendicular then $\mathbf{a}.\mathbf{b} = 0$.

Conversely, if the scalar product of two non-zero vectors is zero, they are perpendicular.

Example 1.2

A, B and C have position vectors $-4\mathbf{i} + 17\mathbf{j} - \mathbf{k}$, $5\mathbf{i} + 2\mathbf{j} - 2\mathbf{k}$ and $\mathbf{i} + 5\mathbf{j} - 9\mathbf{k}$ respectively.

(i) Prove that triangle ABC is right-angled.

(ii) Find the area of ABC correct to three significant figures.

Solution

(i) $\overrightarrow{AB} = \begin{pmatrix} 5 \\ 2 \\ -2 \end{pmatrix} - \begin{pmatrix} -4 \\ 17 \\ -1 \end{pmatrix} = \begin{pmatrix} 9 \\ -15 \\ -1 \end{pmatrix}$

$\overrightarrow{BC} = \begin{pmatrix} 1 \\ 5 \\ -9 \end{pmatrix} - \begin{pmatrix} 5 \\ 2 \\ -2 \end{pmatrix} = \begin{pmatrix} -4 \\ 3 \\ -7 \end{pmatrix}$

$\overrightarrow{AC} = \begin{pmatrix} 1 \\ 5 \\ -9 \end{pmatrix} - \begin{pmatrix} -4 \\ 17 \\ -1 \end{pmatrix} = \begin{pmatrix} 5 \\ -12 \\ -8 \end{pmatrix}$

> The scalar product of vectors $\overrightarrow{AB}$ and $\overrightarrow{BC}$ is not zero, so these two vectors are not perpendicular.

$\overrightarrow{AB}.\overrightarrow{BC} = \begin{pmatrix} 9 \\ -15 \\ -1 \end{pmatrix}.\begin{pmatrix} -4 \\ 3 \\ -7 \end{pmatrix} = (9 \times -4) + (-15 \times 3) + (-1 \times -7) = -74$

$\overrightarrow{BC}.\overrightarrow{AC} = \begin{pmatrix} -4 \\ 3 \\ -7 \end{pmatrix}.\begin{pmatrix} 5 \\ -12 \\ -8 \end{pmatrix} = (-4 \times 5) + (3 \times -12) + (-7 \times -8) = 0$

So vectors $\overrightarrow{BC}$ and $\overrightarrow{AC}$ are perpendicular and angle C is $90°$.

(ii) The area of triangle ABC is $\frac{1}{2} \times |\overrightarrow{BC}| \times |\overrightarrow{AC}|$

$= \frac{1}{2} \times \sqrt{(-4)^2 + 3^2 + (-7)^2} \times \sqrt{5^2 + (-12)^2 + (-8)^2}$

$= \frac{1}{2} \times \sqrt{74} \times \sqrt{233}$

$= 65.7$ square units (to 3 significant figures)

The equation of a plane

You can write the equation of a plane in either vector or cartesian form.

A vector which is at right angles to every straight line in a plane is called a **normal** to the plane and is often denoted $\mathbf{n} = \begin{pmatrix} n_1 \\ n_2 \\ n_3 \end{pmatrix}$.

In Figure 1.4 the point A is on the plane and the vector $\mathbf{n}$ is perpendicular to the plane.

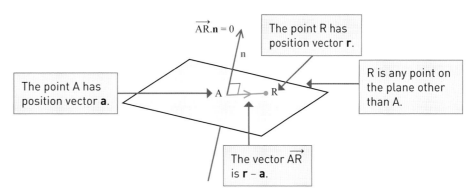

$\overrightarrow{AR}.\mathbf{n} = 0$

The point R has position vector $\mathbf{r}$.

R is any point on the plane other than A.

The point A has position vector $\mathbf{a}$.

The vector $\overrightarrow{AR}$ is $\mathbf{r} - \mathbf{a}$.

Figure 1.4

The vector $\overrightarrow{AR}$ is a line in the plane and can be written $\overrightarrow{AR} = \mathbf{r} - \mathbf{a}$.
As $\overrightarrow{AR}$ is at right angles to the direction $\mathbf{n}$:

$$(\mathbf{r} - \mathbf{a}).\mathbf{n} = 0$$

This is the **vector equation of the plane**.

Expanding the brackets gives an alternative form:

$$\mathbf{r.n} - \mathbf{a.n} = 0 \quad \longleftarrow \quad \boxed{\text{This can also be written as } \mathbf{r.n} = \mathbf{a.n}.}$$

Writing the normal vector $\mathbf{n}$ as $\begin{pmatrix} n_1 \\ n_2 \\ n_3 \end{pmatrix}$, the position vector of A as $\mathbf{a} = \begin{pmatrix} a_1 \\ a_2 \\ a_3 \end{pmatrix}$

and the position vector of the general point R as $\mathbf{r} = \begin{pmatrix} x \\ y \\ z \end{pmatrix}$ and substituting

> Notice that d is a constant and is a scalar quantity.

into the equation $\mathbf{r.n} - \mathbf{a.n} = 0$ gives

$$n_1 x + n_2 y + n_3 z + d = 0 \quad \text{where} \quad d = -(a_1 n_1 + a_2 n_2 + a_3 n_3).$$

This is known as the **cartesian equation of the plane**. Notice that the coefficients of x, y and z are the three elements of the normal vector.

In general the vector $\begin{pmatrix} n_1 \\ n_2 \\ n_3 \end{pmatrix}$ is perpendicular to all planes of the form

$n_1x + n_2y + n_3z + d = 0$ whatever the value of d (see Figure 1.5).

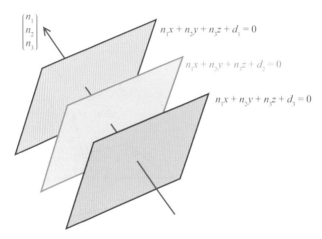

Figure 1.5

Consequently, all planes of that form are parallel; the coefficients of x, y and z determine the direction of the plane, the value of d its location.

Example 1.3

The point A($-1, 2, 4$) lies on a plane. The vector $\mathbf{n} = \begin{pmatrix} -3 \\ 1 \\ -2 \end{pmatrix}$ is perpendicular to the plane.

Find the vector and cartesian equations of the plane.

Solution

The general vector equation of a plane is given by $(\mathbf{r} - \mathbf{a}).\mathbf{n} = 0$.

The vector equation of this plane can therefore be written as

$$\left(\mathbf{r} - \begin{pmatrix} -1 \\ 2 \\ 4 \end{pmatrix} \right) . \begin{pmatrix} -3 \\ 1 \\ -2 \end{pmatrix} = 0$$

The general cartesian equation of a plane is given by $n_1x + n_2y + n_3z + d = 0$.

$$\mathbf{n} = \begin{pmatrix} -3 \\ 1 \\ -2 \end{pmatrix} \text{ so } n_1 = -3, \ n_2 = 1 \text{ and } n_3 = -2.$$

The equation of the plane is $-3x + y - 2z + d = 0$, where d is a constant to be found.

You can use either the cartesian equation or the vector equation to find d.

Using the cartesian equation

The point A is $(-1, 2, 4)$.

Substituting for x, y and z in $-3x + y - 2z + d = 0$ gives $(-3 \times -1) + 2 - (2 \times 4) + d = 0$ and so $d = 3$.

Using the vector equation

$$d = -\mathbf{a}.\mathbf{n}$$

where $\mathbf{a}$ is the position vector of A$(-1, 2, 4)$.

$$\text{so } \mathbf{a} = \begin{pmatrix} -1 \\ 2 \\ 4 \end{pmatrix} \quad \text{and} \quad \mathbf{n} = \begin{pmatrix} -3 \\ 1 \\ -2 \end{pmatrix}$$

Then

$$d = - \begin{pmatrix} -1 \\ 2 \\ 4 \end{pmatrix} . \begin{pmatrix} -3 \\ 1 \\ -2 \end{pmatrix}$$

$$= -[(-1 \times -3) + (2 \times 1) + (4 \times -2)]$$

$$= 3$$

Using either method the cartesian equation of the plane is $-3x + y - 2z + 3 = 0$.

The angle between planes

The angle between two planes can be found by using the scalar product. As Figures 1.6 and 1.7 show, the acute angle between planes π_1 and π_2 is the same as the acute angle between their normal vectors, $\mathbf{n}_1$ and $\mathbf{n}_2$.

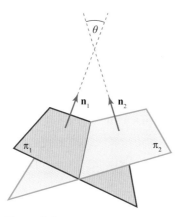

Figure 1.6

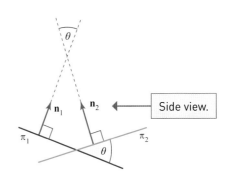

Figure 1.7

Example 1.4

Find, to one decimal place, the acute angles between the planes $-x + 4y - 2z = 3$ and $2x + 5y + z = 9$.

Solution

The planes have normal $\mathbf{n}_1 = \begin{pmatrix} -1 \\ 4 \\ -2 \end{pmatrix}$ and $\mathbf{n}_2 = \begin{pmatrix} 2 \\ 5 \\ 1 \end{pmatrix}$ so the angle

between the planes is satisfied by $\mathbf{n}_1.\mathbf{n}_2 = |\mathbf{n}_1||\mathbf{n}_2|\cos\theta$.

$$\Rightarrow \begin{pmatrix} -1 \\ 4 \\ -2 \end{pmatrix} \cdot \begin{pmatrix} 2 \\ 5 \\ 1 \end{pmatrix} = \sqrt{1 + 16 + 4}\sqrt{4 + 25 + 1}\cos\theta$$

$$\Rightarrow -2 + 20 - 2 = \sqrt{21}\sqrt{30}\cos\theta$$

$$\Rightarrow \theta = \cos^{-1}\left(\frac{16}{\sqrt{21}\sqrt{30}}\right) = 50.4°$$

So the acute angle between the planes is 50.4°.

Review exercise

① Find, to one decimal place, the angle between each pair of vectors.

(i) $\begin{pmatrix} 2 \\ 5 \end{pmatrix}$ and $\begin{pmatrix} -1 \\ 3 \end{pmatrix}$ (ii) $\begin{pmatrix} 4 \\ -3 \end{pmatrix}$ and $\begin{pmatrix} 3 \\ 4 \end{pmatrix}$

(iii) $\mathbf{i} + 7\mathbf{j}$ and $2\mathbf{i} + 3\mathbf{j}$

② Find, to one decimal place, the angle between each pair of vectors.

(i) $\begin{pmatrix} 2 \\ 3 \\ -1 \end{pmatrix}$ and $\begin{pmatrix} 1 \\ 2 \\ 8 \end{pmatrix}$ (ii) $\begin{pmatrix} 4 \\ 0 \\ -1 \end{pmatrix}$ and $\begin{pmatrix} -3 \\ 1 \\ 5 \end{pmatrix}$

(iii) $\mathbf{i} + \mathbf{j} + \mathbf{k}$ and $2\mathbf{i} + 2\mathbf{j} + 2\mathbf{k}$

③ Find the vector and cartesian equations of the plane which is perpendicular to the vector $\mathbf{n} = \begin{pmatrix} 3 \\ -2 \\ 1 \end{pmatrix}$ and passes through the point A(5, −2, 0).

④ Find the cartesian equation of the plane which is parallel to the plane
$$\left(\mathbf{r} - (4\mathbf{i} - \mathbf{j} + 3\mathbf{k})\right) \cdot (2\mathbf{i} + \mathbf{j} - 7\mathbf{k}) = 0$$
and which passes through the point A(−5, 0, 3).

⑤ Find, to one decimal place, the acute angle between the planes:

(i) $\mathbf{r} \cdot \begin{pmatrix} 4 \\ -1 \\ 2 \end{pmatrix} = 11$ and $\mathbf{r} \cdot \begin{pmatrix} -3 \\ 2 \\ 8 \end{pmatrix} = -4$

(ii) $2x - y - 3z = 9$ and $5x + 3y + 3z = 7$.

⑥ The planes $\pi_1: kx - 5y + 3z = 2$ and $\pi_2: kx + ky + 2 = 1$ are perpendicular.

(i) Find the possible values of k.

A third plane π_3 is parallel to π_1 and passes through the point A(2, 1, 2).

(ii) Find the possible equations of plane π_3.

⑦ Find the equation of the plane π which is perpendicular to the planes
$$-2x + y - z + 6 = 0$$
$$3x + 2y + z + 2 = 0$$
and which passes through the point P(−3, 0, 3).

⑧ The points A, B and C have coordinates $(2, 0, -1)$, $(4, 1, 1)$ and $(-3, -1, 0)$.

(i) Write down the vectors $\overrightarrow{AB}$ and $\overrightarrow{AC}$.

(ii) Show that $\overrightarrow{AB} . \begin{pmatrix} 1 \\ -4 \\ 1 \end{pmatrix} = \overrightarrow{AC} . \begin{pmatrix} 1 \\ -4 \\ 1 \end{pmatrix} = 0.$

(iii) Find the equation of the plane containing the points A, B and C.

⑨ Find an equation of the plane that passes through the three points A$(-4, 0, 1)$, B$(-2, 2, -2)$ and C$(5, -2, 1)$.

⑩ Three planes have equations.

$$\pi_1: ax + 2y + z = 5$$
$$\pi_2: 2x + ay + z = 1$$
$$\pi_3: x + y + az = -2$$

Given that the angle between planes π_1 and π_2 is equal to the angle between the planes π_2 and π_3, show that a must satisfy the equation:

$$12a^6 + 69a^4 - 24a^3 + 27a^2 - 120a - 90 = 0$$

⑪ Four planes are given by the equations:

$$\pi_1: 3x + y + 2z - 2 = 0$$
$$\pi_2: x + 3y - 3z + 4 = 0$$
$$\pi_3: -6x - 2y - 4z - 4 = 0$$
$$\pi_4: -x + y + z - 5 = 0$$

Determine whether each *pair* of planes are parallel, perpendicular or neither.

⑫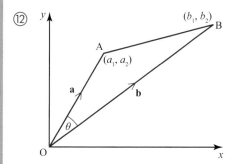

Figure 1.8

(i) Write down the compound angle formula for $\cos(A - B)$.

(ii) Hence show that $\cos\theta = \dfrac{a_1 b_1 + a_2 b_2}{|\mathbf{a}||\mathbf{b}|}$.

How would you usually write the numerator of the right hand side?

1 The vector equation of a line

Lines in two dimensions

Before looking at the equation of a line in three dimensions, Activity 1.1 looks at a new format for the equation of a line in two dimensions. This is called the vector equation of the line.

ACTIVITY 1.1

The position vector of a set of points is given by

$$\mathbf{r} = \begin{pmatrix} 2 \\ -1 \end{pmatrix} + \lambda \begin{pmatrix} 2 \\ 4 \end{pmatrix}$$

where λ is a parameter that can take any value and A is the point $(2, -1)$.

(i)   Show that $\lambda = 1$ corresponds to the point B with position vector $\begin{pmatrix} 4 \\ 3 \end{pmatrix}$.

(ii) Find the position vectors of the points corresponding to values of λ of

  $-2, -1, 0, \dfrac{1}{2}, \dfrac{3}{4}, 2, 3.$

(iii) Plot the points from parts (i) and (ii) on a sheet of graph paper and show they can be joined to form a straight line.

(iv) What can you say about the position of the point if:

  (a) $0 < \lambda < 1$

  (b) $\lambda > 1$

  (c) $\lambda < 0$?

This activity should have convinced you that $\mathbf{r} = \begin{pmatrix} 2 \\ -1 \end{pmatrix} + \lambda \begin{pmatrix} 2 \\ 4 \end{pmatrix}$ is the

equation of a straight line passing through the point $(2, -1)$. The vector $\begin{pmatrix} 2 \\ 4 \end{pmatrix}$

determines the direction of the line. You might find it helpful to think of this as shown in Figure 1.9.

Starting from the origin, you can 'step' on to the line at a given point A. All other points on the line can then be reached by taking 'steps' of different sizes (λ) in the direction of a given vector, called the **direction vector**.

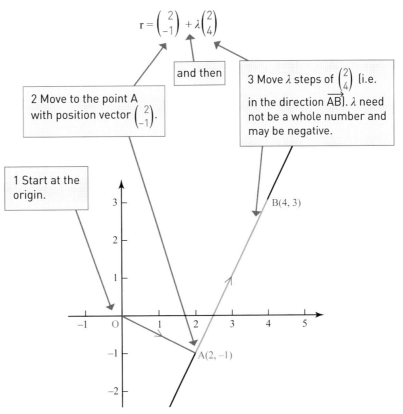

$$\mathbf{r} = \begin{pmatrix} 2 \\ -1 \end{pmatrix} + \lambda \begin{pmatrix} 2 \\ 4 \end{pmatrix}$$

and then

3 Move λ steps of $\begin{pmatrix} 2 \\ 4 \end{pmatrix}$ (i.e. in the direction $\overrightarrow{AB}$). λ need not be a whole number and may be negative.

2 Move to the point A with position vector $\begin{pmatrix} 2 \\ -1 \end{pmatrix}$.

1 Start at the origin.

Figure 1.9

You should also have noticed that:

$\lambda = 0$	corresponds to the point A
$\lambda = 1$	corresponds to the point B, one 'step' along the line away from A
$0 < \lambda < 1$	corresponds to points lying between A and B
$\lambda > 1$	corresponds to points lying beyond B
$\lambda < 0$	corresponds to points beyond A, in the opposite direction to B.

This is shown by the green part of the line in Figure 1.9.

The vector equation of a line is not unique. In this case, any vector parallel, or in the opposite direction to, $\begin{pmatrix} 2 \\ 4 \end{pmatrix}$ could be used as the direction vector, for example, $\begin{pmatrix} 1 \\ 2 \end{pmatrix}$, $\begin{pmatrix} 3 \\ 6 \end{pmatrix}$ or $\begin{pmatrix} 20 \\ 40 \end{pmatrix}$. Similarly, you can 'step' on to the line at any point, such as B(4, 3).

So the line $\mathbf{r} = \begin{pmatrix} 2 \\ -1 \end{pmatrix} + \lambda \begin{pmatrix} 2 \\ 4 \end{pmatrix}$ could also have equation

$$\mathbf{r} = \begin{pmatrix} 2 \\ -1 \end{pmatrix} + \lambda \begin{pmatrix} 1 \\ 2 \end{pmatrix} \quad \text{or} \quad \mathbf{r} = \begin{pmatrix} 4 \\ 3 \end{pmatrix} + \lambda \begin{pmatrix} 2 \\ 4 \end{pmatrix}, \text{ for example.}$$

In general, the **vector equation of a line** in two dimensions is given by:

$$\mathbf{r} = \mathbf{a} + \lambda \mathbf{d}$$

In two dimensions the equation of a line usually looks easier in cartesian form than in vector form. However, as you are about to see, the opposite is the case in three dimensions where the vector form is much easier to work with.

where $\mathbf{a}$ is the position vector of a point A on the line and $\mathbf{d}$ is the direction vector of the line. Sometimes a different letter, such as μ or t is used as the parameter instead of λ.

ACTIVITY 1.2

The vector equation of a line

$$\mathbf{r} = \mathbf{a} + \lambda\mathbf{d}$$

is written in the form

$$\begin{pmatrix} x \\ y \end{pmatrix} = \begin{pmatrix} a_1 \\ a_2 \end{pmatrix} + \lambda \begin{pmatrix} d_1 \\ d_2 \end{pmatrix}.$$

(i) Write down expressions for x and y in terms of λ.

(ii) Rearrange the two expressions from part (i) to make λ the subject.

By equating these two expressions, show that the vector equation of the line can be written in the form

$$y = mx + c$$

where m and c are constants.

Activity 1.2 shows that the vector and cartesian equations of a line are equivalent.

Lines in three dimensions

The same form for the vector equation of a line can be used in three dimensions. For example:

$$\mathbf{r} = \begin{pmatrix} 3 \\ 4 \\ 1 \end{pmatrix} + \lambda \begin{pmatrix} 2 \\ 3 \\ 6 \end{pmatrix}$$

represents a line through the point with position vector $\begin{pmatrix} 3 \\ 4 \\ 1 \end{pmatrix}$ with direction vector $\begin{pmatrix} 2 \\ 3 \\ 6 \end{pmatrix}$.

Writing $\mathbf{r}$ as $\begin{pmatrix} x \\ y \\ z \end{pmatrix}$ gives $\begin{pmatrix} x \\ y \\ z \end{pmatrix} = \begin{pmatrix} 3 \\ 4 \\ 1 \end{pmatrix} + \lambda \begin{pmatrix} 2 \\ 3 \\ 6 \end{pmatrix}$. This equation contains the three relationships

$$x = 3 + 2\lambda \qquad y = 4 + 3\lambda \qquad z = 1 + 6\lambda$$

Making λ the subject of each of these gives:

$$\lambda = \frac{x-3}{2} = \frac{y-4}{3} = \frac{z-1}{6}$$

> This form is not easy to work with and you will often find that the first step in a problem is to convert the cartesian form into vector form.

This is the **cartesian equation of a line** in three dimensions.

Generally, a line with direction vector $\mathbf{d} = \begin{pmatrix} d_1 \\ d_2 \\ d_3 \end{pmatrix}$ passing through the point A

with position vector $\mathbf{a} = \begin{pmatrix} a_1 \\ a_2 \\ a_3 \end{pmatrix}$ has the cartesian equation:

> The direction vector of the line can be read from the denominators of the three expressions in this equation; the point A can be determined from the three numerators.

$$\frac{x - a_1}{d_1} = \frac{y - a_2}{d_2} = \frac{z - a_3}{d_3}$$

Special cases of the cartesian equation of a line

In the equation

$$\frac{x - a_1}{d_1} = \frac{y - a_2}{d_2} = \frac{z - a_3}{d_3}$$

it is possible that one or two of the values of d_i might equal zero. In such cases the equation of the line needs to be written differently.

For example, the line through $(7, 2, 3)$ in the direction $\begin{pmatrix} 0 \\ 5 \\ 2 \end{pmatrix}$ would have equation:

$$\lambda = \frac{x - 7}{0} = \frac{y - 2}{5} = \frac{z - 3}{2}$$

The first fraction involves division by zero, which is undefined. The expression

$\dfrac{x - 7}{0}$ comes from the rearrangement of $x - 7 = 0\lambda$ and so you can write this as $x - 7 = 0$ or $x = 7$.

So the equation of this line would be written:

$$x = 7 \quad \text{and} \quad \lambda = \frac{y - 2}{5} = \frac{z - 3}{2}$$

Note that all three of the d_i could not equal zero as it is not possible for a line to have a zero direction vector.

| Example 1.5 | Write the equation of this line in vector form: |

$$\frac{x - 4}{3} = \frac{y - 3}{-2} = \frac{z + 6}{4}$$

Solution

$$\lambda = \frac{x-4}{3} = \frac{y-3}{-2} = \frac{z+6}{4}$$

$$\lambda = \frac{x-4}{3} \Rightarrow x = 3\lambda + 4$$

$$\lambda = \frac{y-3}{-2} \Rightarrow y = -2\lambda + 3$$

$$\lambda = \frac{z+6}{4} \Rightarrow z = 4\lambda - 6$$

$$\text{So } \mathbf{r} = \begin{pmatrix} x \\ y \\ z \end{pmatrix} = \begin{pmatrix} 3\lambda + 4 \\ -2\lambda + 3 \\ 4\lambda - 6 \end{pmatrix} \Rightarrow \mathbf{r} = \begin{pmatrix} 4 \\ 3 \\ -6 \end{pmatrix} + \lambda \begin{pmatrix} 3 \\ -2 \\ 4 \end{pmatrix}$$

Example 1.6

Find the cartesian form of the equation of the line through the point A(7, −12, 4) in the direction $2\mathbf{i} - 5\mathbf{j} - 3\mathbf{k}$.

Solution

The line has vector form $\mathbf{r} = \mathbf{a} + \lambda\mathbf{d} = \begin{pmatrix} 7 \\ -12 \\ 4 \end{pmatrix} + \lambda \begin{pmatrix} 2 \\ -5 \\ -3 \end{pmatrix}$.

This leads to the equations

$$x = 7 + 2\lambda \qquad y = -12 - 5\lambda \qquad z = 4 - 3\lambda$$

which can be rearranged to give the cartesian equation

$$\lambda = \frac{x-7}{2} = \frac{y+12}{-5} = \frac{z-4}{-3}$$

The intersection of straight lines in two dimensions

You already know how to find the point of intersection of two straight lines given in cartesian form, by using simultaneous equations.

You can also use vector methods to find the position vector of the point where two lines intersect.

Example 1.7

Find the position vector of the point where the following lines intersect.

$$\mathbf{r} = \begin{pmatrix} 2 \\ 3 \end{pmatrix} + \lambda \begin{pmatrix} 1 \\ 2 \end{pmatrix} \quad \text{and} \quad \mathbf{r} = \begin{pmatrix} 6 \\ 1 \end{pmatrix} + \mu \begin{pmatrix} 1 \\ -3 \end{pmatrix}$$

Notice that different letters are used for the parameters in the two equations to avoid confusion.

Solution

When the lines intersect, the position vector is the same for each of them.

$$\mathbf{r} = \begin{pmatrix} x \\ y \end{pmatrix} = \begin{pmatrix} 2 \\ 3 \end{pmatrix} + \lambda \begin{pmatrix} 1 \\ 2 \end{pmatrix} = \begin{pmatrix} 6 \\ 1 \end{pmatrix} + \mu \begin{pmatrix} 1 \\ -3 \end{pmatrix}$$

This gives two simultaneous equations for λ and μ.

$$2 + \lambda = 6 + \mu$$

$$3 + 2\lambda = 1 - 3\mu$$

Solving these gives $\lambda = 2$ and $\mu = -2$. Substituting in either equation gives

$$\mathbf{r} = \begin{pmatrix} 4 \\ 7 \end{pmatrix}$$

which is the position vector of the point of intersection.

The intersection of straight lines in three dimensions

Hold a pen and a pencil to represent two distinct straight lines as follows:

■ hold them to represent two parallel lines;

■ hold them to represent two lines intersecting at a unique point;

■ hold them to represent lines which are not parallel and which do not intersect even if you were to extend them.

In three-dimensional space two or more straight lines which are not parallel and which do not meet are known as **skew** lines. In two dimensions, two distinct lines are either parallel or intersecting but in three dimensions there are three possibilities: the lines are either parallel, intersecting or skew. This is illustrated in the following examples.

Example 1.8

The lines l_1 and l_2 are represented by the equations

$$l_1: \frac{x-1}{1} = \frac{y+6}{2} = \frac{z+1}{3} \qquad l_2: \frac{x-9}{2} = \frac{y-7}{3} = \frac{z-2}{-1}$$

(i) Write these lines in vector form.

(ii) Hence find whether the lines meet and if so, the coordinates of their point of intersection.

Solution

(i) The equation of l_1 is $\mathbf{r} = \begin{pmatrix} 1 \\ -6 \\ -1 \end{pmatrix} + \lambda \begin{pmatrix} 1 \\ 2 \\ 3 \end{pmatrix}$

The equation of l_2 is $\mathbf{r} = \begin{pmatrix} 9 \\ 7 \\ 2 \end{pmatrix} + \mu \begin{pmatrix} 2 \\ 3 \\ -1 \end{pmatrix}$

(ii)   If there is a point $\begin{pmatrix} X \\ Y \\ Z \end{pmatrix}$ that is common to both lines then

$$\begin{pmatrix} X \\ Y \\ Z \end{pmatrix} = \begin{pmatrix} 1 \\ -6 \\ -1 \end{pmatrix} + \lambda \begin{pmatrix} 1 \\ 2 \\ 3 \end{pmatrix} = \begin{pmatrix} 9 \\ 7 \\ 2 \end{pmatrix} + \mu \begin{pmatrix} 2 \\ 3 \\ -1 \end{pmatrix}$$

for some parameters λ and μ.

This gives the three equations

$$X = \lambda + 1 = 2\mu + 9 \quad \text{①}$$
$$Y = 2\lambda - 6 = 3\mu + 7 \quad \text{②}$$
$$Z = 3\lambda - 1 = -\mu + 2 \quad \text{③}$$

Using ① and ②

Now solve any two of the three equations simultaneously.

$$\left. \begin{matrix} \lambda - 2\mu = 8 \\ 2\lambda - 3\mu = 13 \end{matrix} \right\} \Leftrightarrow \left. \begin{matrix} 2\lambda - 4\mu = 16 \\ 2\lambda - 3\mu = 13 \end{matrix} \right\} \Leftrightarrow \mu = -3, \lambda = 2$$

If these values for λ and μ also satisfy equation ③, then the lines meet.

Using equation ③, when $\lambda = 2$, $Z = 6 - 1 = 5$ and when $\mu = -3$, $Z = 3 + 2 = 5$.

As both values of Z are equal this proves the lines intersect.

Using either $\lambda = 2$ or $\mu = -3$ in equations ①, ② and ③ gives $X = 3$, $Y = -2$, $Z = 5$ so the lines meet at the point $(3, -2, 5)$.

Example 1.9

Prove that the lines l_1 and l_2 are skew, where:

$$l_1: \begin{pmatrix} 1 \\ -6 \\ -1 \end{pmatrix} + \lambda \begin{pmatrix} 1 \\ 2 \\ 3 \end{pmatrix}$$

$$l_2: \begin{pmatrix} 9 \\ 8 \\ 2 \end{pmatrix} + \mu \begin{pmatrix} 2 \\ 3 \\ -1 \end{pmatrix}$$

Solution

If there is a point (X, Y, Z) common to both lines then

$$\begin{pmatrix} X \\ Y \\ Z \end{pmatrix} = \begin{pmatrix} 1 \\ -6 \\ -1 \end{pmatrix} + \lambda \begin{pmatrix} 1 \\ 2 \\ 3 \end{pmatrix} = \begin{pmatrix} 9 \\ 8 \\ 2 \end{pmatrix} + \mu \begin{pmatrix} 2 \\ 3 \\ -1 \end{pmatrix}$$

for some parameters λ and μ.

$$X = \lambda + 1 = 2\mu + 9 \quad \text{①}$$
$$Y = 2\lambda - 6 = 3\mu + 8 \quad \text{②}$$
$$Z = 3\lambda - 1 = -\mu + 2 \quad \text{③}$$

Solving equations ① and ② simultaneously

$$\left. \begin{array}{r} \lambda - 2\mu = 8 \\ 2\lambda - 3\mu = 14 \end{array} \right\} \Leftrightarrow \left. \begin{array}{r} 2\lambda - 4\mu = 16 \\ 2\lambda - 3\mu = 14 \end{array} \right\} \Leftrightarrow \mu = -2, \lambda = 4$$

When $\lambda = 4$, $Z = 12 - 1 = 11$
and when $\mu = -2$, $Z = 2 + 2 = 4$.

Substitute these values into equation ③

Therefore the values $\mu = -2$, $\lambda = 4$ do not satisfy the third equation and so the lines do not meet. As the lines are distinct, the only other alternatives are that the lines are parallel or skew.

Look at the direction vectors of the lines: $\begin{pmatrix} 1 \\ 2 \\ 3 \end{pmatrix}$ and $\begin{pmatrix} 2 \\ 3 \\ -1 \end{pmatrix}$. Neither of these is a multiple of the other so they are not parallel and hence the two lines are not parallel. So, lines l_1 and l_2 are skew.

Finding the angle between two lines

Figure 1.10 shows two lines in two dimensions, with their equations given in vector form.

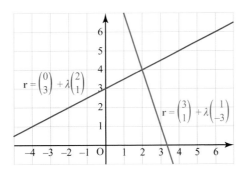

Figure 1.10

The angle between the two lines is the same as the angle between their direction vectors, $\begin{pmatrix} 2 \\ 1 \end{pmatrix}$ and $\begin{pmatrix} 1 \\ -3 \end{pmatrix}$. So you can use the scalar product to find the angle between the two lines.

Example 1.10

Find the acute angle between the lines $\mathbf{r} = \begin{pmatrix} 0 \\ 3 \end{pmatrix} + \lambda \begin{pmatrix} 2 \\ 1 \end{pmatrix}$ and

$\mathbf{r} = \begin{pmatrix} 3 \\ 1 \end{pmatrix} + \mu \begin{pmatrix} 1 \\ -3 \end{pmatrix}$.

Solution

$$\begin{pmatrix} 2 \\ 1 \end{pmatrix} \cdot \begin{pmatrix} 1 \\ -3 \end{pmatrix} = (2 \times 1) + (1 \times -3) = 2 - 3 = -1$$

$$\left| \begin{pmatrix} 2 \\ 1 \end{pmatrix} \right| = \sqrt{2^2 + 1^2} = \sqrt{5} \text{ and } \left| \begin{pmatrix} 1 \\ -3 \end{pmatrix} \right| = \sqrt{1^2 + (-3)^2} = \sqrt{10}$$

$$\cos\theta = \frac{-1}{\sqrt{5}\sqrt{10}}$$

$$\theta = 98.1°$$

So the acute angle between the lines is $180° - 98.1° = 81.9°$.

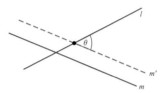

ACTIVITY 1.3

Find the cartesian forms of the two equations in Example 1.10. How can you find the angle between them without using vectors?

The same method can be used for lines in three dimensions. Even if the lines do not meet, the angle between them is still the angle between their direction vectors.

The lines l and m shown in Figure 1.11 are skew. The angle between them is shown in the diagram by the angle θ between the lines l and m', where m' is a translation of the line m to a position where it intersects the line l.

Figure 1.11

Example 1.11

Find the angle between the lines $\mathbf{r} = \begin{pmatrix} 1 \\ 0 \\ 4 \end{pmatrix} + \lambda \begin{pmatrix} 2 \\ -1 \\ -1 \end{pmatrix}$ and

$$\mathbf{r} = \begin{pmatrix} 2 \\ -1 \\ 3 \end{pmatrix} + \mu \begin{pmatrix} 3 \\ 0 \\ 1 \end{pmatrix}.$$

Solution

The angle between the lines is the angle between their direction vectors

$$\begin{pmatrix} 2 \\ -1 \\ -1 \end{pmatrix} \text{ and } \begin{pmatrix} 3 \\ 0 \\ 1 \end{pmatrix}.$$

Using $\mathbf{a}.\mathbf{b} = |\mathbf{a}||\mathbf{b}|\cos\theta$,

$$\begin{pmatrix} 2 \\ -1 \\ -1 \end{pmatrix} \cdot \begin{pmatrix} 3 \\ 0 \\ 1 \end{pmatrix} = \sqrt{2^2 + (-1)^2 + (-1)^2} \sqrt{3^2 + 0^2 + 1^2} \cos\theta$$

$$\Rightarrow 6 + 0 - 1 = \sqrt{6}\sqrt{10} \cos\theta$$

$$\Rightarrow \theta = \arccos\left(\frac{5}{\sqrt{6}\sqrt{10}}\right) = 49.8°$$

Exercise 1.1

① Find the equation of the following lines in vector form:

(i) through $(3, 1)$ in the direction $\begin{pmatrix} 5 \\ -2 \end{pmatrix}$

(ii) through $(5, -1)$ in the direction $\begin{pmatrix} 0 \\ 4 \end{pmatrix}$

(iii) through $(-2, 4)$ and $(3, 9)$

(iv) through $(0, 8)$ and $(-2, -3)$

② Find the equation of the following lines in vector form:

(i) through $(2, 4, -1)$ in the direction $\begin{pmatrix} 3 \\ 6 \\ 4 \end{pmatrix}$

(ii) through $(1, 0, -1)$ in the direction $\begin{pmatrix} 1 \\ 0 \\ 0 \end{pmatrix}$

(iii) through $(1, 0, 4)$ and $(6, 3, -2)$

(iv) through $(0, 0, 1)$ and $(2, 1, 4)$

③ Write the equations of the following lines in cartesian form.

(i) $\mathbf{r} = \begin{pmatrix} 2 \\ 4 \\ -1 \end{pmatrix} + t \begin{pmatrix} 3 \\ 6 \\ 4 \end{pmatrix}$
(ii) $\mathbf{r} = \begin{pmatrix} 1 \\ 0 \\ -1 \end{pmatrix} + t \begin{pmatrix} 1 \\ 3 \\ 4 \end{pmatrix}$

(iii) $\mathbf{r} = \begin{pmatrix} 3 \\ 0 \\ 4 \end{pmatrix} + t \begin{pmatrix} 1 \\ 0 \\ 2 \end{pmatrix}$
(iv) $\mathbf{r} = \begin{pmatrix} 0 \\ 4 \\ 1 \end{pmatrix} + t \begin{pmatrix} 2 \\ 0 \\ 4 \end{pmatrix}$

④ Write the equations of the following lines in vector form.

(i) $\dfrac{x - 3}{5} = \dfrac{y + 2}{3} = \dfrac{z - 1}{4}$
(ii) $x = \dfrac{y}{2} = \dfrac{z + 1}{3}$

(iii) $x = y = z$
(iv) $x = 2$ and $y = z$

⑤ Write down the vector and cartesian equations of the line through the point $(3, -5, 2)$ which is parallel to the y-axis.

⑥ Find the position vector of the point of intersection of each of these pairs of lines.

(i) $\mathbf{r} = \begin{pmatrix} 2 \\ 1 \end{pmatrix} + \lambda \begin{pmatrix} 1 \\ 0 \end{pmatrix}$ and $\mathbf{r} = \begin{pmatrix} 3 \\ 0 \end{pmatrix} + \mu \begin{pmatrix} 1 \\ 1 \end{pmatrix}$

(ii) $\mathbf{r} = \begin{pmatrix} 2 \\ -1 \end{pmatrix} + \lambda \begin{pmatrix} 1 \\ 2 \end{pmatrix}$ and $\mathbf{r} = \mu \begin{pmatrix} 1 \\ 1 \end{pmatrix}$

(iii) $\mathbf{r} = \begin{pmatrix} -2 \\ -3 \end{pmatrix} + \lambda \begin{pmatrix} -1 \\ 3 \end{pmatrix}$ and $\mathbf{r} = \begin{pmatrix} 1 \\ 3 \end{pmatrix} + \mu \begin{pmatrix} 2 \\ -1 \end{pmatrix}$

⑦ Decide whether the following pairs of lines intersect or not. If they do intersect, find the point of intersection; if not, state whether the lines are parallel or skew.

(i) $L_1: \dfrac{x - 6}{1} = \dfrac{y + 4}{-2} = \dfrac{z - 2}{5}$ $\qquad$ $L_2: \dfrac{x - 1}{1} = \dfrac{y - 4}{-1} = \dfrac{z + 17}{2}$

(ii) $L_1: \dfrac{x}{5} = \dfrac{y + 1}{3} = \dfrac{z - 4}{-3}$ $\qquad$ $L_2: \dfrac{x - 2}{4} = \dfrac{y - 5}{-3} = \dfrac{z + 1}{2}$

(iii) $\mathbf{r}_1 = \begin{pmatrix} 2 \\ 0 \\ 1 \end{pmatrix} + \lambda \begin{pmatrix} 3 \\ 2 \\ 1 \end{pmatrix}$ $\qquad$ $\mathbf{r}_2 = \begin{pmatrix} 4 \\ 9 \\ -1 \end{pmatrix} + \mu \begin{pmatrix} -6 \\ -4 \\ -2 \end{pmatrix}$

(iv) $\mathbf{r}_1 = \begin{pmatrix} 9 \\ 3 \\ -4 \end{pmatrix} + \lambda \begin{pmatrix} 1 \\ 2 \\ -3 \end{pmatrix}$ $\qquad$ $\mathbf{r}_2 = \begin{pmatrix} 1 \\ -4 \\ 5 \end{pmatrix} + \mu \begin{pmatrix} 1 \\ -1 \\ 2 \end{pmatrix}$

(v) $\mathbf{r}_1 = \begin{pmatrix} 2 \\ 3 \\ 1 \end{pmatrix} + \lambda \begin{pmatrix} 1 \\ 1 \\ -2 \end{pmatrix}$ $\qquad$ $\mathbf{r}_2 = \begin{pmatrix} -1 \\ -3 \\ -1 \end{pmatrix} + \mu \begin{pmatrix} 1 \\ 3 \\ 2 \end{pmatrix}$

⑧ Find the acute angle between these pairs of lines:

(i) $\mathbf{r} = \begin{pmatrix} 2 \\ 5 \end{pmatrix} + \lambda \begin{pmatrix} 1 \\ 2 \end{pmatrix}$ and $\mathbf{r} = \begin{pmatrix} 1 \\ 2 \end{pmatrix} + \mu \begin{pmatrix} -1 \\ 3 \end{pmatrix}$

(ii) $\mathbf{r} = \begin{pmatrix} 0 \\ 3 \end{pmatrix} + \lambda \begin{pmatrix} -5 \\ 1 \end{pmatrix}$ and $\mathbf{r} = \begin{pmatrix} 2 \\ -1 \end{pmatrix} + \mu \begin{pmatrix} 1 \\ 1 \end{pmatrix}$

(iii) $\mathbf{r} = \begin{pmatrix} 2 \\ 1 \\ 3 \end{pmatrix} + \lambda \begin{pmatrix} 1 \\ 4 \\ 0 \end{pmatrix}$ and $\mathbf{r} = \begin{pmatrix} 6 \\ 10 \\ 4 \end{pmatrix} + \mu \begin{pmatrix} 2 \\ 1 \\ 1 \end{pmatrix}$

(iv) $\mathbf{r} = \lambda \begin{pmatrix} 4 \\ 1 \\ 4 \end{pmatrix}$ and $\mathbf{r} = \begin{pmatrix} 7 \\ 0 \\ -3 \end{pmatrix} + \mu \begin{pmatrix} 1 \\ 2 \\ -1 \end{pmatrix}$

(v) $\dfrac{x - 4}{3} = \dfrac{y - 2}{7} = \dfrac{z + 1}{-4}$ and $\dfrac{x - 5}{2} = \dfrac{y - 1}{8} = \dfrac{z}{-5}$

⑨ To support a tree damaged in a gale a tree surgeon attaches wire ropes to four of the branches, as shown in Figure 1.12.

He joins $(2, 0, 3)$ to $(-1, 2, 6)$ and $(0, 3, 5)$ to $(-2, -2, 4)$.

Do the ropes, assumed to be straight, meet?

Figure 1.12

⑩ Show that the lines

$$L_1: \frac{x+7}{4} = \frac{y-24}{-7} = \frac{z+4}{4}$$

$$L_2: \frac{x-3}{2} = \frac{y+10}{2} = \frac{z-15}{-1}$$

$$L_3: \frac{x+3}{8} = \frac{y-6}{-3} = \frac{z-6}{2}$$

form a triangle and find the length of its sides.

⑪ Figure 1.13 shows a music stand, consisting of a rectangle DEFG with a vertical support OA.

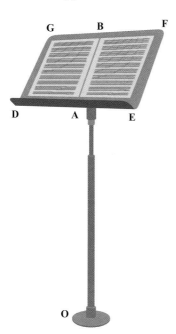

Relative to axes through the origin O, which is on the floor, the coordinates of various points are given, with dimensions in metres, as A(0, 0, 1), D(−0.25, 0, 1) and F(0.25, 0.15, 1.3).

DE and GF are horizontal. A is the midpoint of DE and B is the midpoint of GF.

C is on AB so that AC = $\frac{1}{3}$ AB.

Figure 1.13

(i) Write down the vector $\overrightarrow{AD}$ and show that $\overrightarrow{EF}$ is $\begin{pmatrix} 0 \\ 0.15 \\ 0.3 \end{pmatrix}$.

(ii) Calculate the coordinates of C.

(iii) Find the equations of the lines DE and EF in vector form.

⑫ The point A(2, 3, −1) lies on the plane $7x + 8y + 5z = d$.

(i) Find the value of d.

(ii) Show that the point B(3, −1, 4) lies on the plane.

(iii) Write down the vector equation of the line L that passes through the point A and is perpendicular to the plane.

(iv) The point C(−5, p, q) lies on the line L. Find the values of p and q and the distance AC.

(v) Find the exact area of the triangle ABC.

⑬ Figure 1.14 illustrates the flight path of a helicopter H taking off from an airport.

The origin O is situated at the base of the airport control tower, the x-axis is due east, the y-axis due north and the z-axis vertical.

The units of distance are kilometres.

The helicopter takes off from the point G.

The position vector $\mathbf{r}$ of the helicopter t minutes after take-off is given by:
$$\mathbf{r} = (1 + t)\mathbf{i} + (0.5 + 2t)\mathbf{j} + 2t\mathbf{k}$$

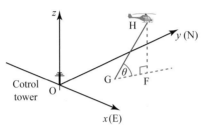

Figure 1.14

(i) Write down the coordinates of G.

(ii) Find the angle the flight path makes with the horizontal (this is shown as angle θ in the diagram).

(iii) Find the bearing of the flight path (i.e. the bearing of the line GF).

The helicopter enters a cloud at a height of 2 km.

(iv) Find the coordinates of the point where the helicopter enters the cloud.

A mountain top is situated at M(5, 4.5, 3).

(v) Find the value of t when HM is perpendicular to the flight path GH.

Find the distance from the helicopter to the mountain top at this time.

2 Lines and planes

The point of intersection of a line and a plane

The next example shows how you can find the point at which a line intersects a plane.

Example 1.12

Find the point of intersection of the line $\mathbf{r} = \begin{pmatrix} 2 \\ 3 \\ 4 \end{pmatrix} + \lambda \begin{pmatrix} 1 \\ 2 \\ -1 \end{pmatrix}$ and the plane $5x + y - z = 1$.

Solution

The line is $\mathbf{r} = \begin{pmatrix} x \\ y \\ z \end{pmatrix} = \begin{pmatrix} 2 \\ 3 \\ 4 \end{pmatrix} + \lambda \begin{pmatrix} 1 \\ 2 \\ -1 \end{pmatrix}$ and so for any point on the line

$$x = 2 + \lambda \quad y = 3 + 2\lambda \quad z = 4 - \lambda$$

Substituting these into the equation of the plane $5x + y - z = 1$ gives

$$5(2 + \lambda) + (3 + 2\lambda) - (4 - \lambda) = 1$$
$$8\lambda = -8$$
$$\lambda = -1$$

Substituting $\lambda = -1$ into the equation of the line gives

$$\mathbf{r} = \begin{pmatrix} x \\ y \\ z \end{pmatrix} = \begin{pmatrix} 2 \\ 3 \\ 4 \end{pmatrix} - \begin{pmatrix} 1 \\ 2 \\ -1 \end{pmatrix} = \begin{pmatrix} 1 \\ 1 \\ 5 \end{pmatrix}$$

So the point of intersection is $(1, 1, 5)$.

The angle between a line and a plane

If they are not perpendicular, the acute angle between a line and a plane is the acute angle θ between the line and its **orthogonal projection** onto the plane, shown by the dotted line AB in Figure 1.15.

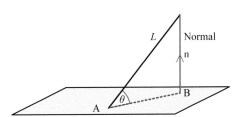

Figure 1.15

You can find the angle θ by first finding the angle α between the direction vector $\mathbf{d}$ of the straight line L and a normal vector $\mathbf{n}$ to the plane, as shown in Figure 1.16.

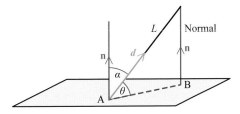

Figure 1.16

The angle θ can then be found by calculating $90 - \alpha$.

This method is illustrated in the following example.

Example 1.13

Find the angle between the line $\mathbf{r} = \begin{pmatrix} 2 \\ 0 \\ -3 \end{pmatrix} + \lambda \begin{pmatrix} 1 \\ 3 \\ 2 \end{pmatrix}$ and the plane

$3x - y + z = 4$.

Solution

For this line, the direction vector $\mathbf{d} = \begin{pmatrix} 1 \\ 3 \\ 2 \end{pmatrix}$ and a normal to the plane is

$\mathbf{n} = \begin{pmatrix} 3 \\ -1 \\ 1 \end{pmatrix}$.

The angle α between the normal vector and the direction vector satisfies

$\mathbf{d}.\mathbf{n} = |\mathbf{d}||\mathbf{n}| \cos\alpha$.

$$\Rightarrow \begin{pmatrix} 1 \\ 3 \\ 2 \end{pmatrix} . \begin{pmatrix} 3 \\ -1 \\ 1 \end{pmatrix} = \sqrt{14}\sqrt{11}\cos\theta$$

$$\Rightarrow \cos\alpha = \frac{2}{\sqrt{14}\sqrt{11}}$$

$$\Rightarrow \alpha = 80.7°$$

So the angle between the line and the plane $\theta = 90° - 80.7° = 9.3°$.

Exercise 1.2

① Show that the point of intersection of the line $\mathbf{r} = \begin{pmatrix} 1 \\ 3 \\ 0 \end{pmatrix} + \lambda \begin{pmatrix} 2 \\ -1 \\ 4 \end{pmatrix}$ and

the plane $2x + y + z = 26$ is $(7, 0, 12)$.

② For each of the following, find the point of intersection of the line and the plane. Find also the angle between the line and the plane.

(i) $x + 2y + 3z = 11$ $\mathbf{r} = \begin{pmatrix} 1 \\ 2 \\ 4 \end{pmatrix} + \lambda \begin{pmatrix} 1 \\ 1 \\ 1 \end{pmatrix}$

(ii) $2x + 3y - 4z = 1$ $\dfrac{x+2}{3} = \dfrac{y+3}{4} = \dfrac{z+4}{5}$

(iii) $3x - 2y - z = 14$ $\mathbf{r} = \begin{pmatrix} 8 \\ 4 \\ 2 \end{pmatrix} + \lambda \begin{pmatrix} 1 \\ 2 \\ 1 \end{pmatrix}$

(iv) $x + y + z = 0$ $\mathbf{r} = \lambda \begin{pmatrix} 1 \\ 1 \\ 2 \end{pmatrix}$

③ (i) Find the equation of the line L passing through A(4, 1, 3) and B(6, 4, 8).

(ii) Find the point of intersection of L with the plane $x + 2y - z + 3 = 0$.

(iii) Find the angle between the line L and the plane.

④ (i) Find the equation of the line through $(13, 5, 0)$ parallel to the line

$$\mathbf{r} = \begin{pmatrix} 2 \\ -1 \\ 4 \end{pmatrix} + \lambda \begin{pmatrix} 3 \\ 1 \\ -2 \end{pmatrix}$$

(ii) Where does this line meet the plane $3x + y - 2z = 2$?

(iii) How far is the point of intersection from $(13, 5, 0)$?

⑤ A plane passes through the points A(2, 3, −1), B(4, 0, 1) and C(−3, 5, −2).

(i) Find the vectors $\overrightarrow{AB}$ and $\overrightarrow{AC}$.

Hence confirm that the vector $\begin{pmatrix} 1 \\ 8 \\ 11 \end{pmatrix}$ is a normal to the plane and find the equation of the plane.

(ii) Find the points of intersection, P and Q, of the lines

$$L_1: \begin{pmatrix} 4 \\ -1 \\ 3 \end{pmatrix} + \lambda \begin{pmatrix} 2 \\ 0 \\ 1 \end{pmatrix}$$

$$L_2: \begin{pmatrix} 7 \\ -2 \\ 1 \end{pmatrix} + \mu \begin{pmatrix} -1 \\ 1 \\ 3 \end{pmatrix}$$

with the plane.

(iii) Determine the point of intersection R of the lines L_1 and L_2.

(iv) Find the angle between the vectors $\overrightarrow{PR}$ and $\overrightarrow{QR}$.

(v) Find the area of the triangle PQR.

⑥ A laser beam ABC is fired from the point A(1, 2, 4) and is reflected at B off the plane with equation $x + 2y - 3z = 0$, as shown in Figure 1.17.

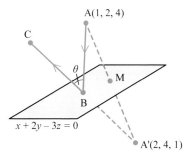

Figure 1.17

A′ is the point (2, 4, 1) and M is the midpoint of AA′.

(i) Show that AA′ is perpendicular to the plane $x + 2y - 3z = 0$ and that M lies in the plane.

The vector equation of the line AB is $\mathbf{r} = \begin{pmatrix} 1 \\ 2 \\ 4 \end{pmatrix} + \lambda \begin{pmatrix} 1 \\ -1 \\ 2 \end{pmatrix}$.

(ii) Find the coordinates of B and a vector equation of the line A′B.

(iii) Given that A′BC is a straight line, find the angle θ.

⑦ Figure 1.18 shows the tetrahedron ABCD. The coordinates of the vertices are A(−3, 0, 0), B(2, 0, −2), C(0, 4, 0) and D(0, 4, 5).

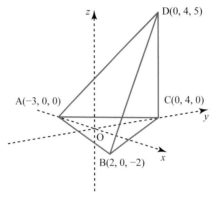

Figure 1.18

(i) Find the lengths of the edges AB and AC, and the size of the angle CAB. Hence calculate the area of the triangle ABC.

(ii) (a) Verify that $\begin{pmatrix} 4 \\ -3 \\ 10 \end{pmatrix}$ is normal to the plane ABC.

(b) Find the equation of this plane.

(iii) Write down the vector equation of the line through D that is perpendicular to the plane ABC. Find the point of intersection of this line with the plane ABC.

The area of a tetrahedron is given by $\frac{1}{3} \times$ base area $\times$ height.

(iv) Find the volume of the tetrahedron ABCD.

LEARNING OUTCOMES

When you have completed this chapter you should be able to:

➤ form the equation of a line in three dimensions in vector or cartesian form

➤ find the angle between two lines

➤ know the different ways in which two lines can intersect or not in three-dimensional space

➤ find out whether two lines in three dimensions are parallel, skew or intersect, and find the point of intersection if there is one

➤ find the point of intersection of a line and a plane

➤ find the angle between a line and a plane.

KEY POINTS

1   The vector equation of a line is given by:

$$\mathbf{r} = \mathbf{a} + \lambda \mathbf{d}$$

where $\mathbf{a}$ is the position vector of a point A on the line and $\mathbf{d}$ is the direction vector of the line. Sometimes a different letter, such as μ or t is used as the parameter instead of λ.

2   The line with direction vector $\mathbf{d} = \begin{pmatrix} d_1 \\ d_2 \\ d_3 \end{pmatrix}$ passing through the point A with

position vector $\mathbf{a} = \begin{pmatrix} a_1 \\ a_2 \\ a_3 \end{pmatrix}$ has the cartesian equation:

$$\lambda = \frac{x - a_1}{d_1} = \frac{y - a_2}{d_2} = \frac{z - a_3}{d_3}$$

3   If two straight lines have equations

$$\mathbf{r}_1 = \mathbf{a}_1 + \lambda \mathbf{d}_1$$
$$\mathbf{r}_2 = \mathbf{a}_2 + \mu \mathbf{d}_2$$

the angle between the lines is found by calculating the scalar product $\mathbf{d}_1 . \mathbf{d}_2$.

4   In three dimensions there are three possibilities for the arrangement of the lines. They are either parallel, intersecting or skew.

5   If they are not perpendicular, the acute angle between a line and a plane is the acute angle θ between the line and its **orthogonal projection** onto the plane.

FUTURE USES

■   You will learn more about working with lines and planes in Vectors 2.

Review: Matrices and transformations

1 Matrices

What is a matrix?

A matrix is an array of numbers (the plural is matrices), usually written inside curved brackets, for example:

$$\mathbf{M} = \begin{pmatrix} 2 & 3 \\ 0 & -1 \\ 4 & 5 \end{pmatrix}$$

It is usual to represent matrices by capital letters, often in bold print.

A matrix consists of rows and columns, and the entries in the various cells are known as **elements**. **M** has 6 elements which are arranged in 3 rows and 2 columns, so **M** is called a 3×2 matrix. The matrix

$$\mathbf{N} = \begin{pmatrix} 3 & 1 & -4 \\ -2 & 5 & 0 \end{pmatrix}$$

also has 6 elements and is described as a 2×3 matrix as it has 2 rows and 3 columns. The number of rows and columns is called the **order** of the matrix – a matrix with r rows and c columns has order $r \times c$.

Some matrices are described by special names which relate to the number of rows and columns or the nature of the elements.

- Matrices such as $\begin{pmatrix} 4 & 2 \\ 1 & 0 \end{pmatrix}$ and $\begin{pmatrix} 3 & 5 & 1 \\ 2 & 0 & -4 \\ 1 & 7 & 3 \end{pmatrix}$ which have the same

  number of rows as columns are called **square matrices**.

- The matrix $\begin{pmatrix} 1 & 0 \\ 0 & 1 \end{pmatrix}$ is called the 2×2 **identity matrix** or **unit matrix**,

  and similarly $\begin{pmatrix} 1 & 0 & 0 \\ 0 & 1 & 0 \\ 0 & 0 & 1 \end{pmatrix}$ is called the 3×3 identity matrix. Identity

  matrices must be square and are usually denoted by **I**.

- The matrix $\mathbf{O} = \begin{pmatrix} 0 & 0 \\ 0 & 0 \end{pmatrix}$ is called the 2×2 **zero matrix**. Zero matrices

  can be of any order.

Two matrices are said to be **equal** if and only if they have the same order and each element in one matrix is equal to the corresponding element in the other matrix. So, for example, the matrices **A** and **C** below are equal, but **B** and **D** are not equal to any of the other matrices.

$$\mathbf{A} = \begin{pmatrix} 2 & 3 \\ -1 & 0 \\ 4 & 4 \end{pmatrix} \quad \mathbf{B} = \begin{pmatrix} 2 & 3 \\ -1 & 0 \end{pmatrix} \quad \mathbf{C} = \begin{pmatrix} 2 & 3 \\ -1 & 0 \\ 4 & 4 \end{pmatrix} \quad \mathbf{D} = \begin{pmatrix} 2 & 3 & 4 \\ -1 & 0 & 4 \end{pmatrix}$$

Working with matrices

Matrices can be added or subtracted if they are of the same order.

Add the elements in corresponding positions.

$$\begin{pmatrix} 1 & 2 & -2 \\ 11 & 2 & -5 \end{pmatrix} + \begin{pmatrix} -5 & 6 & 4 \\ -5 & 0 & -5 \end{pmatrix} = \begin{pmatrix} -4 & 8 & 2 \\ 6 & 2 & -10 \end{pmatrix}$$

$$\begin{pmatrix} -7 & -2 \\ 3 & 12 \end{pmatrix} - \begin{pmatrix} 4 & -7 \\ -1 & 2 \end{pmatrix} = \begin{pmatrix} -11 & 5 \\ 4 & 10 \end{pmatrix}$$

Subtract the elements in corresponding positions.

But $\begin{pmatrix} 1 & -2 & 7 \\ 0 & 4 & 5 \end{pmatrix} + \begin{pmatrix} -5 & -3 \\ -2 & 1 \end{pmatrix}$ cannot be evaluated because the matrices are

not of the same order. These matrices are **non-conformable for addition**.

You can also multiply a matrix by a number, or **scalar**:

Multiply each of the elements by 3.

$$3 \begin{pmatrix} 4 & -2 \\ -5 & 1 \end{pmatrix} = \begin{pmatrix} 12 & -6 \\ -15 & 3 \end{pmatrix}$$

Matrix addition is **associative** and **commutative** as it is always true that:

Commutative as the addition can take place in any order to obtain the same answer.

$$\mathbf{A} + (\mathbf{B} + \mathbf{C}) = (\mathbf{A} + \mathbf{B}) + \mathbf{C}$$
$$\mathbf{A} + \mathbf{B} = \mathbf{B} + \mathbf{A}$$

Associative as the matrices can be grouped in different ways to obtain the same answer

Matrix subtraction is neither associative nor commutative as it is generally not true that:

$$\mathbf{A} - (\mathbf{B} - \mathbf{C}) = (\mathbf{A} - \mathbf{B}) - \mathbf{C}$$
$$\mathbf{A} - \mathbf{B} = \mathbf{B} - \mathbf{A}$$

Matrix multiplication

Two matrices can be multiplied together if their orders satisfy the following rule:

The two 'middle' numbers, in this case 4, must be the same for it to be possible to multiply two matrices. If two matrices can be multiplied, they are **conformable for multiplication**.

$$2 \times 4 \quad \times \quad 4 \times 1$$

The two 'outside' numbers give you the order of the product matrix, in this case 2 × 1.

Figure R.1

So, for example, the product $\begin{pmatrix} 1 & 0 & -1 \\ 3 & 2 & 5 \end{pmatrix}\begin{pmatrix} 4 \\ -1 \\ 5 \end{pmatrix}$ can be calculated because

the orders of the matrices are 2×3 and 3×1; the resulting product would be a

2×1 matrix. However the product $\begin{pmatrix} 4 \\ -1 \\ 5 \end{pmatrix}\begin{pmatrix} 1 & 0 & -1 \\ 3 & 2 & 5 \end{pmatrix}$ is not possible as

the orders 3×1 and 2×3 do not have the same 'middle' numbers. This also illustrates that, generally, matrix multiplication is not commutative.

The example which follows shows how you multiply two matrices which are conformable.

Example R.1

Find $\begin{pmatrix} 6 & -2 \\ 4 & 3 \end{pmatrix}\begin{pmatrix} 8 \\ -1 \end{pmatrix}$.

Solution

The matrices have orders 2×2 and 2×1, so the matrices are conformable and the product will have order 2×1.

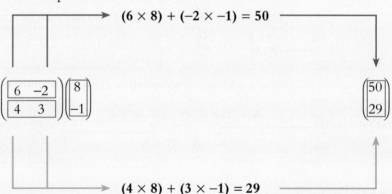

TECHNOLOGY

In addition to using this method, you should check that you are able to multiply conformable matrices using your calculator.

$(6 \times 8) + (-2 \times -1) = 50$

$(4 \times 8) + (3 \times -1) = 29$

Figure R.2

Note that when a square matrix **A** is multiplied by an identity matrix **I** of the same size:

AI = IA = A

Matrix multiplication is **associative**. This means that for three conformable matrices **A**, **B** and **C**:

This result is important for later work on matrices.

$(\mathbf{AB})\mathbf{C} = \mathbf{A}(\mathbf{BC})$.

① (i) Write down the orders of the matrices:

$$\mathbf{A} = \begin{pmatrix} 2 & 3 \\ 0 & 7 \end{pmatrix} \qquad \mathbf{B} = \begin{pmatrix} 5 & -2 & 4 \end{pmatrix}$$

$$\mathbf{C} = \begin{pmatrix} 1 \\ -4 \end{pmatrix} \qquad \mathbf{D} = \begin{pmatrix} 2 & -5 & 1 \\ 0 & 2 & 1 \end{pmatrix}$$

$$\mathbf{E} = \begin{pmatrix} 3 & 0 \\ -1 & -7 \\ 3 & 2 \end{pmatrix} \qquad \mathbf{F} = \begin{pmatrix} 5 & -2 & 1 \\ 3 & -2 & 1 \\ 0 & 5 & 0 \end{pmatrix}$$

$$\mathbf{G} = (-4) \qquad \mathbf{H} = \begin{pmatrix} 1 & -2 & 3 & -4 & 5 \end{pmatrix}$$

(ii) Without a calculator find, where possible, the matrix products:

   (a) **AC**   (b) **EC**   (c) **BF**   (d) **FB**

   (e) **GH**   (f) **CD**   (g) **DF**

(iii) Given also the matrices:

$$\mathbf{I} = \begin{pmatrix} -6 & 0 \\ 1 & 3 \end{pmatrix} \qquad \mathbf{J} = \begin{pmatrix} 2 & 5 & -3 \\ 0 & 1 & 1 \\ 2 & 0 & -1 \end{pmatrix} \qquad \mathbf{K} = \begin{pmatrix} 2 \\ -3 \end{pmatrix}$$

find, where possible:

   (a) $\mathbf{A} + \mathbf{I}$ (b) $\mathbf{A} - \mathbf{I}$ (c) $2\mathbf{C} + 3\mathbf{K}$ (d) $\mathbf{I} + \mathbf{D}$

   (e) $\mathbf{F} - 2\mathbf{J}$ (f) $\mathbf{K} - 3\mathbf{G}$ (g) $\mathbf{J} + \dfrac{1}{2}\mathbf{E}$

② Given the matrices $\mathbf{M} = \begin{pmatrix} 2 & 1 \\ -3 & 0 \end{pmatrix}$ and $\mathbf{N} = \begin{pmatrix} 1 & -1 \\ 4 & 2 \end{pmatrix}$

(i) find **MN** and **NM**

(ii) What property of matrices does your result from (i) illustrate?

③ Given the matrices $\mathbf{P} = \begin{pmatrix} 2 & 1 \\ 0 & -1 \end{pmatrix}$, $\mathbf{Q} = \begin{pmatrix} 0 & 1 & -2 \\ 3 & -2 & 1 \end{pmatrix}$ and

$$\mathbf{R} = \begin{pmatrix} 1 & 3 & 0 \\ -2 & 1 & 1 \\ 0 & 2 & -1 \end{pmatrix}$$

(i) find **PQ** and (**PQ**)**R**

(ii) find **QR** and **P**(**QR**)

(iii) What property of matrices do your answers to (i) and (ii) illustrate?

④ Find the possible values of a and b such that:

$$\begin{pmatrix} 5a^2 & 4 \\ 0 & 18 \end{pmatrix} - \begin{pmatrix} 7(2a+1) & 7 \\ 5 & b^2 \end{pmatrix} = \begin{pmatrix} -4 & -3 \\ -5 & 2 \end{pmatrix}$$

⑤ For the matrix $\mathbf{A} = \begin{pmatrix} 5 & 1 \\ 0 & 1 \end{pmatrix}$ find, without using a calculator:

(i) $\mathbf{A}^2$

(ii) $\mathbf{A}^3$

(iii) $\mathbf{A}^4$

(iv) Suggest a general form for the matrix $\mathbf{A}^n$ in terms of n.

(v) Find $\mathbf{A}^6$ on your calculator and confirm that it gives the same answer as using (iv).

(vi) Use proof by induction to prove the result you found in (iv).

⑥ For the matrices $\mathbf{A} = \begin{pmatrix} 5 & x & 0 \\ -1 & -1 & 3 \end{pmatrix}$ and $\mathbf{B} = \begin{pmatrix} 2 & -3 \\ 3 & x \\ x & 5 \end{pmatrix}$:

> Proof by induction was covered in A Level Further Mathematics Year 1 and will be revised in Chapter 3.

(i) Find the product $\mathbf{AB}$ in terms of x.

> A **symmetric** matrix is one in which the entries are symmetrical about the leading diagonal,
>
> for example $\begin{pmatrix} 2 & 5 \\ 5 & 0 \end{pmatrix}$ and $\begin{pmatrix} 3 & 4 & -6 \\ 4 & 2 & 5 \\ -6 & 5 & 1 \end{pmatrix}$.

(ii) Given that the matrix $\mathbf{AB}$ is symmetric, find the possible values of x.

(iii) Write down the possible matrices $\mathbf{AB}$.

⑦ If $\mathbf{A} = \begin{pmatrix} a & 0 & -b \\ 2 & a & 3 \end{pmatrix}$, $\mathbf{B} = \begin{pmatrix} a & 6 \\ 1 & 2a \\ b & 4 \end{pmatrix}$ and $\mathbf{AB} = \begin{pmatrix} 45 & -50 \\ -15 & 122 \end{pmatrix}$, find the possible values of a and b.

2 Using matrices to represent transformations

A transformation maps an object according to a rule and can be represented by a matrix.

Figure R.3 and Figure R.4 show two possible transformations of a triangle OAB (shown in red). The vertices of the image (shown in blue) are denoted by the same letters with a dash e.g. A′, B′.

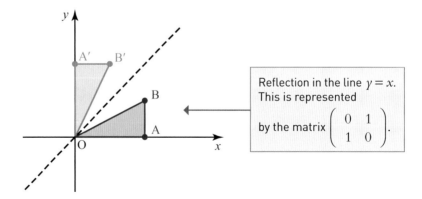

Reflection in the line $y = x$. This is represented by the matrix $\begin{pmatrix} 0 & 1 \\ 1 & 0 \end{pmatrix}$.

Figure R.3

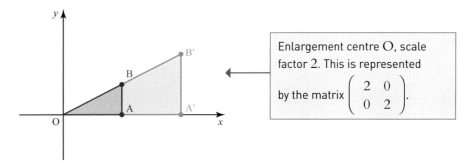

Enlargement centre O, scale factor 2. This is represented by the matrix $\begin{pmatrix} 2 & 0 \\ 0 & 2 \end{pmatrix}$.

Figure R.4

The transformation represented by a matrix can be found by looking at the effect of multiplying the unit vectors $\mathbf{i} = \begin{pmatrix} 1 \\ 0 \end{pmatrix}$ and $\mathbf{j} = \begin{pmatrix} 0 \\ 1 \end{pmatrix}$ by the matrix. Figure R.3 shows reflection in the line $y = x$ which is represented by the matrix $\begin{pmatrix} 0 & 1 \\ 1 & 0 \end{pmatrix}$. The images of the unit vectors $\mathbf{i}$ and $\mathbf{j}$ are:

$$\begin{pmatrix} 0 & 1 \\ 1 & 0 \end{pmatrix}\begin{pmatrix} 1 \\ 0 \end{pmatrix} = \begin{pmatrix} 0 \\ 1 \end{pmatrix}$$

The image of the unit vector $\mathbf{i}$ is $\mathbf{j}$.

$$\begin{pmatrix} 0 & 1 \\ 1 & 0 \end{pmatrix}\begin{pmatrix} 0 \\ 1 \end{pmatrix} = \begin{pmatrix} 1 \\ 0 \end{pmatrix}$$

The image of the unit vector $\mathbf{j}$ is $\mathbf{i}$.

The interchange of the unit vectors $\mathbf{i}$ and $\mathbf{j}$ can also be determined by thinking about the transformation geometrically. Notice that $\begin{pmatrix} 0 \\ 1 \end{pmatrix}$ and $\begin{pmatrix} 1 \\ 0 \end{pmatrix}$ form the columns of the transformation matrix.

Figure R.4 shows an enlargement centre O scale factor 2 which is represented by the matrix $\begin{pmatrix} 2 & 0 \\ 0 & 2 \end{pmatrix}$. The images of the unit vectors $\mathbf{i}$ and $\mathbf{j}$ are:

$$\begin{pmatrix} 2 & 0 \\ 0 & 2 \end{pmatrix}\begin{pmatrix} 1 \\ 0 \end{pmatrix} = \begin{pmatrix} 2 \\ 0 \end{pmatrix}$$

The image of the unit vector $\mathbf{i}$ is $\begin{pmatrix} 2 \\ 0 \end{pmatrix}$.

The image of the unit vector $\mathbf{j}$ is $\begin{pmatrix} 0 \\ 2 \end{pmatrix}$.

$$\begin{pmatrix} 2 & 0 \\ 0 & 2 \end{pmatrix}\begin{pmatrix} 0 \\ 1 \end{pmatrix} = \begin{pmatrix} 0 \\ 2 \end{pmatrix}$$

Again $\begin{pmatrix} 2 \\ 0 \end{pmatrix}$ and $\begin{pmatrix} 0 \\ 2 \end{pmatrix}$ form the columns of the transformation matrix.

This connection between the images of the unit vectors $\mathbf{i}$ and $\mathbf{j}$ and the matrix representing the transformation provides a quick method for finding the matrix representing a transformation.

You may find it easier to see what the transformation is when you use a shape, like the unit square, rather than points or lines.

It is common to use the unit square with coordinates $O(0, 0)$, $I(1, 0)$, $P(1, 1)$ and $J(0, 1)$. You can think about the images of the points I and J, and from this you can write down the images of the unit vectors $\mathbf{i}$ and $\mathbf{j}$.
This is done in the next example.

Example R.2

By drawing a diagram to show the image of the unit square, find the matrices which represent each of the following transformations:

(i) a rotation of 90° clockwise about the origin

(ii) a stretch scale factor 3 parallel to the y-axis.

Solution

(i) Figure R.5 shows the effect of a rotation of 90° clockwise about the origin on the unit square.

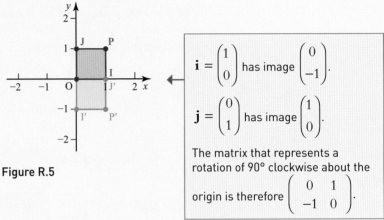

Figure R.5

$\mathbf{i} = \begin{pmatrix} 1 \\ 0 \end{pmatrix}$ has image $\begin{pmatrix} 0 \\ -1 \end{pmatrix}$.

$\mathbf{j} = \begin{pmatrix} 0 \\ 1 \end{pmatrix}$ has image $\begin{pmatrix} 1 \\ 0 \end{pmatrix}$.

The matrix that represents a rotation of 90° clockwise about the origin is therefore $\begin{pmatrix} 0 & 1 \\ -1 & 0 \end{pmatrix}$.

(ii) Figure R.6 shows the effect of a stretch scale factor 3 parallel to the y-axis.

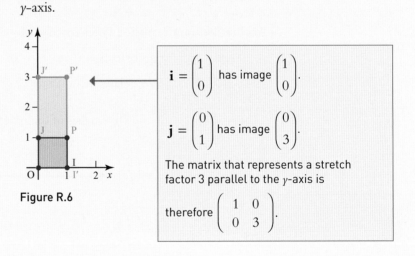

Figure R.6

$\mathbf{i} = \begin{pmatrix} 1 \\ 0 \end{pmatrix}$ has image $\begin{pmatrix} 1 \\ 0 \end{pmatrix}$.

$\mathbf{j} = \begin{pmatrix} 0 \\ 1 \end{pmatrix}$ has image $\begin{pmatrix} 0 \\ 3 \end{pmatrix}$.

The matrix that represents a stretch factor 3 parallel to the y-axis is therefore $\begin{pmatrix} 1 & 0 \\ 0 & 3 \end{pmatrix}$.

In summary:

- The matrix $\begin{pmatrix} k & 0 \\ 0 & k \end{pmatrix}$ represents an enlargement of scale factor k, centre the origin.

- The matrix $\begin{pmatrix} m & 0 \\ 0 & 1 \end{pmatrix}$ represents a stretch of scale factor m parallel to the x-axis.

- The matrix $\begin{pmatrix} 1 & 0 \\ 0 & n \end{pmatrix}$ represents a stretch of scale factor n parallel to the y-axis.

- The matrix $\begin{pmatrix} 0 & 1 \\ -1 & 0 \end{pmatrix}$ represents a rotation of 90° clockwise about the origin. This is a special case of the matrix $\begin{pmatrix} \cos\theta & -\sin\theta \\ \sin\theta & \cos\theta \end{pmatrix}$ which represents a rotation through angle θ degrees about the origin.

- The matrices $\begin{pmatrix} 1 & 0 \\ 0 & -1 \end{pmatrix}$ and $\begin{pmatrix} -1 & 0 \\ 0 & 1 \end{pmatrix}$ represent reflections in the x-axis and y-axis respectively.

- The matrix $\begin{pmatrix} 0 & 1 \\ 1 & 0 \end{pmatrix}$ represents a reflection in the line $y = x$.

Figure R.7 shows the unit square and its image under the transformation represented by the matrix $\begin{pmatrix} 1 & 5 \\ 0 & 1 \end{pmatrix}$ on the unit square. The matrix $\begin{pmatrix} 1 & 5 \\ 0 & 1 \end{pmatrix}$ transforms the unit vector $\mathbf{i} = \begin{pmatrix} 1 \\ 0 \end{pmatrix}$ to the vector $\begin{pmatrix} 1 \\ 0 \end{pmatrix}$ and transforms the unit vector $\mathbf{j} = \begin{pmatrix} 0 \\ 1 \end{pmatrix}$ to the vector $\begin{pmatrix} 5 \\ 1 \end{pmatrix}$.

The point with position vector $\begin{pmatrix} 1 \\ 1 \end{pmatrix}$ is transformed to the point with position vector $\begin{pmatrix} 6 \\ 1 \end{pmatrix}$. $\longleftarrow$ As $\begin{pmatrix} 1 & 5 \\ 0 & 1 \end{pmatrix}\begin{pmatrix} 1 \\ 1 \end{pmatrix} = \begin{pmatrix} 6 \\ 1 \end{pmatrix}$.

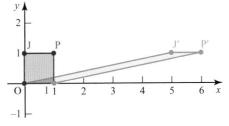

Figure R.7

This transformation is called a **shear**. The points on the x-axis stay the same and the points J and P move parallel to the x-axis to the right.

This shear can be described fully by saying that the x-axis is fixed, and giving the image of one point that is not on the x-axis, e.g. (0, 1) is mapped to (5, 1).

Generally, a shear with the x-axis fixed has the form $\begin{pmatrix} 1 & k \\ 0 & 1 \end{pmatrix}$

and a shear with the y-axis fixed has the form $\begin{pmatrix} 1 & 0 \\ k & 1 \end{pmatrix}$.

For each point, calculating the quantity

$$\frac{\text{distance between the point and its image}}{\text{distance of original point from } x\text{-axis}}$$

produces the same numerical value, which is the same as the number in the top right of the matrix. This is called the **shear factor** for the shear.

 There are different conventions about the sign of a shear factor and, for this reason, shear factors are not used to define a shear in this book. It is possible to show the effect of matrix transformations using some geometrical computer software packages. You might find that some packages use different approaches towards shears and define them in different ways.

Transformations in three dimensions

A plane is an infinite two-dimensional flat surface with no thickness. Figure R.8 below illustrates some common planes in three dimensions – the XY plane, the XZ plane and the YZ plane. The plane XY can also be referred to as $z = 0$, since the z coordinate would be zero for all points on the XY plane. Similarly, the XZ plane is referred to as $y = 0$ and the YZ plane as $x = 0$.

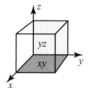

Figure R.8

As in two dimensions, the matrix can be found algebraically or by considering the effect of the transformation on the three unit vectors

$$\mathbf{i} = \begin{pmatrix} 1 \\ 0 \\ 0 \end{pmatrix}, \mathbf{j} = \begin{pmatrix} 0 \\ 1 \\ 0 \end{pmatrix} \text{ and } \mathbf{k} = \begin{pmatrix} 0 \\ 0 \\ 1 \end{pmatrix}.$$

So, for example, Figure R.9 shows the effect of a reflection in the plane $y = 0$.

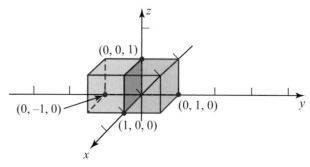

Figure R.9

$$\mathbf{i} = \begin{pmatrix} 1 \\ 0 \\ 0 \end{pmatrix} \text{ maps to } \begin{pmatrix} 1 \\ 0 \\ 0 \end{pmatrix}, \ \mathbf{j} = \begin{pmatrix} 0 \\ 1 \\ 0 \end{pmatrix} \text{ maps to } \begin{pmatrix} 0 \\ -1 \\ 0 \end{pmatrix} \text{ and } \mathbf{k} = \begin{pmatrix} 0 \\ 0 \\ 1 \end{pmatrix} \text{ maps to } \begin{pmatrix} 0 \\ 0 \\ 1 \end{pmatrix}.$$

The images of $\mathbf{i}$, $\mathbf{j}$ and $\mathbf{k}$ form the columns of the 3×3 transformation matrix.

$$\begin{pmatrix} 1 & 0 & 0 \\ 0 & -1 & 0 \\ 0 & 0 & 1 \end{pmatrix}$$

When rotating an object about an axis, the rotation is taken to be anticlockwise about the axis of rotation when looking along the axis from the positive end towards the origin. Figure R.10 shows a rotation of 90° anticlockwise about the x-axis.

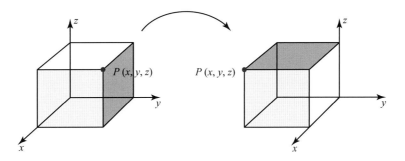

Figure R.10

Looking at the effect of the transformation on the unit vectors $\mathbf{i}$, $\mathbf{j}$ and $\mathbf{k}$ shows

that $\mathbf{i} = \begin{pmatrix} 1 \\ 0 \\ 0 \end{pmatrix}$ maps to $\begin{pmatrix} 1 \\ 0 \\ 0 \end{pmatrix}$, $\mathbf{j} = \begin{pmatrix} 0 \\ 1 \\ 0 \end{pmatrix}$ maps to $\begin{pmatrix} 0 \\ 0 \\ 1 \end{pmatrix}$ and $\mathbf{k} = \begin{pmatrix} 0 \\ 0 \\ 1 \end{pmatrix}$ maps to $\begin{pmatrix} 0 \\ -1 \\ 0 \end{pmatrix}$.

The images of $\mathbf{i}$, $\mathbf{j}$ and $\mathbf{k}$ form the columns of the 3×3 transformation

matrix $\begin{pmatrix} 1 & 0 & 0 \\ 0 & 0 & -1 \\ 0 & 1 & 0 \end{pmatrix}$.

Successive transformations

Figure R.11 shows the effect of two successive transformations on a triangle. The transformation A represents a reflection in the x-axis. A maps the point P to the point A(P).

The transformation B represents a rotation of 90° anticlockwise about O. When you apply B to the image formed by A, the point A(P) is mapped to the point B(A(P)). This is abbreviated to BA(P).

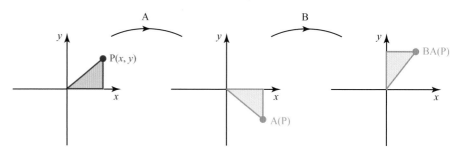

Figure R.11

Notice that a transformation written as BA means 'carry out A, then carry out B'.

This process is sometimes called **composition of transformations**.

In general, the matrix for a composite transformation is found by multiplying the matrices of the individual transformations in reverse order. So, for two transformations the matrix representing the first transformation is on the right and the matrix for the second transformation is on the left. For n transformations $T_1, T_2, \ldots, T_{n-1}, T_n$ the matrix product would be $\mathbf{A}_n \mathbf{A}_{n-1} \ldots \mathbf{A}_2 \mathbf{A}_1$.

Exercise R.2

① A triangle has vertices at the origin, A(3,0) and B(3,1).

For each of the transformations below, draw a diagram to show the triangle OAB and its image OA′B′ and find the matrix which represents the transformation.

(i) Reflection in the line $y = x$

(ii) Rotation 180° anticlockwise about O

(iii) Enlargement scale factor 4, centre O

(iv) Shear with the x-axis fixed, which maps the point (0,1) to (2,1)

② Describe the geometrical transformations represented by these matrices:

(i) $\begin{pmatrix} 0 & -1 \\ -1 & 0 \end{pmatrix}$ (ii) $\begin{pmatrix} 4 & 0 \\ 0 & 1 \end{pmatrix}$ (iii) $\begin{pmatrix} 4 & 0 \\ 0 & 4 \end{pmatrix}$

(iv) $\begin{pmatrix} 1 & 0 \\ 0 & -1 \end{pmatrix}$ (v) $\begin{pmatrix} 0 & 1 \\ -1 & 0 \end{pmatrix}$

③ Figure R.12 shows a square with vertices at the points A(1, 1), B(1, −1), C(−1, −1) and D(−1, 1).

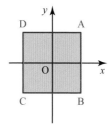

Figure R.12

(i) Draw a diagram to show the image of this square of the transformation matrix $\mathbf{M} = \begin{pmatrix} 1 & 0 \\ 5 & 1 \end{pmatrix}$.

(ii) Describe fully the transformation represented by the matrix $\mathbf{M}$. State the fixed line and the image of the point A.

④ Find the matrix that represents each of the following transformations in three dimensions:

(i) Rotation of 270° anticlockwise about the x-axis

(ii) Reflection in the plane $z = 0$

(iii) Rotation of 180° about the y-axis

⑤ Describe the transformations represented by these matrices:

(i) $\begin{pmatrix} 2 & 0 & 0 \\ 0 & 2 & 0 \\ 0 & 0 & 2 \end{pmatrix}$ (ii) $\begin{pmatrix} 1 & 0 & 0 \\ 0 & 1 & 0 \\ 0 & 0 & -1 \end{pmatrix}$

(iii) $\begin{pmatrix} 1 & 0 & 0 \\ 0 & 0 & 1 \\ 0 & -1 & 0 \end{pmatrix}$ (iv) $\begin{pmatrix} 4 & 0 & 0 \\ 0 & \frac{1}{2} & 0 \\ 0 & 0 & 3 \end{pmatrix}$

⑥ The 2×2 matrix $\mathbf{P}$ represents a rotation of 90° anticlockwise about the origin.

The 2×2 matrix $\mathbf{Q}$ represents a reflection in the line $y = -x$.

(i) Write down the matrices $\mathbf{P}$ and $\mathbf{Q}$.

(ii) Find the matrix $\mathbf{PQ}$ and describe the transformation it represents.

(iii) Find the matrix $\mathbf{QP}$ and describe the transformation it represents.

(iv) Show algebraically that there is only one point $r = \begin{pmatrix} x \\ y \end{pmatrix}$ which has the same image under the transformations represented by $\mathbf{PQ}$ and $\mathbf{QP}$ and state this point.

⑦ (i) Write down the matrix $\mathbf{A}$ which represents an anticlockwise rotation of 120° about the origin.

(ii) Write down the matrices $\mathbf{B}$ and $\mathbf{C}$ which represent anticlockwise rotations of 30° and 90° respectively about the origin. Find the matrix $\mathbf{BC}$ and verify that $\mathbf{A} = \mathbf{BC}$.

(iii) Calculate the matrix $\mathbf{B}^3$ and comment on your answer.

⑧ In three dimensions, the four matrices $\mathbf{J}, \mathbf{K}, \mathbf{L}$ and $\mathbf{M}$ represent transformations as follows:

$\mathbf{J}$ represents a reflection in the plane $y = 0$.

$\mathbf{K}$ represents an anticlockwise rotation of 90° about the z-axis.

$\mathbf{L}$ represents a reflection in the plane $x = 0$.

$\mathbf{M}$ represents a rotation of 180° about the x-axis.

(i) Write down the matrices $\mathbf{J}, \mathbf{K}, \mathbf{L}$ and $\mathbf{M}$.

(ii) Write down matrix products which would represent the single transformations obtained by each of the following combinations of transformations. Find the matrix in each case:

(a) a reflection in the plane $y = 0$ followed by a reflection in the plane $x = 0$

(b) a reflection in the plane $y = 0$ followed by an anticlockwise rotation of 90° about the z-axis

(c) an anticlockwise rotation of 90° about the z-axis followed by a second anticlockwise rotation of 90° about the z-axis

(d) a rotation of 180° about the x axis followed by a reflection in the plane $x = 0$ followed by a reflection in the plane $y = 0$

⑨ (i) Write down the matrix **P** which represents a stretch of scale factor 3 parallel to the y-axis.

(ii) The matrix $\mathbf{Q} = \begin{pmatrix} 2 & 0 \\ 0 & -1 \end{pmatrix}$. Write down the two single transformations which are represented by the matrix **Q**.

(iii) Find the matrix **PQ**. Write a list of the three transformations which are represented by the matrix **PQ**. In how many different orders could the three transformations occur?

(iv) Find the matrix **R** for which the matrix product **RPQ** would transform an object to its original position.

⑩ (i) Write down the matrix **A** representing a rotation about the origin through angle θ, and the matrix **B** representing a rotation about the origin through angle ϕ.

(ii) Find the matrix **BA**, representing a rotation about the origin through angle θ, followed by a rotation about the origin through angle ϕ.

(iii) Write down the matrix **C** representing a rotation about the origin through angle $\theta + \phi$.

(iv) By equating **C** to **BA**, write down expressions for $\sin(\theta + \phi)$ and $\cos(\theta + \phi)$.

(v) Explain why **BA** = **AB** in this case.

⑪ The matrix **R** represents a reflection in the line $y = mx$.

Show that $\mathbf{R}^2 = \begin{pmatrix} 1 & 0 \\ 0 & 1 \end{pmatrix}$ and explain geometrically why this is the case.

⑫ The matrix **A** is $\begin{pmatrix} 0 & -1 \\ 1 & 0 \end{pmatrix}$.

(i) Explain in terms of transformations why $\mathbf{A}^4 = \mathbf{I}$.

(ii) Using geometrical considerations, find the matrix **B** such that **BA** = **I**.

(iii) Write down the matrix **C** which represents a rotation of 60° anticlockwise about the origin.

(iv) Write down the smallest positive integers m and n such that $\mathbf{A}^m = \mathbf{C}^n$, explaining your answer in terms of transformations.

(v) Find **AC** and explain in terms of transformations why **AC** = **CA**.

3 Invariance

Points which map to themselves under a transformation are called **invariant points**. The origin is always an invariant point under a transformation that can be represented by a matrix, as the following statement is always true:

$$\begin{pmatrix} a & b \\ c & d \end{pmatrix}\begin{pmatrix} 0 \\ 0 \end{pmatrix} = \begin{pmatrix} 0 \\ 0 \end{pmatrix}$$

More generally, a point (x, y) is invariant if it satisfies the matrix equation:

$$\begin{pmatrix} a & b \\ c & d \end{pmatrix}\begin{pmatrix} x \\ y \end{pmatrix} = \begin{pmatrix} x \\ y \end{pmatrix}$$

For example, the point $(-2, 2)$ is invariant under the transformation represented by the matrix $\begin{pmatrix} 6 & 5 \\ 2 & 3 \end{pmatrix}$:

$$\begin{pmatrix} 6 & 5 \\ 2 & 3 \end{pmatrix}\begin{pmatrix} -2 \\ 2 \end{pmatrix} = \begin{pmatrix} -2 \\ 2 \end{pmatrix}$$

Example R.3

M is the matrix $\begin{pmatrix} 10 & -3 \\ 3 & 0 \end{pmatrix}$.

(i) Show that $(-2, -6)$ is an invariant point under the transformation represented by **M**.

(ii) What can you say about the invariant points under this transformation?

Solution

(i) $$\begin{pmatrix} 10 & -3 \\ 3 & 0 \end{pmatrix}\begin{pmatrix} -2 \\ -6 \end{pmatrix} = \begin{pmatrix} -2 \\ -6 \end{pmatrix}$$

so $(-2, -6)$ is an invariant point under the transformation represented by **M**.

(ii) Suppose the point $\begin{pmatrix} x \\ y \end{pmatrix}$ maps to itself. Then:

$$\begin{pmatrix} 10 & -3 \\ 3 & 0 \end{pmatrix}\begin{pmatrix} x \\ y \end{pmatrix} = \begin{pmatrix} x \\ y \end{pmatrix}$$

$$\begin{pmatrix} 10x - 3y \\ 3x \end{pmatrix} = \begin{pmatrix} x \\ y \end{pmatrix}$$

$\Leftrightarrow 10x - 3y = x$ and $3x = y$.

> These points all have the form $\lambda, 3\lambda$. The point $(-2, -6)$ is just one of the points on this line.

> Both equations simplify to $y = 3x$.

So the invariant points of the transformation are all the points on the line $y = 3x$.

The simultaneous equations in Example R.3 were equivalent and so all the invariant points were on a straight line. Generally, any matrix equation set up to find the invariant points will lead to two equations of the form $ax + by = 0$, which can also be expressed in the form $y = -\dfrac{ax}{b}$. These equations may be equivalent, in which case this is a line of invariant points. If the two equations are not equivalent, the origin is the only point which satisfies both equations, and so this is the only invariant point.

Invariant lines

A line AB is known as an **invariant line** under a transformation if the image of every point on AB is also on AB. It is important to note that it is not necessary for each of the points to map to itself; it can map to itself or to some other point on the line AB.

Sometimes it is easy to spot which lines are invariant. For example, in Figure R.13 the position of the points A − F and their images A′ − F′ show that the transformation is a reflection in the line *l*. So every point on *l* maps onto itself and *l* is a line of invariant points.

Look at the lines perpendicular to the mirror line in Figure R.13 , for example the line ABB′A′. Any point on one of these lines maps onto another point on the same line. Such a line is invariant but it is not a line of invariant points.

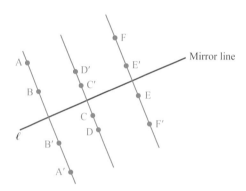

Figure R.13

Example R.4

Find the invariant lines of the transformation given by the matrix

$$\mathbf{M} = \begin{pmatrix} 4 & -1 \\ -2 & 1 \end{pmatrix}.$$

Solution

Suppose the invariant line has the form $y = mx + c$.

> Let the original point be (x, y) and the image point be (x', y').

$$\begin{pmatrix} x' \\ y' \end{pmatrix} = \begin{pmatrix} 4 & -1 \\ -2 & 1 \end{pmatrix} \begin{pmatrix} x \\ y \end{pmatrix} \Leftrightarrow x' = 4x - y \text{ and } y' = -2x + y.$$

$$\Leftrightarrow \begin{cases} x' = 4x - (mx + c) = (4 - m)x - c \quad \longleftarrow \boxed{\text{Using } y = mx + c} \\ \\ y' = -2x + (mx + c) = (-2 + m)x + c \end{cases}$$

As the line is invariant, (x', y') also lies on the line, so $y' = mx' + c$.

Therefore:

$$(-2 + m)x + c = m\big[(4 - m)x - c\big] + c$$

$$\Leftrightarrow 0 = \big(m^2 - 3m - 2\big)x + mc$$

For the right hand side to equal zero, both $m^2 - 3m - 2 = 0$ and $mc = 0$.

$$(m - 1)(m - 2) = 0 \Leftrightarrow m = 1 \text{ or } m = 2.$$

and

$$mc = 0 \Leftrightarrow m = 0 \text{ or } c = 0. \longleftarrow$$

> $m = 0$ is not a viable solution as $m^2 - 3m - 2 \neq 0$.

So, there are two possible solutions for the invariant line:

$$m = 1, c = 0 \Rightarrow y = x$$
$$\text{or } m = 2, c = 0 \Rightarrow y = 2x$$

Exercise R.3

① Find the invariant points under the transformations represented by the following matrices:

(i) $\begin{pmatrix} 3 & 3 \\ 1 & 1 \end{pmatrix}$ (ii) $\begin{pmatrix} 2 & 3 \\ 1 & 2 \end{pmatrix}$ (iii) $\begin{pmatrix} 3 & 2 \\ 1 & 2 \end{pmatrix}$ (iv) $\begin{pmatrix} 7 & -4 \\ 3 & -1 \end{pmatrix}$

② What lines, if any, are invariant under the following transformations?

(i)   Enlargement, centre the origin

(ii) Rotation through 180° about the origin

(iii) Rotation through 90° about the origin

(iv) Reflection in the line $y = x$

(v)   Reflection in the line $y = -x$

(vi) Shear, x-axis fixed

③ For the matrix $\mathbf{M} = \begin{pmatrix} 2 & 7 \\ 7 & 2 \end{pmatrix}$:

(i) show that the origin is the only invariant point

(ii) find the invariant lines of the transformation represented by $\mathbf{M}$

④ For the matrix $\mathbf{M} = \begin{pmatrix} 1 & 0 \\ 3 & -1 \end{pmatrix}$:

(i)   find the line of invariant points of the transformation given by $\mathbf{M}$

(ii) find the invariant lines of the transformation

(iii) draw a diagram to show the effect of the transformation on the unit square

⑤ A reflection in a line l is represented by the matrix $\mathbf{A} = \begin{pmatrix} -0.6 & 0.8 \\ 0.8 & 0.6 \end{pmatrix}$.

(i)   Find the image of the point $(3, 6)$ and hence write down the equation of the mirror line l.

(ii) The matrix $\mathbf{B} = \begin{pmatrix} 0 & -1 \\ 1 & 0 \end{pmatrix}$ represents a rotation. By considering the point $(3, 2)$, find the centre and angle of rotation.

(iii) Find the matrix $\mathbf{BA}$.

(iv) Show that under the transformation $\mathbf{BA}$ the point $(1, -3)$ is invariant. Hence state the equation of the line of invariant points under the transformation $\mathbf{BA}$.

⑥ The matrix $\begin{pmatrix} \dfrac{1-m^2}{1+m^2} & \dfrac{2m}{1+m^2} \\[2mm] \dfrac{2m}{1+m^2} & \dfrac{m^2-1}{1+m^2} \end{pmatrix}$ represents a reflection in the line $y = mx$.

Prove that the line $y = mx$ is a line of invariant points.

KEY POINTS

1 A matrix is a rectangular array of numbers or letters.

2 The shape of a matrix is described by its order. A matrix with r rows and c columns has order $r \times c$.

3 A matrix with the same number of rows and columns is called a **square matrix**.

4 The matrix $\mathbf{O} = \begin{pmatrix} 0 & 0 \\ 0 & 0 \end{pmatrix}$ is known as the 2×2 **zero matrix**. Zero matrices can be of any order.

5 A matrix of the form $\mathbf{I} = \begin{pmatrix} 1 & 0 \\ 0 & 1 \end{pmatrix}$ is known as an **identity matrix**. All identity matrices are square, with 1's on the leading diagonal and zeros elsewhere.

6 Matrices can be added or subtracted if they have the same order.

7 Two matrices **A** and **B** can be multiplied to give matrix **AB** if their orders are of the form $p \times q$ and $q \times r$ respectively. The resulting matrix will have the order $p \times r$.

8 Matrix multiplication

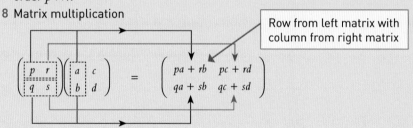

> Row from left matrix with column from right matrix

Figure R.14

9 Matrix addition and multiplication are **associative**.
$$\mathbf{A} + (\mathbf{B} + \mathbf{C}) = (\mathbf{A} + \mathbf{B}) + \mathbf{C}$$
$$\mathbf{A}(\mathbf{BC}) = (\mathbf{AB})\mathbf{C}$$

10 Matrix addition is **commutative** but matrix multiplication is generally not commutative.
$$\mathbf{A} + \mathbf{B} = \mathbf{B} + \mathbf{A}$$
$$\mathbf{AB} \neq \mathbf{BA}$$

11 The matrix $\mathbf{M} = \begin{pmatrix} a & b \\ c & d \end{pmatrix}$ represents the transformation which maps the point with position vector $\begin{pmatrix} x \\ y \end{pmatrix}$ to the point with position vector $\begin{pmatrix} ax + by \\ cx + dy \end{pmatrix}$.

12 Under the transformation represented by $\mathbf{M}$, the image of $\mathbf{i} = \begin{pmatrix} 1 \\ 0 \end{pmatrix}$ is the first

column of $\mathbf{M}$ and the image of $\mathbf{j} = \begin{pmatrix} 0 \\ 1 \end{pmatrix}$ is the second column of $\mathbf{M}$.

Similarly, in three dimensions the images of the unit vectors

$\mathbf{i} = \begin{pmatrix} 1 \\ 0 \\ 0 \end{pmatrix}$, $\mathbf{j} = \begin{pmatrix} 0 \\ 1 \\ 0 \end{pmatrix}$ and $\mathbf{k} = \begin{pmatrix} 0 \\ 0 \\ 1 \end{pmatrix}$ are the first, second and third columns of the

transformation matrix.

13 Summary of transformations in two dimensions

$\begin{pmatrix} 1 & 0 \\ 0 & -1 \end{pmatrix}$ Reflection in the x-axis

$\begin{pmatrix} -1 & 0 \\ 0 & 1 \end{pmatrix}$ Reflection in the y-axis

$\begin{pmatrix} 0 & 1 \\ 1 & 0 \end{pmatrix}$ Reflection in the line $y = x$

$\begin{pmatrix} 0 & -1 \\ -1 & 0 \end{pmatrix}$ Reflection in the line $y = -x$

$\begin{pmatrix} \cos\theta & -\sin\theta \\ \sin\theta & \cos\theta \end{pmatrix}$ Rotation anticlockwise about the origin through angle θ

$\begin{pmatrix} k & 0 \\ 0 & k \end{pmatrix}$ Enlargement centre the origin, scale factor k

$\begin{pmatrix} k & 0 \\ 0 & 1 \end{pmatrix}$ Stretch parallel to the x-axis, scale factor k

$\begin{pmatrix} 1 & 0 \\ 0 & k \end{pmatrix}$ Stretch parallel to the y-axis, scale factor k

$\begin{pmatrix} 1 & k \\ 0 & 1 \end{pmatrix}$ Shear, x-axis invariant, with $(0,1)$ mapped to $(k, 1)$

$\begin{pmatrix} 1 & 0 \\ k & 1 \end{pmatrix}$ Shear, y-axis invariant, with $(1, 0)$ mapped to $(1, k)$

14 Examples of transformations in three dimensions

$\begin{pmatrix} -1 & 0 & 0 \\ 0 & 1 & 0 \\ 0 & 0 & 1 \end{pmatrix}$ Reflection in plane $x = 0$

$\begin{pmatrix} 1 & 0 & 0 \\ 0 & -1 & 0 \\ 0 & 0 & 1 \end{pmatrix}$ Reflection in plane $y = 0$

$$\begin{pmatrix} 1 & 0 & 0 \\ 0 & 1 & 0 \\ 0 & 0 & -1 \end{pmatrix}$$ Reflection in plane $z = 0$

$$\begin{pmatrix} 1 & 0 & 0 \\ 0 & 0 & -1 \\ 0 & 1 & 0 \end{pmatrix}$$ Rotation of 90° about the x-axis

$$\begin{pmatrix} 0 & 0 & 1 \\ 0 & 1 & 0 \\ -1 & 0 & 0 \end{pmatrix}$$ Rotation of 90° about the y-axis

$$\begin{pmatrix} 0 & -1 & 0 \\ 1 & 0 & 0 \\ 0 & 0 & 1 \end{pmatrix}$$ Rotation of 90° about the z-axis

When rotating an object about an axis, the rotation is taken to be anticlockwise about the axis of rotation when looking along the axis from the positive end towards the origin.

15 The composite of the transformation represented by **M** followed by that represented by **N** is represented by the matrix product **NM**.

16 If (x, y) is an **invariant point** under a transformation represented by the matrix **M** then $\mathbf{M} \begin{pmatrix} x \\ y \end{pmatrix} = \begin{pmatrix} x \\ y \end{pmatrix}$.

17 A line AB is known as an **invariant line** under a transformation if the image of every point on AB is also on AB.

2 Matrices

> *I do not hesitate to maintain, that what we are conscious of is constructed out of what we are not conscious of, that our whole knowledge, in fact, is made up of the unknown and the incognisable.*
>
> William Hamilton (1788–1856)

Search engines' algorithms determine which order to display results by solving huge systems of simultaneous equations. In practice this happens using matrices. Matrices are also used extensively in computer games – particularly to apply rotations and translations to cameras and objects whose positions and orientations will be described by coordinates and vectors in three dimensional worlds.

Review: The determinant of a 2 × 2 matrix

Figure 2.1 shows the parallelogram produced when the unit square is transformed by the matrix $\begin{pmatrix} a & b \\ c & d \end{pmatrix}$.

> You can find the area of the red parallelogram by subtracting the areas of the yellow rectangles and the green and blue triangles from the area of the large rectangle.

Figure 2.1

47

The area of the parallelogram is $(ad - bc)$ units^2.

Since the area of the unit square is one unit, the quantity $(ad - bc)$ is the area scale factor associated with the transformation matrix $\begin{pmatrix} a & b \\ c & d \end{pmatrix}$. It is called the **determinant** of the matrix and is denoted by det **M** or |**M**|. Another notation that is sometimes used as an alternative is Δ.

Example 2.1

For each of the matrices $\mathbf{A} = \begin{pmatrix} 3 & 4 \\ 1 & 2 \end{pmatrix}$ and $\mathbf{B} = \begin{pmatrix} 3 & 4 \\ 2 & 1 \end{pmatrix}$:

(i)   Draw a diagram to show the image of the unit square OIPJ under the transformation represented by the matrix.

(ii)   Find the determinant of the matrix.

(iii)   Use your answer to (ii) to find the area of the transformed shape.

Solution

(a)   (i)

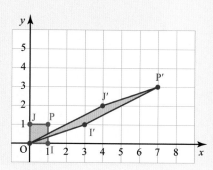

Figure 2.2

(ii) $\det \mathbf{A} = (3 \times 2) - (1 \times 4) = 2$

(iii)   Area of quadrilateral OI′P′J′ is $1 \times 2 = 2$.

(b)   (i)

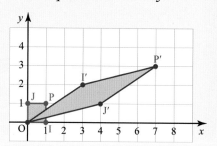

Figure 2.3

(ii) $\det \mathbf{B} = (3 \times 1) - (4 \times 2) = -5$

> Notice that the determinant is negative. Since area cannot be negative, the area of the transformed shape is 5 square units.

(iii)   Area of quadrilateral OI′P′J′ is $1 \times 5 = 5$.

In Example 2.1 the sign of the determinant has significance. In part (i), if you move anticlockwise around the original unit square you come to vertices O, I, P, J in that order. Moving anticlockwise around the image gives O, I′, P′, J′, i.e. the order is unchanged.

However, in part (ii) moving anticlockwise about the image reverses the order of the vertices, i.e. O, J′, P′, I′. This reversal in the order of the vertices produces the negative determinant.

The same principle applies in three dimensions.

Important results about determinants

1. For square matrices $\mathbf{P}$ and $\mathbf{Q}$, $\det \mathbf{PQ} = \det \mathbf{P} \times \det \mathbf{Q}$.

A special case is the zero matrix which maps all points to the origin.

→ 2. If a matrix $\mathbf{M}$ has zero determinant, the area scale factor of the transformation is zero, so all points are mapped to a shape with zero area. In fact the matrix maps all points in the plane to a straight line.

Finding the inverse of a 2 × 2 matrix

The inverse of a 2 × 2 matrix $\mathbf{M} = \begin{pmatrix} a & b \\ c & d \end{pmatrix}$ is $\mathbf{M}^{-1} = \dfrac{1}{ad - bc} \begin{pmatrix} d & -b \\ -c & a \end{pmatrix}$.

If the determinant is zero then the inverse matrix does not exist and the matrix is said to be **singular**. If $\det \mathbf{M} \neq 0$ the matrix is said to be **non-singular**.

If a matrix is singular, then it maps an infinite number of points in the plane to the same point on the straight line. It is therefore not possible to find the inverse of the transformation, because it would need to map a point on that straight line to just one other point, not to an infinite number of them.

Example 2.2

$$\mathbf{A} = \begin{pmatrix} 5 & -3 \\ 9 & 1 \end{pmatrix}$$

(i)   Find $\mathbf{A}^{-1}$.

(ii)   The point P is mapped to the point Q(−22, −14) under the transformation represented by $\mathbf{A}$. Find the coordinates of P.

Solution

(i) $\det \mathbf{A} = (5 \times 1) - (-3 \times 9) = 32$

$$\mathbf{A}^{-1} = \frac{1}{32} \begin{pmatrix} 1 & 3 \\ -9 & 5 \end{pmatrix}$$

$\boxed{\text{A maps P to Q, so } \mathbf{A}^{-1} \text{ maps Q to P.}}$

(ii) $\mathbf{A}^{-1} \begin{pmatrix} -22 \\ -14 \end{pmatrix} = \frac{1}{32} \begin{pmatrix} 1 & 3 \\ -9 & 5 \end{pmatrix} \begin{pmatrix} -22 \\ -14 \end{pmatrix} = \frac{1}{32} \begin{pmatrix} -64 \\ 128 \end{pmatrix} = \begin{pmatrix} -2 \\ 4 \end{pmatrix}$

So P has coordinates (−2, 4).

An important result about a matrix and its inverse is that $\mathbf{MM}^{-1} = \mathbf{M}^{-1}\mathbf{M} = \mathbf{I}$. This is true for all square matrices, not just 2 × 2 matrices.

The inverse of a product of matrices

Suppose you want to find the inverse of the product $\mathbf{MN}$, where $\mathbf{M}$ and $\mathbf{N}$ are non-singular matrices. This means that you need to find a matrix $\mathbf{X}$ such that $\mathbf{X(MN)} = \mathbf{I}$.

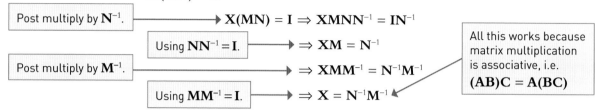

Post multiply by $\mathbf{N}^{-1}$. → $\mathbf{X(MN)} = \mathbf{I} \Rightarrow \mathbf{XMNN}^{-1} = \mathbf{IN}^{-1}$

Using $\mathbf{NN}^{-1} = \mathbf{I}$. → $\Rightarrow \mathbf{XM} = \mathbf{N}^{-1}$

Post multiply by $\mathbf{M}^{-1}$. → $\Rightarrow \mathbf{XMM}^{-1} = \mathbf{N}^{-1}\mathbf{M}^{-1}$

Using $\mathbf{MM}^{-1} = \mathbf{I}$. → $\Rightarrow \mathbf{X} = \mathbf{N}^{-1}\mathbf{M}^{-1}$

All this works because matrix multiplication is associative, i.e.
$\mathbf{(AB)C} = \mathbf{A(BC)}$

So $\mathbf{(MN)}^{-1} = \mathbf{N}^{-1}\mathbf{M}^{-1}$ for matrices $\mathbf{M}$ and $\mathbf{N}$ of the same order. This means that when working backwards, you must reverse the second transformation before reversing the first transformation.

Using matrices to solve simultaneous equations

There are a number of methods to solve a pair of linear simultaneous equations of the form

$$3x + 4y = 7$$
$$2x - y = 12$$

such as elimination, substitution or graphical methods.

Another method involves the use of inverse matrices. This method has the advantage that it can more easily be extended to solving a set of n equations in n variables.

Example 2.3

Use a matrix method to solve the simultaneous equations

$$3x + 4y = 7$$
$$2x - y = 12$$

Solution

$$\begin{pmatrix} 3 & 4 \\ 2 & -1 \end{pmatrix}\begin{pmatrix} x \\ y \end{pmatrix} = \begin{pmatrix} 7 \\ 12 \end{pmatrix}$$

Write the equations in matrix form.

The inverse of the matrix $\begin{pmatrix} 3 & 4 \\ 2 & -1 \end{pmatrix}$ is $-\dfrac{1}{11}\begin{pmatrix} -1 & -4 \\ -2 & 3 \end{pmatrix} = \dfrac{1}{11}\begin{pmatrix} 1 & 4 \\ 2 & -3 \end{pmatrix}$.

Pre-multiply both sides of the matrix equation by the inverse matrix.

$$\frac{1}{11}\begin{pmatrix} 1 & 4 \\ 2 & -3 \end{pmatrix}\begin{pmatrix} 3 & 4 \\ 2 & -1 \end{pmatrix}\begin{pmatrix} x \\ y \end{pmatrix} = \frac{1}{11}\begin{pmatrix} 1 & 4 \\ 2 & -3 \end{pmatrix}\begin{pmatrix} 7 \\ 12 \end{pmatrix}$$

$$\begin{pmatrix} x \\ y \end{pmatrix} = \frac{1}{11}\begin{pmatrix} 55 \\ -22 \end{pmatrix} = \begin{pmatrix} 5 \\ -2 \end{pmatrix}$$

As $\mathbf{M}^{-1}\mathbf{M} = \mathbf{I}$ the left hand side simplifies to $\begin{pmatrix} x \\ y \end{pmatrix}$.

The solution is $x = 5$, $y = -2$.

Geometrical interpretation in two dimensions

Two equations in two unknowns can be represented in a plane by two straight lines. The number of points of intersection of the lines determines the number of solutions to the equations.

There are three different possibilities.

Case 1

Example 2.3 shows that two simultaneous equations can have a unique solution. Graphically, this is represented by a single point of intersection, as shown in Figure 2.4.

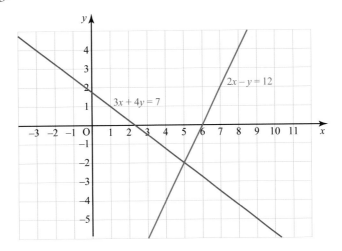

Figure 2.4

This is the case where $\det \mathbf{M} \neq 0$ and so the inverse matrix $\mathbf{M}^{-1}$ exists, allowing the equations to be solved.

Case 2

If two lines are parallel they do not have a point of intersection. For example, the lines

$$x + 2y = 10$$
$$x + 2y = 4$$

> The equations can be written in matrix form as $\begin{pmatrix} 1 & 2 \\ 1 & 2 \end{pmatrix} \begin{pmatrix} x \\ y \end{pmatrix} = \begin{pmatrix} 10 \\ 4 \end{pmatrix}$.

are parallel (see Figure 2.5).

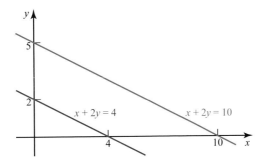

Figure 2.5

The matrix $\mathbf{M} = \begin{pmatrix} 1 & 2 \\ 1 & 2 \end{pmatrix}$ has determinant zero and hence the inverse matrix does not exist.

Case 3

More than one solution is possible in cases where the lines are coincident, i.e. lie on top of each other. For example, the two lines

$$x + 2y = 10$$
$$3x + 6y = 30$$

are coincident (see Figure 2.6). You can see this because the equations are multiples of each other.

The equations can be written in matrix form as

$$\begin{pmatrix} 1 & 2 \\ 3 & 6 \end{pmatrix}\begin{pmatrix} x \\ y \end{pmatrix} = \begin{pmatrix} 10 \\ 30 \end{pmatrix}.$$

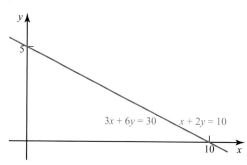

Figure 2.6

In this case the matrix **M** is $\begin{pmatrix} 1 & 2 \\ 3 & 6 \end{pmatrix}$ and $\det \mathbf{M} = 0$.

There are infinitely many solutions to these equations.

Review

① The matrix $\begin{pmatrix} 2x - 7 & 2 \\ 9 - 8x & x + 1 \end{pmatrix}$ has determinant -4.

Find the possible values of x.

② (i) Write down the matrices **A** and **B** which represent:

A - a reflection in the y-axis

B - a reflection in the line $y = -x$

(ii) Show that the matrices **A** and **B** each have determinant of -1.

(iii) Draw diagrams for each of the transformations **A** and **B** to demonstrate that the images of the vertices labelled anticlockwise on the unit square OIPJ are reversed to a clockwise labelling.

③ Figure 2.7 shows the unit square transformed by a shear.

(i) Write down the matrix which represents this transformation.

(ii) Show that under this transformation the area of the image is always equal to the area of the object.

④ For the matrix $\begin{pmatrix} 3 & -5 \\ 1 & 5 \end{pmatrix}$:

(i) find the image of the point $(2, -1)$

(ii) find the inverse matrix

(iii) find the point which maps to the image $(3, 1)$

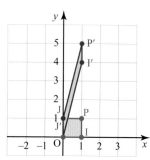

Figure 2.7

⑤ The matrix $\begin{pmatrix} 3-k & k \\ 2 & 2-k \end{pmatrix}$ is singular.

Find the possible values of k.

⑥ $\mathbf{M} = \begin{pmatrix} -2 & 3 \\ -1 & 6 \end{pmatrix}$ and $\mathbf{N} = \begin{pmatrix} 4 & 7 \\ 9 & -1 \end{pmatrix}$.

(i)   Find the determinants of $\mathbf{M}$ and $\mathbf{N}$.

(ii) Find the matrix $\mathbf{MN}$ and show that $\det(\mathbf{MN}) = \det \mathbf{M} \times \det \mathbf{N}$.

⑦ Given that $\mathbf{M} = \begin{pmatrix} 1 & -3 \\ 2 & 1 \end{pmatrix}$ and $\mathbf{MN} = \begin{pmatrix} 6 & -1 & 3 & 6 \\ 5 & -2 & 6 & 12 \end{pmatrix}$, find the matrix $\mathbf{N}$.

⑧ The plane is transformed by the matrix $\mathbf{M} = \begin{pmatrix} 3 & -9 \\ -2 & 6 \end{pmatrix}$.

(i)   Draw a diagram to show the image of the unit square under the transformation represented by $\mathbf{M}$.

(ii) Describe the effect of the transformation and explain this with reference to the determinant of $\mathbf{M}$.

⑨ Use matrices to solve the following pairs of simultaneous equations:

(i) $5x - 3y = 13$
      $2x - y = 5$

(ii) $4x - y = -16$
      $x - 3y = -15$

⑩ Find the two values of k for which the equations

$2x + ky = 3$
$kx + 8y = 6$

do not have a unique solution.

How many solutions are there in each case?

⑪ $\mathbf{M} = \begin{pmatrix} a & b \\ c & d \end{pmatrix}$ is a singular matrix.

(i)   Show that $\mathbf{M}^2 = (a+d)\mathbf{M}$.

(ii) Find a formula which expresses $\mathbf{M}^n$ in terms of $\mathbf{M}$, where n is a positive integer.

(iii) Prove the formula you found in part (ii) by induction.

⑫ The plane is transformed using the matrix $\begin{pmatrix} a & b \\ c & d \end{pmatrix}$ where $ad - bc = 0$.

Prove that the general point $\mathrm{P}(x, y)$ maps to P' on the line $cx - ay = 0$.

⑬ Triangle T has vertices at $(1, 0)$, $(0, 2)$ and $(-3, 0)$.

It is transformed to triangle T' by the matrix $\mathbf{M} = \begin{pmatrix} 4 & 1 \\ 1 & 1 \end{pmatrix}$.

(i)   Find the coordinates of the vertices of T'.
      Show the triangles T and T' on a single diagram.

(ii) Find the ratio of the area of T' to the area of T.
      Comment on your answer in relation to the matrix $\mathbf{M}$.

(iii) Find $\mathbf{M}^{-1}$ and verify that this matrix maps the vertices of T′ to the vertices of T.

⑭ (i) The matrix $\mathbf{S} = \begin{pmatrix} -1 & 2 \\ -3 & 4 \end{pmatrix}$ represents a transformation.

    (a) Show that the point (1, 1) is invariant under this transformation.

    (b) Calculate $\mathbf{S}^{-1}$.

    (c) Verify that (1, 1) is also invariant under the transformation represented by $\mathbf{S}^{-1}$.

(ii) Part (i) can be generalised as follows:

If (x, y) is an invariant point under a transformation represented by the non-singular matrix $\mathbf{T}$, it is also invariant under the transformation represented by $\mathbf{T}^{-1}$.

Starting with $\mathbf{T}\begin{pmatrix} x \\ y \end{pmatrix} = \begin{pmatrix} x \\ y \end{pmatrix}$ prove this result algebraically.

⑮ The simultaneous equations

$$2x - y = 1$$
$$3x + ky = b$$

are represented by the matrix $\mathbf{M}\begin{pmatrix} x \\ y \end{pmatrix} = \begin{pmatrix} 1 \\ b \end{pmatrix}$.

(i) Write down the matrix $\mathbf{M}$.

(ii) State the value of k for which $\mathbf{M}^{-1}$ does not exist and find $\mathbf{M}^{-1}$ in terms of k when $\mathbf{M}^{-1}$ exists.

Use $\mathbf{M}^{-1}$ to solve the simultaneous equations when $k = 5$, $b = 21$.

(iii) What can you say about the solutions of the equations when $k = -\dfrac{3}{2}$?

(iv) The two equations can be interpreted as representing two lines in the x–y plane.

Describe the relationship between the two lines when:

    (a) $k = 5$, $b = 21$

    (b) $k = -\dfrac{3}{2}$, $b = 1$

    (c) $k = -\dfrac{3}{2}$, $b = \dfrac{3}{2}$

⑯ Matrices $\mathbf{M}$ and $\mathbf{N}$ are given by $\mathbf{M} = \begin{pmatrix} 3 & 2 \\ 0 & 1 \end{pmatrix}$ and $\mathbf{N} = \begin{pmatrix} 1 & -3 \\ 1 & 4 \end{pmatrix}$.

(i) Find $\mathbf{M}^{-1}$ and $\mathbf{N}^{-1}$.

(ii) Find $\mathbf{MN}$ and $(\mathbf{MN})^{-1}$. Verify that $(\mathbf{MN})^{-1} = \mathbf{N}^{-1}\mathbf{M}^{-1}$.

(iii) The result $(\mathbf{PQ})^{-1} = \mathbf{Q}^{-1}\mathbf{P}^{-1}$ is true for any two 2 × 2 non-singular matrices $\mathbf{P}$ and $\mathbf{Q}$.

The first two lines of a proof of this general result are given below.

Beginning with these two lines, complete the general proof.

$$(\mathbf{PQ})^{-1}\mathbf{PQ} = \mathbf{I}$$

$$(\mathbf{PQ})^{-1}\mathbf{PQQ}^{-1} = \mathbf{IQ}^{-1}$$

1 Finding the inverse of a 3 × 3 matrix

The determinant of a 3 × 3 matrix is sometimes denoted $|\mathbf{a}\ \mathbf{b}\ \mathbf{c}|$.

→ In this section you will find the determinant and inverse of 3 × 3 matrices using the calculator facility and also using a non-calculator method.

Finding the inverse of a 3 × 3 matrix using a calculator

> **ACTIVITY 2.1**
>
> Using a calculator, find the determinant and inverse of the matrix
>
> $$\mathbf{A} = \begin{pmatrix} 3 & -2 & 1 \\ 0 & 1 & 2 \\ 4 & 0 & 1 \end{pmatrix}.$$
>
> Still using a calculator, find out which of the following matrices are non-singular and find the inverse in each of these cases.
>
> $$\mathbf{B} = \begin{pmatrix} 5 & 5 & 5 \\ 2 & 2 & 2 \\ 2 & 4 & -3 \end{pmatrix} \qquad \mathbf{C} = \begin{pmatrix} 1 & 3 & 2 \\ -1 & 0 & 1 \\ 2 & 1 & 4 \end{pmatrix} \qquad \mathbf{D} = \begin{pmatrix} 0 & 3 & -2 \\ 1 & -1 & 2 \\ 3 & 0 & 3 \end{pmatrix}$$

Finding the determinant of a 3 × 3 matrix without using a calculator

It is also possible to find the determinant and inverse of a 3 × 3 matrix without using a calculator. This is useful in cases where some of the elements of the matrix are algebraic rather than numerical.

If $\mathbf{M}$ is the 3 × 3 matrix $\begin{pmatrix} a_1 & b_1 & c_1 \\ a_2 & b_2 & c_2 \\ a_3 & b_3 & c_3 \end{pmatrix}$ then the determinant of $\mathbf{M}$ is defined by

$$\det \mathbf{M} = a_1 \begin{vmatrix} b_2 & c_2 \\ b_3 & c_3 \end{vmatrix} - a_2 \begin{vmatrix} b_1 & c_1 \\ b_3 & c_3 \end{vmatrix} + a_3 \begin{vmatrix} b_1 & c_1 \\ b_2 & c_2 \end{vmatrix},$$

which is sometimes referred to as the **expansion of the determinant by the first column**.

Notice that you do not really need to calculate $\begin{vmatrix} -2 & 1 \\ 0 & 1 \end{vmatrix}$ as it is going to be multiplied by zero. Keeping an eye open for helpful zeros can reduce the number of calculations needed.

For example, to find the determinant of the matrix $\mathbf{A} = \begin{pmatrix} 3 & -2 & 1 \\ 0 & 1 & 2 \\ 4 & 0 & 1 \end{pmatrix}$ from Activity 2.1:

$$\begin{aligned} \det \mathbf{A} &= 3 \begin{vmatrix} 1 & 2 \\ 0 & 1 \end{vmatrix} - 0 \begin{vmatrix} -2 & 1 \\ 0 & 1 \end{vmatrix} + 4 \begin{vmatrix} -2 & 1 \\ 1 & 2 \end{vmatrix} \\ &= 3(1 - 0) - 0(-2 - 0) + 4(-4 - 1) \\ &= 3 - 20 \\ &= -17 \end{aligned}$$

This is the same answer as you will have obtained earlier using your calculator.

The 2 × 2 determinant $\begin{vmatrix} b_2 & c_2 \\ b_3 & c_3 \end{vmatrix}$ is called the **minor** of the

element a_1. It is obtained by deleting the row and column containing a_1:

$$\begin{vmatrix} a_1 & b_1 & c_1 \\ a_2 & b_2 & c_2 \\ a_3 & b_3 & c_3 \end{vmatrix}$$

Other minors are defined in the same way, for example the minor of a_2 is

$$\begin{vmatrix} a_1 & b_1 & c_1 \\ a_2 & b_2 & c_2 \\ a_3 & b_3 & c_3 \end{vmatrix} = \begin{vmatrix} b_1 & c_1 \\ b_3 & c_3 \end{vmatrix}$$

> ### Note
>
> As an alternative to using the first column, you could use the **expansion of the determinant by the second column**:
>
> $$\det \mathbf{M} = -b_1 \begin{vmatrix} a_2 & c_2 \\ a_3 & c_3 \end{vmatrix} + b_2 \begin{vmatrix} a_1 & c_1 \\ a_3 & c_3 \end{vmatrix} - b_3 \begin{vmatrix} a_1 & c_1 \\ a_2 & c_2 \end{vmatrix},$$
>
> or the **expansion of the determinant by the third column**:
>
> $$\det \mathbf{M} = c_1 \begin{vmatrix} a_2 & b_2 \\ a_3 & b_3 \end{vmatrix} - c_2 \begin{vmatrix} a_1 & b_1 \\ a_3 & b_3 \end{vmatrix} + c_3 \begin{vmatrix} a_1 & b_1 \\ a_2 & b_2 \end{vmatrix}.$$
>
> It is fairly easy to show that all three expressions above for $\det \mathbf{M}$ simplify to:
> $$a_1 b_2 c_3 + a_2 b_3 c_1 + a_3 b_1 c_2 - a_3 b_2 c_1 - a_1 b_3 c_2 - a_2 b_1 c_3$$

You may have noticed that in the expansions of the determinant, the signs on the minors alternate as shown:

$$\begin{vmatrix} + & - & + \\ - & + & - \\ + & - & + \end{vmatrix}$$

A minor, together with its correct sign, is known as a **cofactor** and is denoted by the corresponding capital letter; for example, the cofactor of a_3 is A_3. This means that the expansion by the first column, say, can be written as

$$a_1 A_1 + a_2 A_2 + a_3 A_3.$$

Example 2.4

Find the determinant of the matrix $\mathbf{M} = \begin{pmatrix} 3 & 0 & -4 \\ 7 & 2 & -1 \\ -2 & 1 & 3 \end{pmatrix}$.

Solution

> To find the determinant you can also expand by rows. So, for example, expanding by the top row would give:
>
> $$3\begin{vmatrix} 2 & -1 \\ 1 & 3 \end{vmatrix} - 0\begin{vmatrix} 7 & -1 \\ -2 & 3 \end{vmatrix} + (-4)\begin{vmatrix} 7 & 2 \\ -2 & 1 \end{vmatrix}$$
>
> which also gives the answer −23.

Expanding by the first column using the expression:

$$\det \mathbf{M} = a_1 \begin{vmatrix} b_2 & c_2 \\ b_3 & c_3 \end{vmatrix} - a_2 \begin{vmatrix} b_1 & c_1 \\ b_3 & c_3 \end{vmatrix} + a_3 \begin{vmatrix} b_1 & c_1 \\ b_2 & c_2 \end{vmatrix}$$

gives:

$$\det \mathbf{M} = 3 \begin{vmatrix} 2 & -1 \\ 1 & 3 \end{vmatrix} - 7 \begin{vmatrix} 0 & -4 \\ 1 & 3 \end{vmatrix} + (-2) \begin{vmatrix} 0 & -4 \\ 2 & -1 \end{vmatrix}$$

$$= 3(6 - (-1)) - 7(0 - (-4)) - 2(0 - (-8))$$
$$= 21 - 28 - 16$$
$$= -23$$

> Notice that expanding by the top row would be quicker here as it has a zero element.

Earlier you saw that the determinant of a 2×2 matrix represents the area scale factor of the transformation represented by the matrix. In the case of a 3×3 matrix the determinant represents the volume scale factor. For example, the

matrix $\begin{pmatrix} 2 & 0 & 0 \\ 0 & 2 & 0 \\ 0 & 0 & 2 \end{pmatrix}$ has determinant 8; this matrix represents an enlargement of

scale factor 2, centre the origin, so the volume scale factor of the transformation is $2 \times 2 \times 2 = 8$.

Finding the inverse of a 3 × 3 matrix without using a calculator

> Recall that a minor, together with its correct sign, is known as a cofactor and is denoted by the corresponding capital letter; for example the cofactor of a_3 is A_3.

The matrix $\begin{pmatrix} A_1 & A_2 & A_3 \\ B_1 & B_2 & B_3 \\ C_1 & C_2 & C_3 \end{pmatrix}$ is known as the **adjugate** or **adjoint** of $\mathbf{M}$,

denoted **adj M**.

The adjugate of $\mathbf{M}$ is formed by
- replacing each element of $\mathbf{M}$ by its cofactor;
- then transposing the matrix (i.e. changing rows into columns and columns into rows).

The unique inverse of a 3×3 matrix can be calculated as follows:

$$\mathbf{M}^{-1} = \frac{1}{\det \mathbf{M}} \text{adj } \mathbf{M} = \frac{1}{\det \mathbf{M}} \begin{pmatrix} A_1 & A_2 & A_3 \\ B_1 & B_2 & B_3 \\ C_1 & C_2 & C_3 \end{pmatrix}, \det \mathbf{M} \neq 0$$

The steps involved in the method are shown in the following example.

Example 2.5

Find the inverse of the matrix $\mathbf{M}$ without using a calculator, where

$$\mathbf{M} = \begin{pmatrix} 2 & 3 & 4 \\ 2 & -5 & 2 \\ -3 & 6 & -3 \end{pmatrix}.$$

Solution

Step 1: Find the determinant Δ and check $\Delta \neq 0$

Expanding by the first column

$$\Delta = 2 \begin{vmatrix} -5 & 2 \\ 6 & -3 \end{vmatrix} - 2 \begin{vmatrix} 3 & 4 \\ 6 & -3 \end{vmatrix} + (-3) \begin{vmatrix} 3 & 4 \\ -5 & 2 \end{vmatrix}$$

$$= (2 \times 3) - (2 \times -33) - (3 \times 26) = -6$$

Therefore the inverse matrix exists.

Step 2: Evaluate the cofactors

> You can evaluate the determinant Δ using these cofactors to check your earlier arithmetic is correct:
>
> 2nd column:
> $\Delta = 3B_1 - 5B_2 + 6B_3$
> $= (3 \times 0) - (5 \times 6)$
> $\quad + (6 \times 4) = -6$
>
> 3rd column:
> $\Delta = 4C_1 + 2C_2 - 3C_3$
> $= (4 \times -3) + (2 \times -21)$
> $\quad - (3 \times -16) = -6$

$$A_1 = \begin{vmatrix} -5 & 2 \\ 6 & -3 \end{vmatrix} = 3 \qquad B_1 = -\begin{vmatrix} 2 & 2 \\ -3 & -3 \end{vmatrix} = 0 \qquad C_1 = \begin{vmatrix} 2 & -5 \\ -3 & 6 \end{vmatrix} = -3$$

$$A_2 = -\begin{vmatrix} 3 & 4 \\ 6 & -3 \end{vmatrix} = 33 \qquad B_2 = \begin{vmatrix} 2 & 4 \\ -3 & -3 \end{vmatrix} = 6 \qquad C_2 = -\begin{vmatrix} 2 & 3 \\ -3 & 6 \end{vmatrix} = -21$$

$$A_3 = \begin{vmatrix} 3 & 4 \\ -5 & 2 \end{vmatrix} = 26 \qquad B_3 = -\begin{vmatrix} 2 & 4 \\ 2 & 2 \end{vmatrix} = 4 \qquad C_3 = \begin{vmatrix} 2 & 3 \\ 2 & -5 \end{vmatrix} = -16$$

Step 3: Form the matrix of cofactors and transpose it, then multiply by $\dfrac{1}{\Delta}$

$$\mathbf{M}^{-1} = \frac{1}{-6} \begin{pmatrix} 3 & 0 & -3 \\ 33 & 6 & -21 \\ 26 & 4 & -16 \end{pmatrix}^T$$

> Multiply by $\dfrac{1}{\Delta}$.

> The capital T indicates the matrix is to be transposed.

> Matrix of cofactors.

$$= \frac{1}{-6} \begin{pmatrix} 3 & 33 & 26 \\ 0 & 6 & 4 \\ -3 & -21 & -16 \end{pmatrix}$$

$$= \frac{1}{6} \begin{pmatrix} -3 & -33 & -26 \\ 0 & -6 & -4 \\ 3 & 21 & 16 \end{pmatrix}$$

The final matrix could then be simplified and written as

$$\mathbf{M}^{-1} = \begin{pmatrix} -\dfrac{1}{2} & -\dfrac{11}{2} & -\dfrac{13}{3} \\ 0 & -1 & -\dfrac{2}{3} \\ \dfrac{1}{2} & \dfrac{7}{2} & \dfrac{8}{3} \end{pmatrix}$$

Check: $\mathbf{M}\mathbf{M}^{-1} = \begin{pmatrix} 2 & 3 & 4 \\ 2 & -5 & 2 \\ -3 & 6 & -3 \end{pmatrix} \dfrac{1}{6} \begin{pmatrix} -3 & -33 & -26 \\ 0 & -6 & -4 \\ 3 & 21 & 16 \end{pmatrix}$

$$= \frac{1}{6} \begin{pmatrix} 6 & 0 & 0 \\ 0 & 6 & 0 \\ 0 & 0 & 6 \end{pmatrix} = \begin{pmatrix} 1 & 0 & 0 \\ 0 & 1 & 0 \\ 0 & 0 & 1 \end{pmatrix}$$

This adjugate method for finding the inverse of a 3 × 3 matrix is reasonably straightforward but it is important to check your arithmetic as you go along, as it is very easy to make mistakes. You can use your calculator to check that you have calculated the inverse correctly.

As shown in the example above, you might also multiply the inverse by the original matrix and check that you obtain the 3 × 3 identity matrix.

Exercise 2.1

① Evaluate these determinants without using a calculator. Check your answers using your calculator.

(i) (a) $\begin{vmatrix} 1 & 1 & 3 \\ -1 & 0 & 2 \\ 3 & 1 & 4 \end{vmatrix}$ (b) $\begin{vmatrix} 1 & -1 & 3 \\ 1 & 0 & 1 \\ 3 & 2 & 4 \end{vmatrix}$

(ii) (a) $\begin{vmatrix} 1 & -5 & -4 \\ 2 & 3 & 3 \\ -2 & 1 & 0 \end{vmatrix}$ (b) $\begin{vmatrix} 1 & 2 & -2 \\ -5 & 3 & 1 \\ -4 & 3 & 0 \end{vmatrix}$

(iii) (a) $\begin{vmatrix} 2 & 1 & 2 \\ 3 & 5 & 3 \\ 1 & -1 & 1 \end{vmatrix}$ (b) $\begin{vmatrix} 1 & 5 & 0 \\ 1 & 5 & 0 \\ 2 & 1 & -2 \end{vmatrix}$

What do you notice about the determinants?

② Find the inverses of the following matrices, if they exist, without using a calculator.

(i) $\begin{pmatrix} 1 & 2 & 4 \\ 2 & 4 & 5 \\ 0 & 1 & 2 \end{pmatrix}$ (ii) $\begin{pmatrix} 3 & 2 & 6 \\ 5 & 3 & 11 \\ 7 & 4 & 16 \end{pmatrix}$

(iii) $\begin{pmatrix} 5 & 5 & -5 \\ -9 & 3 & -5 \\ -4 & -6 & 8 \end{pmatrix}$ (iv) $\begin{pmatrix} 6 & 5 & 6 \\ -5 & 2 & -4 \\ -4 & -6 & -5 \end{pmatrix}$

③ Find the inverse of the matrix $\begin{pmatrix} 4 & -5 & 3 \\ 3 & 3 & -4 \\ 5 & 4 & -6 \end{pmatrix}$ and hence solve the simultaneous equations:

$4x - 5y + 3z = 3$

$3x + 3y - 4z = 48$

$5x + 4y - 6z = 74$

④ Find the inverse of the matrix $\mathbf{M} = \begin{pmatrix} 1 & 3 & -2 \\ k & 0 & 4 \\ 2 & -1 & 4 \end{pmatrix}$ where $k \neq 0$.

For what value of k is the matrix $\mathbf{M}$ singular?

⑤ (i) Investigate the relationship between the matrices

$\mathbf{A} = \begin{pmatrix} 0 & 3 & 1 \\ 2 & 4 & 2 \\ -1 & 3 & 5 \end{pmatrix}$ $\mathbf{B} = \begin{pmatrix} 1 & 0 & 3 \\ 2 & 2 & 4 \\ 5 & -1 & 3 \end{pmatrix}$ $\mathbf{C} = \begin{pmatrix} 3 & 1 & 0 \\ 4 & 2 & 2 \\ 3 & 5 & -1 \end{pmatrix}$

(ii) Find $\det \mathbf{A}$, $\det \mathbf{B}$ and $\det \mathbf{C}$ and comment on your answer.

⑥ Show that $x = 1$ is one root of the equation $\begin{vmatrix} 2 & 2 & x \\ 1 & x & 1 \\ x & 1 & 4 \end{vmatrix} = 0$ and find the other roots.

⑦ Find the values of x for which the matrix $\begin{pmatrix} 3 & -1 & 1 \\ 2 & x & 4 \\ x & 1 & 3 \end{pmatrix}$ is singular.

⑧ Given that the matrix $\mathbf{M} = \begin{pmatrix} k & 2 & 1 \\ 0 & -k & 2 \\ 2k & 1 & 3 \end{pmatrix}$ has determinant greater than 5, find the range of possible values for k.

⑨ (i) $\mathbf{P}$ and $\mathbf{Q}$ are non-singular matrices. Prove that $(\mathbf{PQ})^{-1} = \mathbf{Q}^{-1}\mathbf{P}^{-1}$.

(ii) Find the inverses of the matrices $\mathbf{P} = \begin{pmatrix} 0 & 3 & -1 \\ -2 & 2 & 2 \\ -3 & 0 & 1 \end{pmatrix}$ and $\mathbf{Q} = \begin{pmatrix} 2 & 1 & 2 \\ 1 & 0 & 1 \\ 4 & -3 & 2 \end{pmatrix}$.

Using the result from part (i), find $(\mathbf{PQ})^{-1}$.

⑩ (i) Prove that $\begin{vmatrix} ka_1 & b_1 & c_1 \\ ka_2 & b_2 & c_2 \\ ka_3 & b_3 & c_3 \end{vmatrix} = k \begin{vmatrix} a_1 & b_1 & c_1 \\ a_2 & b_2 & c_2 \\ a_3 & b_3 & c_3 \end{vmatrix}$, where k is a constant.

(ii) Explain in terms of volumes why multiplying all the elements in the first column by a constant k multiplies the value of the determinant by k.

(iii) What would happen if you multiplied a different column by k?

⑪ Given that $\begin{vmatrix} 1 & 2 & 3 \\ 6 & 4 & 5 \\ 7 & 5 & 1 \end{vmatrix} = 43$, write down the values of the determinants:

(i) $\begin{vmatrix} 10 & 2 & 3 \\ 60 & 4 & 5 \\ 70 & 5 & 1 \end{vmatrix}$

(ii) $\begin{vmatrix} 4 & 10 & -21 \\ 24 & 20 & -35 \\ 28 & 25 & -7 \end{vmatrix}$

(iii) $\begin{vmatrix} x & 4 & 3y \\ 6x & 8 & 5y \\ 7x & 10 & y \end{vmatrix}$

(iv) $\begin{vmatrix} x^4 & \dfrac{1}{x} & 12y \\ 6x^4 & \dfrac{2}{x} & 20y \\ 7x^4 & \dfrac{5}{2x} & 4y \end{vmatrix}$

2 Intersection of three planes

In most cases, two planes intersect in a straight line. The exception occurs when they are parallel. It is also possible that they are one and the same plane. However in what follows, it is assumed that all the planes being considered are distinct.

In this section you will look at the different possibilities for how *three* planes can be arranged in three-dimensional space.

You saw in Chapter 1 that the cartesian equation of a plane is of the form $ax + by + cz + d = 0$. Solving a system of three equations in three unknowns x, y and z is equivalent to finding where the three planes represented by those equations intersect.

There are five ways in which three distinct planes π_1, π_2 and π_2 can intersect in 3D space.

If two of the planes are parallel, there are two possibilities for the third:

- it can be parallel to the other two (see Figure 2.8); or
- it can cut the other two (see Figure 2.9).

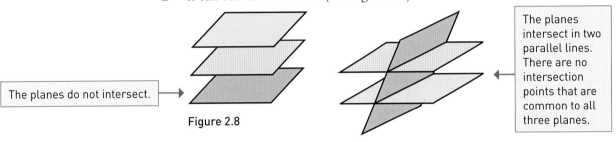

The planes do not intersect.

Figure 2.8

The planes intersect in two parallel lines. There are no intersection points that are common to all three planes.

Figure 2.9

If none of the planes are parallel, there are three possibilities:

- The planes intersect in a single point (see Figure 2.10)

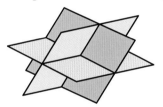

Figure 2.10

- The planes form a **sheaf** (see Figure 2.11)

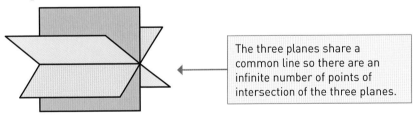

The three planes share a common line so there are an infinite number of points of intersection of the three planes.

Figure 2.11

- The planes form a **triangular prism** (see Figure 2.12)

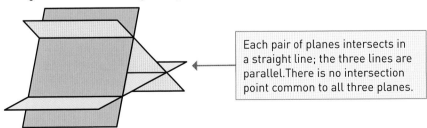

Each pair of planes intersects in a straight line; the three lines are parallel. There is no intersection point common to all three planes.

Figure 2.12

The diagrams above show that three planes either intersect in a unique point, intersect in an infinite number of points or do not have a common intersection point.

Finding the unique point of intersection of three planes

3×3 matrices can be used to find the point of intersection of three planes that intersect in a unique point or to determine that the planes have a different arrangement.

Suppose the three planes have equations

$$a_1 x + b_1 y + c_1 z = d_1$$
$$a_2 x + b_2 y + c_2 z = d_2$$
$$a_3 x + b_3 y + c_3 z = d_3$$

They can be written as $\mathbf{M} \begin{pmatrix} x \\ y \\ z \end{pmatrix} = \begin{pmatrix} d_1 \\ d_2 \\ d_3 \end{pmatrix}$, where $\mathbf{M} = \begin{pmatrix} a_1 & b_1 & c_1 \\ a_2 & b_2 & c_2 \\ a_3 & b_3 & c_3 \end{pmatrix}$.

Figure 2.13 summarises the decisions that need to be made and Example 2.6 shows how these decision are carried out. Note that the trivial situation where the matrix $\mathbf{M}$ consists entirely of zeroes has been ignored, and also cases where two or three of the equations are the same (i.e. two or three of the planes are coincident).

If the determinant of $\mathbf{M}$ is zero then there is no unique solution to the equations i.e. no unique point of intersection of the planes represented by the matrix. In this case one of the other arrangements of the planes would be relevant.

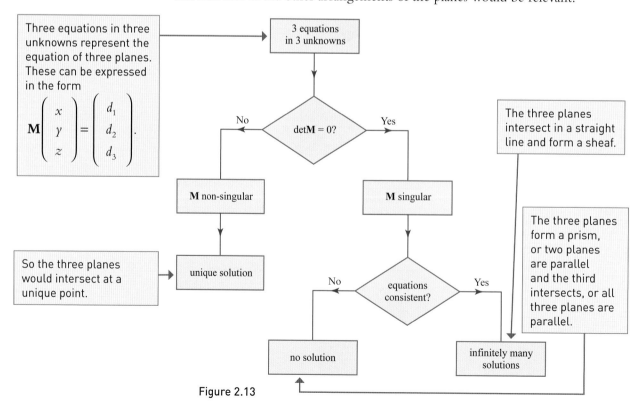

Figure 2.13

To summarise:

If **M** is non-singular, the planes intersect in a unique point.

If **M** is singular, the planes must be arranged in one of the other four possible arrangements:

■ three parallel planes
■ two parallel planes that are cut by the third to form two parallel lines
■ a sheaf of planes that intersect in a common line
■ a prism of planes in which each pair of planes meet in a straight line but there are no common points of intersection between the three planes.

Example 2.6

Three planes have equations

$$\pi_1: x + 3y - 2z = 7 \qquad ①$$
$$\pi_2: 2x - 2y + z = 3 \qquad ②$$
$$\pi_3: 3x + y - z = k \qquad ③$$

(i) Explain how you know that none of the planes are parallel to any of the other planes.

(ii) Investigate the intersection of the planes when:

    (a) $k = 10$ (b) $k = 12$

Solution

You cannot multiply any of the equations by a number to obtain one of the other equations.

→ (i) If planes are parallel they can be written as a scalar multiple of each other. That is not true in this case, so none of the planes are parallel to any of the others.

Writing the equation of the planes in the matrix format $$\mathbf{M}\begin{pmatrix} x \\ y \\ z \end{pmatrix} = \begin{pmatrix} d_1 \\ d_2 \\ d_3 \end{pmatrix}.$$

→ (ii) $$\begin{pmatrix} 1 & 3 & -2 \\ 2 & -2 & 1 \\ 3 & 1 & -1 \end{pmatrix}\begin{pmatrix} x \\ y \\ z \end{pmatrix} = \begin{pmatrix} 7 \\ 3 \\ k \end{pmatrix}$$

Using a calculator, $\det \mathbf{M} = 0$ so the matrix is singular and hence the planes do not intersect at a unique point.

To investigate the intersection of the planes in a case where the matrix is similar, try to solve the equations algebraically.

Eliminating the variable z produces two equations in x and y:

① $- 2 \times$ ③: $-5x + y = 7 - 2k$

② $+$ ③: $5x - y = 3 + k$

For the equations to be consistent, the value of $5x - y$ must be the same in each equation.

These equations are consistent if $2k - 7 = 3 + k$
$$\Leftrightarrow k = 10$$

and in this case both equations reduce to $5x - y = 13$.

(a) The equations are consistent in the case $k = 10$ and so there are infinitely many solutions. Since none of the planes are coincident, they intersect in a straight line and form a sheaf.

(b) When $k = 12$ the two equations in x and y are inconsistent:

$$-5x + y = -17$$
$$5x - y = 15$$

$5x - y$ cannot be equal to both 17 and 15.

Therefore there are no solutions. Since the planes are not parallel they must form a prism of planes.

Exercise 2.2

① (i) Without using a calculator, find the determinant and inverse of the

matrix $\mathbf{M} = \begin{pmatrix} 2 & -3 & 1 \\ 3 & 0 & 4 \\ 1 & -1 & 3 \end{pmatrix}$.

(ii) Using your answer to part (i), show that the planes

$2x - 3y + z = 10$
$3x + 4z = 25$
$x - y + 3z = 20$

intersect in the unique point $(0, -1.25, 6.25)$.

② Without carrying out any calculations, describe the arrangements of the following sets of planes:

(i) $\pi_1: 2x + y - z = 4$
$\pi_2: x + y - z = 2$
$\pi_3: 2x + y - z = 6$

(ii) $\pi_1: 2x + y - z = 4$
$\pi_2: 6x + 3y - 3z = 12$
$\pi_3: 2x + y - z = 6$

(iii) $\pi_1: 2x + y - z = 4$
$\pi_2: 10x + 5y - 5z = 15$
$\pi_3: 2x + y - z = 6$

③ (i) Express the equations of the three planes

$\pi_1: 5x + 3y - 2z = 6$
$\pi_2: 6x + 2y + 3z = 11$
$\pi_3: 7x + y + 8z = 12$

in the form $\mathbf{M} \begin{pmatrix} x \\ y \\ z \end{pmatrix} = \begin{pmatrix} d_1 \\ d_2 \\ d_3 \end{pmatrix}$ and show that $\det \mathbf{M} = 0$.

(ii) Eliminate the variable y from the three equations and hence show that the three planes form a prism of planes.

④ Determine the arrangement of the following planes in three dimensions. You should find the determinant and inverse of matrices without using a calculator.

(i) $\pi_1: x + 2y + 4z = 7$
$\pi_2: 3x + 2y + 5z = 21$
$\pi_3: 4x + y + 2z = 14$

(ii) $\pi_1: x + y + z = 4$
$\pi_2: 2x + 3y - 4z = 3$
$\pi_3: 5x + 8y - 13z = 8$

(iii) $\pi_1: 2x - y = 1$
$\pi_2: 3x + 2z = 13$
$\pi_3: 3y + 4z = 23$

(iv) $\pi_1: 3x + 2y + z = 2$
$\pi_2: 5x + 3y - 4z = 1$
$\pi_3: x + y + 4z = 5$

(v) $\pi_1: 2x + y - z = 5$
$\pi_2: 8x + 4y - 4z = 20$
$\pi_3: -2x - y + z = -5$

⑤ Solve, where possible, the equation

$$\begin{pmatrix} 1 & 3 & -2 \\ -3 & 1 & m \\ -3 & 11 & -4 \end{pmatrix} \begin{pmatrix} x \\ y \\ z \end{pmatrix} = \begin{pmatrix} -2 \\ 6 \\ k \end{pmatrix}$$

in each of these cases:

(i) $m = 2, k = 3$ (ii) $m = 1, k = 3$ (iii) $m = 1, k = 6$

In each case interpret the solution geometrically with reference to three planes in three dimensions.

⑥ (i) Obtain an expression for the inverse of the matrix $\begin{pmatrix} k & -7 & 4 \\ 2 & -2 & 3 \\ 1 & -3 & -2 \end{pmatrix}$ in terms of k.

State the value of k for which the planes

$$kx - 7y + 4z = 3$$
$$2x - 2y + 3z = 7$$
$$x - 3y - 2z = -1$$

would not intersect in a unique point.

(ii) Describe the intersection of the planes

$$4x - 7y + 4z = p$$
$$2x - 2y + 3z = 1$$
$$x - 3y - 2z = 2$$

giving your answer in terms of p.

(iii) Find the value of p for which the planes

$$5x - 7y + 4z = p$$
$$2x - 2y + 3z = 1$$
$$x - 3y - 2z = 2$$

have at least one point of intersection and describe the arrangement of the planes geometrically in three dimensions.

LEARNING OUTCOMES

When you have completed this chapter you should be able to:

➤ find the determinant of a 2×2 and a 3×3 matrix and explain their geometrical significance

➤ find the inverse of a non-singular 2×2 or 3×3 matrix

➤ use matrices to solve simultaneous linear equations

➤ use matrices to determine how three planes intersect in three dimensions.

KEY POINTS

1 If $\mathbf{M} = \begin{pmatrix} a & b \\ c & d \end{pmatrix}$ then the determinant of $\mathbf{M}$, written det $\mathbf{M}$ or $|\mathbf{M}|$ or Δ is given by det $\mathbf{M} = ad - bc$.

2 The determinant of a 2×2 matrix represents the area scale factor of the transformation.

3 If $\mathbf{M} = \begin{pmatrix} a & b \\ c & d \end{pmatrix}$ then $\mathbf{M}^{-1} = \dfrac{1}{ad - bc} \begin{pmatrix} d & -b \\ -c & a \end{pmatrix}$.

4 The determinant of a 3×3 matrix $\mathbf{M} = \begin{pmatrix} a_1 & b_1 & c_1 \\ a_2 & b_2 & c_2 \\ a_3 & b_3 & c_3 \end{pmatrix}$ is given by

$$\det \mathbf{M} = a_1 \begin{vmatrix} b_2 & c_2 \\ b_3 & c_3 \end{vmatrix} - a_2 \begin{vmatrix} b_1 & c_1 \\ b_3 & c_3 \end{vmatrix} + a_3 \begin{vmatrix} b_1 & c_1 \\ b_2 & c_2 \end{vmatrix}.$$

5 The determinant of a 3×3 matrix represents the volume scale factor of the transformation.

6 For a 3×3 matrix $\begin{pmatrix} a_1 & b_1 & c_1 \\ a_2 & b_2 & c_2 \\ a_3 & b_3 & c_3 \end{pmatrix}$ the **minor** of an element is formed by crossing out the row and column containing that element and finding the determinant of the resulting 2×2 matrix.

7 A minor, together with its correct sign, given by the matrix $\begin{vmatrix} + & - & + \\ - & + & - \\ + & - & + \end{vmatrix}$ is known as a **cofactor** and is denoted by the corresponding capital letter; for example the cofactor of a_3 is A_3.

8 The inverse of a 3×3 matrix $\mathbf{M} = \begin{pmatrix} a_1 & b_1 & c_1 \\ a_2 & b_2 & c_2 \\ a_3 & b_3 & c_3 \end{pmatrix}$ can be found using a calculator or using the formula

$$\mathbf{M}^{-1} = \frac{1}{\det \mathbf{M}} \operatorname{adj} \mathbf{M} = \frac{1}{\det \mathbf{M}} \begin{pmatrix} A_1 & A_2 & A_3 \\ B_1 & B_2 & B_3 \\ C_1 & C_2 & C_3 \end{pmatrix}, \Delta \neq 0$$

The matrix $\begin{pmatrix} A_1 & A_2 & A_3 \\ B_1 & B_2 & B_3 \\ C_1 & C_2 & C_3 \end{pmatrix}$ is the **adjoint** or **adjugate** matrix, denoted **adj** $\mathbf{M}$, formed by replacing each element of $\mathbf{M}$ by its cofactor and then transposing (i.e. changing rows into columns and columns into rows).

9 $(\mathbf{MN})^{-1} = \mathbf{N}^{-1}\mathbf{M}^{-1}$

10 A matrix is **singular** if the determinant is zero. If the determinant is non-zero the matrix is said to be **non-singular**.

11 If the determinant of a matrix is zero, all points are mapped to either a straight line (in two dimensions) or to a plane (three dimensions).

12 If $\mathbf{A}$ is a non-singular matrix, $\mathbf{AA}^{-1} = \mathbf{A}^{-1}\mathbf{A} = \mathbf{I}$.

13 When solving 2 simultaneous equations in 2 unknowns, the equations can be written as a matrix equation $\mathbf{M}\begin{pmatrix} x \\ y \end{pmatrix} = \begin{pmatrix} a \\ b \end{pmatrix}$.

When solving 3 simultaneous equations in 3 unknowns, the equations can be written as a matrix equation $\mathbf{M}\begin{pmatrix} x \\ y \\ z \end{pmatrix} = \begin{pmatrix} a \\ b \\ c \end{pmatrix}$.

In both cases, if $\det \mathbf{M} \neq 0$ there is a unique solution to the equations which can be found by pre-multiplying both sides of the equation by the inverse matrix $\mathbf{M}^{-1}$.

If $\det \mathbf{M} = 0$ there is no unique solution to the equations. In this case there is either no solution or an infinite number of solutions.

14 Three distinct planes in three dimensions will be arranged in one of five ways:

■ They meet in a unique point of intersection

■ All three planes are parallel and therefore do not meet

■ Two of the planes are parallel, and these are cut by the third plane to form two parallel lines

■ The planes form a sheaf of planes that intersect in a common line

■ The planes form a prism of planes in which each pair of planes meet in a straight line but there are no common points of intersection between the three planes

15 Three distinct planes

$$a_1 x + b_1 y + c_1 z = d_1$$
$$a_2 x + b_2 y + c_2 z = d_2$$
$$a_3 x + b_3 y + c_3 z = d_3$$

can be expressed in the form

$$\mathbf{M}\begin{pmatrix} x \\ y \\ z \end{pmatrix} = \begin{pmatrix} d_1 \\ d_2 \\ d_3 \end{pmatrix}$$

where $\mathbf{M} = \begin{pmatrix} a_1 & b_1 & c_1 \\ a_2 & b_2 & c_2 \\ a_3 & b_3 & c_3 \end{pmatrix}$.

If $\mathbf{M}$ is non-singular, the unique point of intersection is given by $\mathbf{M}^{-1}\begin{pmatrix} d_1 \\ d_2 \\ d_3 \end{pmatrix}$.

Otherwise, the planes meet in one of the other four possible arrangements. In the case of a sheaf of planes, the equations have an infinite number of possible solutions, and in the other three cases the equations have no solutions.

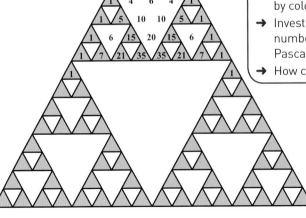

The essence of mathematics is not to make simple things complicated, but to make complicated things simple.

S. Gudder

Figure 3.1

Discussion points

The image of Pascal's triangle shown here has the odd numbers coloured. This results in a pattern similar to the Sierspinksy triangle, which is an example of a fractal.

→ Investigate the patterns produced by colouring multiples of 3, 4, etc.

→ Investigate the sum of the numbers in the first *n* rows of Pascal's triangle.

→ How could you prove your result?

1 Summing series

Review: Terminology and notation

A **sequence** is an ordered set of objects with an underlying rule.

- The terms of a sequence are often written as $a_1, a_2, a_3, \ldots$ or $u_1, u_2, u_3, \ldots$.

- The general term of a sequence may be written as a_r or u_r. (Sometimes the letters k or i are used instead of r.)

- The last term is usually written as a_n or u_n.

A **series** is the sum of the terms of a numerical sequence.

- The sum of the first n terms of a sequence is often denoted by S_n.

- $S_n = a_1 + a_2 + \ldots + a_n = \displaystyle\sum_{r=1}^{n} a_r$.

Types of sequences

- A sequence is **increasing** if each term is greater than the previous term.

- A sequence is **decreasing** if each term is smaller than the previous term.

- In an **oscillating** sequence, the terms lie above and below a middle number.

- In an **arithmetic sequence**, the difference between each term and the next is constant. It is called the **common difference** and denoted by d.

- In a **geometric sequence**, the ratio of each term to the next is constant. It is called the **common ratio** and denoted by r.

- The terms of a **convergent** sequence get closer and closer to a limiting value. A geometric sequence is convergent if $-1 < r < 1$.

Review: Summing series using standard formulae

Many series can be summed by using the standard formulae:

$$\sum_{r=1}^{n} 1 = n$$

> Remember that $\displaystyle\sum_{r=1}^{n} 1 = 1 + 1 + 1 + \ldots + 1$, so this means that 1 is added together n times, giving a total of n.

$$\sum_{r=1}^{n} r = \frac{1}{2}n(n + 1)$$

$$\sum_{r=1}^{n} r^2 = \frac{1}{6}n(n + 1)(2n + 1)$$

$$\sum_{r=1}^{n} r^3 = \frac{1}{4}n^2(n + 1)^2$$

> This result is precisely the square of the result $\displaystyle\sum r$.

ACTIVITY 3.1

Prove that $1 + 2 + 3 + \ldots + n = \frac{1}{2}n(n + 1)$ by using the formula for the sum of an arithmetic series.

You will prove the formulae for $\sum\limits_{r=1}^{n} r^2$ and $\sum\limits_{r=1}^{n} r^3$ later in this chapter.

Find the sum of the series $(2 \times 3) + (3 \times 4) + (4 \times 5) + \ldots + (n + 1)(n + 2)$.

Solution

The series can be written in the form $\sum\limits_{r=1}^{n} (r + 1)(r + 2)$.

> To simplify this expression, look for common factors. It's usually helpful to take out any fractions as factors.

$$\sum\limits_{r=1}^{n} (r + 1)(r + 2) = \sum\limits_{r=1}^{n} r^2 + 3\sum\limits_{r=1}^{n} r + 2\sum\limits_{r=1}^{n} 1$$

$$= \tfrac{1}{6}n(n + 1)(2n + 1) + 3 \times \tfrac{1}{2}n(n + 1) + 2n$$

$$= \tfrac{1}{6}n[(n + 1)(2n + 1) + 9(n + 1) + 12]$$

$$= \tfrac{1}{6}n\left[2n^2 + 3n + 1 + 9n + 9 + 12\right]$$

$$= \tfrac{1}{6}n\left[2n^2 + 12n + 22\right]$$

$$= \tfrac{1}{3}n(n^2 + 6n + 11)$$

Review: Summing series using the method of differences

Sometimes the general term of a sequence can be written so that most of the terms cancel out.

(i)   Show that $(r + 1)^2 - (r - 1)^2 = 4r$.

(ii)   Hence find $\sum\limits_{r=1}^{n} 4r$.

(iii)   Deduce that $\sum\limits_{r=1}^{n} r = \tfrac{1}{2}n(n + 1)$.

Solution

(i) $(r + 1)^2 - (r - 1)^2 = r^2 + 2r - 1 - (r^2 - 2r + 1)$

$$= 4r$$

(ii) $\sum\limits_{r=1}^{n} 4r = \sum\limits_{r=1}^{n} \left[(r + 1)^2 - (r - 1)^2\right]$ ← Using the result of (i).

$$= \cancel{2^2} \quad \boxed{-0^2}$$
$$+\cancel{3^2} \quad -1^2$$
$$+\cancel{4^2} \quad -\cancel{2^2}$$
$$+\ldots$$
$$+\cancel{(n-1)^2} \quad -\cancel{(n-3)^2}$$
$$+\boxed{n^2} \quad -\cancel{(n-2)^2}$$
$$+\boxed{(n+1)^2} \quad -\cancel{(n-1)^2}$$

> Write out the first few terms and the last few terms.

> Most of the terms cancel out in pairs, leaving only two at the start and two at the end.

$$= (n+1)^2 + n^2 - 1^2$$
$$= n^2 + 2n + 1 + n^2 - 1$$
$$= 2n^2 + 2n$$
$$= 2n(n+1)$$

(iii) From part (ii), $\sum_{r=1}^{n} 4r = 2n(n+1)$

so $\sum_{r=1}^{n} r = \frac{1}{4} \times 2n(n+1) = \frac{1}{2}n(n+1)$

> This is the standard formula for the sum of the integers, given on page 69.

Prior Knowledge

You need to know how to use partial fractions. This is covered in Chapter 7 of the Year 2 A Level Mathematics book.

Summing series using partial fractions

If the general term of a sequence can be expressed in partial fractions, this may give the opportunity to use the method of differences.

Example 3.3

Find $\sum_{r=1}^{n} \frac{1}{r(r+1)}$.

Solution

$$\frac{1}{r(r+1)} = \frac{A}{r} + \frac{B}{r+1}$$
$$1 = A(r+1) + Br$$

> First, write the expression using partial fractions.

Let $r = 0 \quad \Rightarrow A = 1$
Let $r = -1 \quad \Rightarrow B = -1$

$$\sum_{r=1}^{n} \frac{1}{r(r+1)} = \sum_{r=1}^{n}\left(\frac{1}{r} - \frac{1}{r+1}\right)$$

$$= \boxed{\tfrac{1}{1}} \quad -\cancel{\tfrac{1}{2}}$$
$$+\cancel{\tfrac{1}{2}} \quad -\cancel{\tfrac{1}{3}}$$
$$+\cancel{\tfrac{1}{3}} \quad -\cancel{\tfrac{1}{4}}$$
$$+\ldots$$
$$+\cancel{\tfrac{1}{n-2}} \quad -\cancel{\tfrac{1}{n-1}}$$
$$+\cancel{\tfrac{1}{n-1}} \quad -\cancel{\tfrac{1}{n}}$$
$$+\cancel{\tfrac{1}{n}} \quad \boxed{-\tfrac{1}{n+1}}$$

> Write out the first few terms and the last few terms. Most of the terms cancel, leaving just one term at the start and one at the end.

71

Discussion point

➜ In the example above, how does the form of the partial fractions tell you that you can use the method of differences to sum the series?

$$= 1 - \frac{1}{n+1}$$

$$= \frac{n+1-1}{n+1}$$

$$= \frac{n}{n+1}$$

Sometimes the partial fractions involve three terms, as in the next example. This means that it is particularly important to lay out your work carefully, so that you can see clearly which terms cancel.

Example 3.4

(i) Find $\displaystyle\sum_{r=1}^{n} \frac{2}{r(r+1)(r+2)}$.

(ii) Hence state the value of $\displaystyle\sum_{r=1}^{\infty} \frac{2}{r(r+1)(r+2)}$.

Solution

(i) $\dfrac{2}{r(r+1)(r+2)} = \dfrac{A}{r} + \dfrac{B}{r+1} + \dfrac{C}{r+2}$

Write the expression using partial fractions.

$2 = A(r+1)(r+2) + Br(r+2) + Cr(r+1)$

Let $r = 0 \quad \Rightarrow 2 = 2A \quad \Rightarrow A = 1$

Let $r = -1 \quad \Rightarrow 2 = -B \quad \Rightarrow B = -2$

Let $r = -2 \quad \Rightarrow 2 = 2C \quad \Rightarrow C = 1$

$$\sum_{r=1}^{n} \frac{2}{r(r+1)(r+2)} = \sum_{r=1}^{n}\left(\frac{1}{r} - \frac{2}{r+1} + \frac{1}{r+2}\right)$$

$$= \frac{1}{1} \qquad -\frac{2}{2} \qquad +\frac{1}{3}$$
$$+\frac{1}{2} \qquad -\frac{2}{3} \qquad +\frac{1}{4}$$
$$+\frac{1}{3} \qquad -\frac{2}{4} \qquad +\frac{1}{5}$$
$$+\dots$$
$$+\frac{1}{n-2} \quad -\frac{2}{n-1} \quad +\frac{1}{n}$$
$$+\frac{1}{n-1} \quad -\frac{2}{n} \quad +\frac{1}{n+1}$$
$$+\frac{1}{n} \qquad -\frac{2}{n+1} \quad +\frac{1}{n+2}$$

Most of the terms cancel out in groups of 3.

There are three terms left at the beginning and three at the end. Notice the symmetrical pattern.

$$= 1 - 1 + \frac{1}{2} + \frac{1}{n+1} - \frac{2}{n+1} + \frac{1}{n+2}$$

$$= \frac{1}{2} - \frac{1}{n+1} + \frac{1}{n+2}$$

(ii) As $n \to \infty$, $\dfrac{1}{n+1} \to 0$ and $\dfrac{1}{n+2} \to 0$

so $\displaystyle\sum_{r=1}^{\infty} \frac{2}{r(r+1)(r+2)} = \frac{1}{2}$

① Find $\displaystyle\sum_{r=1}^{n}(4r-1)$.

② Find $\displaystyle\sum_{r=1}^{n}(3r^2+r)$.

③ Find $\displaystyle\sum_{r=1}^{n}(2r^3+r)$.

④ (i) Write $\dfrac{2}{r(r+2)}$ in the form $\dfrac{A}{r}+\dfrac{B}{r+2}$.

　　(ii) Hence find $\displaystyle\sum_{r=1}^{n}\dfrac{2}{r(r+2)}$.

⑤ (i) Find $\displaystyle\sum_{r=1}^{n}r(r+1)(r+2)$.

　　(ii) Hence find $1\times 2\times 3+2\times 3\times 4+3\times 4\times 5+\ldots+100\times 101\times 102$.

⑥ (i) Show that $\frac{1}{3}(r+1)(r+2)(r+3)-\frac{1}{3}r(r+1)(r+2)=(r+1)(r+2)$.

　　(ii) Using the result from (i) and the method of differences,

　　　　find $\displaystyle\sum_{r=1}^{n}(r+1)(r+2)$.

　　(iii) Use standard results to find $\displaystyle\sum_{r=1}^{n}(r+1)(r+2)$ and show that this is the same as the result from (ii).

⑦ (i) Find $\displaystyle\sum_{r=1}^{n}\dfrac{7r+10}{r(r+1)(r+2)}$.

　　(ii) Hence find $\displaystyle\sum_{r=1}^{\infty}\dfrac{7r+10}{r(r+1)(r+2)}$.

⑧ (i) Find $\displaystyle\sum_{r=1}^{n}\dfrac{12r+2}{(2r-1)(2r+1)(2r+3)}$.

　　(ii) Hence find $\dfrac{7}{1\times 3\times 5}+\dfrac{13}{3\times 5\times 7}+\dfrac{19}{5\times 7\times 9}+\ldots$

⑨ A sum is given by $S=1-2\left(\frac{1}{3}\right)+3\left(\frac{1}{3}\right)^2-4\left(\frac{1}{3}\right)^3+\ldots$

　　(i) Write down an expression for $\frac{1}{3}S$.

　　(ii) Add S and $\frac{1}{3}S$. Describe the resulting series and find its sum to infinity. Hence find the value of S.

　　(iii) Show that S is the binomial expansion of $\left(1+\frac{1}{3}\right)^{-2}$. Use this result to confirm the value of S you found in part (ii).

2 Proof by induction

When you are solving a mathematical problem, you may sometimes make a **conjecture**. You might, for example, find a formula which seems to work in the cases you have investigated. You would then want to prove your conjecture.

Mathematical induction is a very powerful method that can be used to prove a conjecture or a given result, such as for the sum of a series.

The principle of proof by induction is to show that:

> **if** the result is true for the case $n = k$
>
> **then** it must be true for the case $n = k + 1$.

If you also show that it is true for an initial case, say $n = 1$, you can then deduce that it must be true for $n = 2$, and therefore it must be true for $n = 3$, and so on. You can then state that it is true for all positive integer values of n.

Steps in mathematical induction

To prove something by mathematical induction you need to state a conjecture to start with. Then there are five elements needed to try to prove the conjecture is true.

- Proving that it is true for a starting value (e.g. $n = 1$).
- Finding the target expression: using the result for $n = k$ to find the equivalent result for $n = k + 1$. ← To find the target expression you replace k with $k + 1$ in the result for $n = k$.

- Proving that: **if** it is true for $n = k$, **then** it is true for $n = k + 1$.

  This can be done before or after finding the target expression, but you may find it easier to find the target expression first so that you know what you are working towards.

- Arguing that since it is true for $n = 1$, it is also true for $n = 1 + 1 = 2$, and so for $n = 2 + 1 = 2$ and for all subsequent values of n.
- Concluding the argument by writing down the result and stating that it has been proved. ← This ensures the argument is properly rounded off. You will often use the word 'therefore'.

Example 3.5	**Finding the sum of a series**

Prove by induction that, for all positive integers n

$$1 + 3 + 5 + \ldots + (2n - 1) = n^2$$

Solution

When $n = 1$,   L.H.S. $= 1$ R.H.S. $= 1$

So it is true for $n = 1$.

Assume the result is true for $n = k$, so

$$1 + 3 + 5 + \ldots + 2k - 1 = k^2$$

You want to prove that the result is true for $n = k + 1$ (if the assumption is true).

Target expression

$$1 + 3 + 5 + \ldots + (2k - 1) + [2(k + 1) - 1] = (k + 1)^2$$

Look at the L.H.S. of the result you want to prove:

$$1 + 3 + 5 + \ldots + 2k - 1 + [2(k + 1) - 1]$$

Use the assumed result for $n = k$, to replace the first k terms

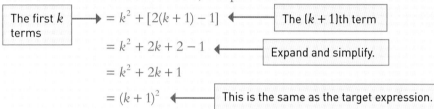

The first k terms $\quad \rightarrow \quad = k^2 + [2(k + 1) - 1] \quad \leftarrow \quad$ The $(k + 1)$th term

$$= k^2 + 2k + 2 - 1 \quad \leftarrow \quad \text{Expand and simplify.}$$

$$= k^2 + 2k + 1$$

$$= (k + 1)^2 \quad \leftarrow \quad \text{This is the same as the target expression.}$$

as required.

If the result is true for $n = k$, *then* it is true for $n = k + 1$.

Since it is true for $n = 1$, it is true for all positive integer values of n. Therefore the result that $1 + 3 + 5 + \ldots + (2n - 1) = n^2$ is true.

The method of proof by induction is often used in the context of the sum of a series, as in the example above. However, it has a number of other applications as well.

Induction can be used in divisibility proofs, as shown in the next example. In proofs like these, there is no 'target expression'; instead your target is to express the result in a form which shows the divisibility property that you are proving.

Example 3.6

Divisibility

Prove that $u_n = 4^n + 6n - 1$ is divisible by 9 for all $n \geqslant 1$.

Solution

When $n = 1$, $u_1 = 4^1 + 6 - 1 = 9$ which is divisible by 9. So it is true for $n = 1$.

Assume the result is true for $n = k$, so

$$u_k = 4^k + 6k - 1 \text{ is divisible by } 9$$

You want to prove that u_{k+1} is divisible by 9 (if the assumption is true).

$$u_{k+1} = 4^{k+1} + 6(k+1) - 1$$
$$= 4 \times 4^k + 6k + 5$$

> You want to express u_{k+1} in terms of u_k.

$$= 4(u_k - 6k + 1) + 6k + 5$$

> Substituting $4^k = u_k - 6k + 1$.

$$= 4u_k - 24k + 4 + 6k + 5$$
$$= 4u_k - 18k + 9$$
$$= 4u_k - 9(2k + 1)$$

> You have assumed that u_k is divisible by 9, and $9(2k+1)$ is divisible by 9, so u_{k+1} is divisible by 9.

If u_k is divisible by 9, then u_{k+1} is divisible by 9.

Since it is true for $n = 1$, it is true for all positive integer values of n.
Therefore the result that $u_n = 4^n + 6n - 1$ is divisible by 9 is true.

Example 3.7

Matrix powers

Given that $\mathbf{A} = \begin{pmatrix} -3 & 8 \\ -2 & 5 \end{pmatrix}$, prove by induction that $\mathbf{A}^n = \begin{pmatrix} 1 - 4n & 8n \\ -2n & 1 + 4n \end{pmatrix}$.

Solution

When $n = 1$, $\mathbf{A}^1 = \begin{pmatrix} 1 - 4 & 8 \\ -2 & 1 + 4 \end{pmatrix} = \begin{pmatrix} -3 & 8 \\ -2 & 5 \end{pmatrix} = \mathbf{A}$

so the result is true for $n = 1$.

Assume the result is true for $n = k$, so

$$\mathbf{A}^k = \begin{pmatrix} 1 - 4k & 8k \\ -2k & 1 + 4k \end{pmatrix}$$

You want to prove that the result is true for $n = k + 1$ (if the assumption is true).

> **Target expression**
> $$\mathbf{A}^{k+1} = \begin{pmatrix} 1 - 4(k+1) & 8(k+1) \\ -2(k+1) & 1 + 4(k+1) \end{pmatrix}$$
> $$= \begin{pmatrix} -3 - 4k & 8k + 8 \\ -2k - 2 & 4k + 5 \end{pmatrix}$$

$$\mathbf{A}^{k+1} = \mathbf{A}^k \mathbf{A}$$

$$= \begin{pmatrix} 1 - 4k & 8k \\ -2k & 1 + 4k \end{pmatrix} \begin{pmatrix} -3 & 8 \\ -2 & 5 \end{pmatrix}$$

> Multiply the assumed result for $\mathbf{A}^k$ by the matrix $\mathbf{A}$.

$$= \begin{pmatrix} -3(1 - 4k) - 16k & 8(1 - 4k) + 40k \\ 6k - 2(1 + 4k) & -16k + 5(1 + 4k) \end{pmatrix}$$

$$= \begin{pmatrix} -3 - 4k & 8k + 8 \\ -2k - 2 & 4k + 5 \end{pmatrix}$$

> This is the same as the target expression.

If the result is true for $n = k$, **then** it is true for $n = k + 1$.

Since it is true for $n = 1$, it is true for all positive integer values of n.

Therefore the result that $\mathbf{A}^n = \begin{pmatrix} 1 - 4n & 8n \\ -2n & 1 + 4n \end{pmatrix}$ is true.

Example 3.8

nth term of a sequence

A sequence is defined by $u_1 = 1$ and $u_{n+1} = 3u_n - 4$.

Prove by induction that $u_n = 2 - 3^{n-1}$.

Solution

When $n = 1$, $u_n = 2 - 3^0 = 2 - 1 = 1$

so the result is true for $n = 1$.

Assume the result is true for $n = k$, so

$$u_k = 2 - 3^{k-1}$$

> Target expression
> $$u_{k+1} = 2 - 3^{(k+1)-1}$$
> $$= 2 - 3^k$$

You want to prove that the result is true for $n = k + 1$
(if the assumption is true).

$$u_{k+1} = 3u_k - 4 \longleftarrow$$
$$= 3(2 - 3^{k-1}) - 4 \longleftarrow$$
$$= 6 - 3 \times 3^{k-1} - 4$$
$$= 2 - 3^k \longleftarrow$$

> Use the given relationship between u_{n+1} and u_n.

> Substitute the assumed result for u_k.

> This is the same as the target expression.

If the result is true for $n = k$, **then** it is true for $n = k + 1$.

Since it is true for $n = 1$, it is true for all positive integer values of n.
Therefore the result that $u_n = 2 - 3^{n-1}$ is true.

Exercise 3.2

① Given that $\mathbf{A} = \begin{pmatrix} -2 & 9 \\ -1 & 4 \end{pmatrix}$, you are going to prove by induction that

$$\mathbf{A}^n = \begin{pmatrix} 1 - 3n & 9n \\ -n & 1 + 3n \end{pmatrix}.$$

(i) Show that the result is true for $n = 1$.

(ii) If the result is true, write down the target expression for $\mathbf{A}^{k+1}$.

(iii) Assuming that the result is true for $n = k$, so $\mathbf{A}^k = \begin{pmatrix} 1 - 3k & 9k \\ -k & 1 + 3k \end{pmatrix}$,

use matrix multiplication to find an expression for $\mathbf{A}^{k+1}$.

(iv) Show that your answers to (ii) and (iii) are the same, and write a conclusion for your proof.

② You are going to prove by induction that $\sum_{r=1}^{n} (3r - 1) = \frac{1}{2}n(3n + 1)$.

(i) Show that the result is true for $n = 1$.

(ii) If the result is true, write down a target expression for $\sum_{r=1}^{k+1} (3r - 1)$.

   (iii) Assuming that the result is true for $n = k$, so $\sum\limits_{r=1}^{k}(3r - 1) = \frac{1}{2}k(3k + 1)$,

         find an expression for $\sum\limits_{r=1}^{k+1}(3r - 1)$ by adding the $(k + 1)$th term to the

         sum of the first k terms.

   (iv) Show that your answers to (ii) and (iii) are the same, and write a conclusion for your proof.

③ A sequence is defined by $u_1 = 3$ and $u_{n+1} = 2u_n + 1$.

    You are going to prove by induction that $u_n = 2^{n+1} - 1$.

   (i)   Show that the result is true for $n = 1$.

   (ii) If the result is true, write down a target expression for u_{k+1}.

   (iii) Assuming that the result is true for $n = k$, so $u_k = 2^{k-1} - 1$, find an expression for u_{k+1} by applying the rule $u_{k+1} = 2u_k + 1$.

   (iv) Show that your answers to (ii) and (iii) are the same, and write a conclusion for your proof.

④ Prove by induction that $1^2 + 2^2 + 3^2 + \ldots + n^2 = \frac{1}{6}n(n + 1)(2n + 1)$.

⑤ Given that $\mathbf{A} = \begin{pmatrix} 6 & 5 \\ -5 & -4 \end{pmatrix}$, prove by induction that

    $\mathbf{A}^n = \begin{pmatrix} 1 + 5n & 5n \\ -5n & 1 - 5n \end{pmatrix}$.

⑥ Prove by induction that $u_n = n^3 + 2n$ is a multiple of 3 for any positive integer n.

⑦ A sequence is defined by $u_1 = 3$ and $u_{n+1} = u_n + 2^n$.

    Prove by induction that $u_n = 2^n + 1$.

⑧ Prove by induction that $u_n = 8^n - 3^n$ is divisible by 5 for any positive integer n.

⑨ Prove by induction that

    $1 \times 3 + 2 \times 5 + 3 \times 7 + \ldots + n(2n + 1) = \frac{1}{6}n(n + 1)(4n + 5)$

⑩ Given that $\mathbf{P} = \begin{pmatrix} 1 & 0 \\ -1 & 2 \end{pmatrix}$, prove by induction that $\mathbf{P}^n = \begin{pmatrix} 1 & 0 \\ 1 - 2^n & 2^n \end{pmatrix}$.

⑪ Prove by induction that $u_n = 4^{4n+1} + 3$ is a multiple of 5 for any positive integer n.

⑫ A sequence is defined by $u_1 = 2$ and $u_{n+1} = 2u_n + 5$.

    Prove by induction that $u_n = 7 \times 2^{n-1} - 5$.

⑬ Prove by induction that $u_n = 11^{n+2} + 12^{2n+1}$ is divisible by 133 for $n \geqslant 0$.

⑭ Given that $\mathbf{M} = \begin{pmatrix} 3 & 2 & -1 \\ 0 & 3 & 0 \\ 0 & 6 & 0 \end{pmatrix}$

   (i)   use a calculator to find $\mathbf{M}^2$, $\mathbf{M}^3$ and $\mathbf{M}^4$

   (ii) make a conjecture about the matrix $\mathbf{M}^n$

   (iii) prove your conjecture by induction.

LEARNING OUTCOMES

When you have finished this chapter you should be able to:

➤ sum a simple series using standard formulae for $\sum r$, $\sum r^2$ and $\sum r^3$
➤ sum a simple series using the method of differences
➤ sum a simple series using partial fractions
➤ construct and present a proof using mathematical induction for given results for a formula for the nth term of a simple sequence, the sum of a simple series, the nth power of a matrix, or a divisibility result.

KEY POINTS

1 Some series can be expressed as combinations of these standard results:

$$\sum_{r=1}^{n} r = \tfrac{1}{2}n(n+1) \qquad \sum_{r=1}^{n} r^2 = \tfrac{1}{6}n(n+1)(2n+1) \qquad \sum_{r=1}^{n} r^3 = \tfrac{1}{4}n^2(n+1)^2$$

2 Some series can be summed by using the method of differences. If the terms of the series can be written as the difference of terms of another series, then many terms may cancel out. This is called a telescoping sum.

3 Some series that involve fractions can be summed by writing the general term as partial fractions, and then using the method of differences.

4 To prove by induction that a statement involving an integer n is true for all $n \geqslant n_0$, you need to:

■ Prove that the result is true for an initial value of n, typically $n = 1$ for $n = n_0$.
■ Find the target expression: use the result for $n = k$ to find the equivalent result for $n = k + 1$.
■ Prove that: **if** it is true for $n = k$, **then** it is true for $n = k + 1$.
■ Argue that since it is true for $n = 1$, it is also true for $n = 1 + 1 = 2$, and so for $n = 2 + 1 = 3$ and for all subsequent values of n.
■ Conclude the argument with a precise statement about what has been proved.

FUTURE USES

■ You will use proof by induction to prove de Moivre's theorem in Chapter 10 Complex numbers.

4 Further calculus

The moving power of mathematics invention is not reasoning, but imagination.

Augustus de Morgan, 1806–71

Discussion point

→ How could you estimate the number of birds in this picture?

Prior knowledge

You should be confident in all the integration methods you have covered previously.

1 Improper integrals

All the definite integrals you have calculated so far have been **proper** integrals. In this section you will meet some examples of **improper integrals**.

Discussion point

→ You have drawn a curve and want to use integration to find the area between the curve and the x-axis. What features of a curve would warn you of possible difficulties?

Example 4.1

(i) Sketch the graph of $y = \dfrac{1}{x^2}$ for $x > 0$ and shade the area represented by the integral $\displaystyle\int_1^\infty \dfrac{1}{x^2}\,dx$. What features of this curve warn you of possible difficulties in evaluating this integral?

(ii) Evaluate $\displaystyle\int_1^\infty \dfrac{1}{x^2}\,dx$.

Solution

(i)

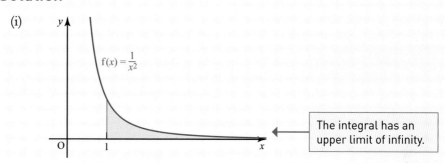

The integral has an upper limit of infinity.

Figure 4.1

(ii) The integral can be rewritten as the limit of an integral with finite limits – replacing the upper limit of infinity by the letter a, and then letting it tend towards infinity.

$$\int_1^\infty \frac{1}{x^2}\,dx = \lim_{a \to \infty}\int_1^a \frac{1}{x^2}\,dx$$

$$= \lim_{a \to \infty}\left[-\frac{1}{x}\right]_1^a$$

$$= \lim_{a \to \infty}\left(-\frac{1}{a} + 1\right)$$

As $a \to \infty$, $\dfrac{1}{a} \to 0$

As a tends to infinity, i.e. a gets very large, $\dfrac{1}{a}$ gets very small, so $\dfrac{1}{a}$ tends to zero.

So

$$\lim_{a \to \infty}\left(-\frac{1}{a} + 1\right) = 1$$

The area under the graph of $\dfrac{1}{x^2}$, from 1 to infinity, is 1 square unit.

The integral in Example 4.1 is said to be **convergent** and it converges to a value of 1. Not all integrals of this type are convergent, as the following example shows.

Example 4.2

(i) Sketch the graph of $y = \dfrac{1}{x}$ for $x > 0$, and shade the area represented by the integral $\displaystyle\int_1^\infty \frac{1}{x}\,\mathrm{d}x$.

(ii) What happens if you try to evaluate $\displaystyle\int_1^\infty \frac{1}{x}\,\mathrm{d}x$?

Solution

(i)

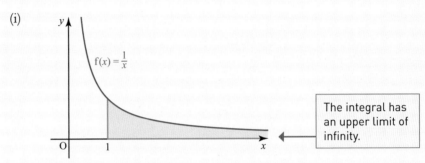

$f(x) = \dfrac{1}{x}$

The integral has an upper limit of infinity.

Figure 4.2

(ii) $\displaystyle\int_1^\infty \frac{1}{x}\,\mathrm{d}x = \lim_{a\to\infty} \int_1^a \frac{1}{x}\,\mathrm{d}x$

As in the last example, replace ∞ by a and let $a \to \infty$.

$\qquad = \lim_{a\to\infty}\big[\ln|x|\big]_1^a$

$\qquad = \lim_{a\to\infty}(\ln a - \ln 1)$

The natural logarithm function does not converge to anything as $a \to \infty$.

$\qquad = \lim_{a\to\infty}(\ln a)$

This is not defined.

Integrals like the one in Example 4.2, where there is no numerical answer, are **divergent**.

Example 4.3

(i) Sketch the graph of $y = \dfrac{1}{\sqrt{x}}$ and shade the area represented by the integral $\displaystyle\int_0^1 \frac{1}{\sqrt{x}}\,\mathrm{d}x$. What feature of the curve warns you of possible difficulties in evaluating this integral?

(ii) Evaluate $\displaystyle\int_0^1 \frac{1}{\sqrt{x}}\,\mathrm{d}x$.

Solution

(i)

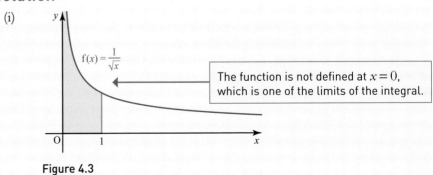

$f(x) = \dfrac{1}{\sqrt{x}}$

The function is not defined at $x = 0$, which is one of the limits of the integral.

Figure 4.3

(ii) $\displaystyle\int_0^1 \frac{1}{\sqrt{x}}\,dx = \lim_{a\to 0}\int_a^1 \frac{1}{\sqrt{x}}\,dx$

> Notice that the variable a tends to 0 this time, not ∞.

$$= \lim_{a\to 0}\left[2x^{\frac{1}{2}}\right]_a^1$$

$$= \lim_{a\to 0}\left(2 - 2a^{\frac{1}{2}}\right)$$

As $a \to 0$, $a^{\frac{1}{2}} \to 0$

So $\displaystyle\lim_{a\to 0}\left(2 - 2a^{\frac{1}{2}}\right) = 2$

The area under the graph of $\dfrac{1}{\sqrt{x}}$, between 0 and 1, converges to 2 square units.

In this case you get the correct answer if you just integrate in the usual way, but it is safer to use the process shown in the example, as often you cannot be sure whether the value of the integral will converge or not.

ACTIVITY 4.1

Karen is trying to work out $\displaystyle\int_1^3 \frac{1}{(x-2)^2}\,dx$. She writes

$$\int_1^3 \frac{1}{(x-2)^2}\,dx = \left[-\frac{1}{x-2}\right]_1^3 = -1 - 1 = -2$$

How do you know that Karen's answer must be wrong?
What is the problem with Karen's working?

Example 4.4

(i) Sketch the graph of $y = \dfrac{1}{(x-2)^2}$ and shade the area represented by the integral $\displaystyle\int_1^3 \frac{1}{(x-2)^2}\,dx$. What feature of the curve warns you of possible difficulties in evaluating this integral?

(ii) What happens when you try to evaluate $\displaystyle\int_1^3 \frac{1}{(x-2)^2}\,dx$?

Solution

(i)

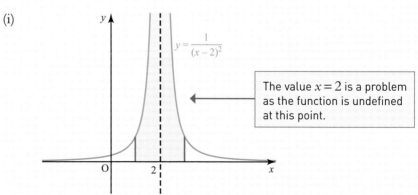

> The value $x = 2$ is a problem as the function is undefined at this point.

Figure 4.4

(ii) $$\int_1^3 \frac{1}{(x-2)^2}\,dx = \int_1^2 \frac{1}{(x-2)^2}\,dx + \int_2^3 \frac{1}{(x-2)^2}\,dx$$

> Split the integral at the point where it is undefined.

$$= \lim_{a\to 2}\int_1^a \frac{1}{(x-2)^2}\,dx + \lim_{a\to 2}\int_a^3 \frac{1}{(x-2)^2}\,dx$$

$$= \lim_{a\to 2}\left[-\frac{1}{x-2}\right]_1^a + \lim_{a\to 2}\left[-\frac{1}{x-2}\right]_a^3$$

> Now remove the problem limits $(x=2)$ in the same way as the previous examples.

$$= \lim_{a\to 2}\left(-\frac{1}{a-2}-1\right) + \lim_{a\to 2}\left(-1+\frac{1}{a-2}\right)$$

As $a \to 2$, $\dfrac{1}{a-2}$ is undefined, so the integral diverges.

The four examples above all involve **improper integrals**.

An improper integral is defined to be a definite integral in which:

- at least one of the limits is infinite
- or the function you wish to integrate approaches infinity at some point in the interval required.

Examples 4.1 and 4.2 both have an infinite limit, Example 4.3 has a function which approaches infinity at $x \to 0$, but 0 is one of the limits, and Example 4.4 includes the value $x = 2$ in the range required, but the function is not defined at that point.

Exercise 4.1

① (i) Sketch the graph of $y = x^{-3}$.

(ii) Evaluate $\displaystyle\int_2^a x^{-3}\,dx$, leaving your answer in terms of a.

(iii) In your answer to part (ii), let $a \to \infty$, and hence state the value of the integral $\displaystyle\int_2^\infty x^{-3}\,dx$.

② The graph below shows the shape of the curve $y = (x-1)^{-\frac{2}{3}}$.

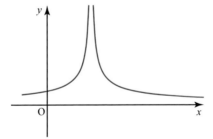

Figure 4.5

(i) Evaluate $\displaystyle\int_0^b (x-1)^{-\frac{2}{3}}\,dx$ and $\displaystyle\int_c^3 (x-1)^{-\frac{2}{3}}\,dx$, leaving your answers in terms of b and c respectively.

(ii) In your answers to part (i), let $b \to 1$ (from below), and let $c \to 1$ (from above). Hence state the value of $\displaystyle\int_0^3 (x-1)^{-\frac{2}{3}}\,dx$.

(iii) Copy the graph of $y = (x-1)^{-\frac{2}{3}}$ above and indicate the area you have evaluated.

③ (i) Sketch the graph of $y = e^{-x}$.

(ii) Evaluate $\int_0^d e^{-x}\,dx$, leaving your answer in terms of d.

(iii) In your answer to part (ii), let $d \to \infty$, and hence state the value of $\int_0^\infty e^{-x}\,dx$.

Show whether the following improper integrals are convergent or divergent, and calculate their value if convergent. In each case show the area represented by the integral on a diagram.

④ $\int_1^\infty x^{-3}\,dx$ ⑤ $\int_0^\infty x^{-3}\,dx$ ⑥ $\int_0^3 \dfrac{1}{x^2}\,dx$ ⑦ $\int_2^\infty \dfrac{1}{x^2}\,dx$

⑧ $\int_{-\infty}^0 e^x\,dx$ ⑨ $\int_0^\infty e^x\,dx$ ⑩ $\int_0^{10} x\left(4 - x^2\right)^{-\frac{2}{3}}\,dx$ ⑪ $\int_0^\infty e^{-2x} - e^{-x}\,dx$

⑫ Evaluate $\int_0^\infty x e^{-x}\,dx$.

⑬ Evaluate $\int_0^\infty \dfrac{1}{x+2} - \dfrac{1}{x+1}\,dx$.

⑭ Evaluate $\int_{-1}^1 \left| x^{\frac{1}{3}} \right|\,dx$.

2 Calculus with inverse trigonometric functions

In this section you will see how the derivatives of the inverse trigonometric functions are very useful in integrating many functions even though they appear to be completely unrelated.

> ### Note
>
> You will often see $\arcsin x$ written as $\sin^{-1} x$. They represent exactly the same function (the inverse sine function) but the second notation has the potential to be somewhat confusing when compared to, for example, $\sin^2 x$, which actually means $(\sin x)^2$. It is vital that you recognise that $\sin^{-1} x$ does NOT mean $(\sin x)^{-1}$ – which is actually $\dfrac{1}{\sin x}$ or $\operatorname{cosec} x$.

Prior knowledge

- You need to be able to differentiate and integrate trigonometric functions such as $\sin x$, $\cos x$ and $\tan x$.
- You should be familiar with the inverse trigonometric functions $\arcsin x$, $\arccos x$ and $\arctan x$ and their domains, ranges and graphs.
- You need to be able to differentiate functions defined implicitly.

Differentiating inverse trigonometric functions

To differentiate the inverse trigonometric functions, you need to use implicit differentiation.

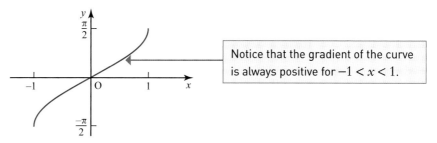

Notice that the gradient of the curve is always positive for $-1 < x < 1$.

Figure 4.6

$$y = \arcsin x$$

$$\sin y = x$$

$$\cos y \frac{dy}{dx} = 1 \quad \longleftarrow$$

> Differentiate implicitly with respect to x.

$$\frac{dy}{dx} = \frac{1}{\cos y}$$

$$= \frac{1}{\pm\sqrt{1 - \sin^2 y}}$$

$$= \frac{1}{\pm\sqrt{1 - x^2}}$$

But $y = \arcsin x$ has a range of $-\frac{\pi}{2} \leqslant y \leqslant \frac{\pi}{2}$, which implies that $\cos x \geqslant 0$, and then you can ignore the $\pm$ symbol since it must be positive in this case.

The conclusion is that:

$$\frac{d}{dx}(\arcsin x) = \frac{1}{\sqrt{1 - x^2}}$$

There are several things to notice with this result:

■ it is positive, and only defined for $-1 < x < 1$

■ it has a minimum at $x = 0$

■ it tends to ∞ as $x \to \pm 1$

and these points are consistent with the graph of $y = \arcsin x$ in Figure 4.6.

ACTIVITY 4.2

Use a similar method to show that:

■ $\dfrac{d}{dx}(\arccos x) = -\dfrac{1}{\sqrt{1 - x^2}}$ ■ $\dfrac{d}{dx}(\arctan x) = \dfrac{1}{1 + x^2}$

Integration using inverse trigonometric substitutions

You can use these results in integration. In practice it is the arcsin x and the arctan x results that are used since the arccos x one is just the negative of the arcsin x one.

$$\frac{d}{dx}(\arcsin x) = \frac{1}{\sqrt{1 - x^2}}$$

$$\frac{d}{dx}(\arctan x) = \frac{1}{1 + x^2}$$

From these results it becomes clear by integrating that:

$$\int \frac{1}{\sqrt{1 - x^2}}\,dx = \arcsin x + c$$

$$\int \frac{1}{1 + x^2}\,dx = \arctan x + c$$

ACTIVITY 4.3

Use the chain rule and the derivatives for arcsin x and arctan x given above, to show that:

- $\dfrac{d}{dx}\left(\arcsin\dfrac{x}{a}\right) = \dfrac{1}{\sqrt{a^2 - x^2}}$

- $\dfrac{d}{dx}\left(\arctan\dfrac{x}{a}\right) = \dfrac{a}{a^2 + x^2}$

The results in Activity 4.3 above lead to the following results:

$$\int \frac{1}{\sqrt{a^2 - x^2}}\, dx = \arcsin\frac{x}{a} + c$$

$$\int \frac{1}{a^2 + x^2}\, dx = \frac{1}{a}\arctan\frac{x}{a} + c$$

These results can be quoted for use in integration.

Example 4.5

Calculate the value of the indefinite integral $\displaystyle\int \frac{1}{\sqrt{9 - x^2}}\, dx$

Solution

$$\int \frac{1}{\sqrt{9 - x^2}}\, dx = \arcsin\frac{x}{3} + c$$

> You can use the standard result with $a = 3$.

Notice that in the standard results, the coefficient of x^2 is 1. If you need to integrate a function of this form in which the coefficient of x^2 is not 1, then you need to first rewrite it in the standard form. This is shown in the next example.

Example 4.6

Find $\displaystyle\int \frac{1}{\sqrt{16 - 3x^2}}\, dx$

Discussion point

→ In the example on the right, why can't you simply use the standard integral, using $a = 4$ and replacing x with $x\sqrt{3}$?

> First factorise out the 3, which becomes $\sqrt{3}$ when it leaves the square root.

Solution

$$\int \frac{1}{\sqrt{16 - 3x^2}}\, dx = \frac{1}{\sqrt{3}}\int \frac{1}{\sqrt{\frac{16}{3} - x^2}}\, dx$$

> This is now in the standard form, with $a = \dfrac{4}{\sqrt{3}}$.

$$= \frac{1}{\sqrt{3}}\arcsin\left(\frac{x\sqrt{3}}{4}\right) + c$$

The arctan x result can also be quoted for use in integration.

Example 4.7

Evaluate the definite integral:

$$\int_0^2 \frac{1}{4+x^2}\,dx$$

Solution

$$\int_0^2 \frac{1}{4+x^2}\,dx = \left[\frac{1}{2}\arctan\left(\frac{x}{2}\right)\right]_0^2 \quad\longleftarrow\quad \boxed{\text{Using the standard result with } a=2.}$$

$$= \frac{1}{2}(\arctan 1 - \arctan 0)$$

$$= \frac{\pi}{8}$$

Exercise 4.2

① State the domain and range of the arcsin, arccos and arctan functions.

② Use the standard results to evaluate the following indefinite integrals:

(i) $\displaystyle\int \frac{1}{\sqrt{25-x^2}}\,dx$

(ii) $\displaystyle\int \frac{1}{16+t^2}\,dt$

③ Use the standard results to evaluate the following definite integrals:

(i) $\displaystyle\int_0^1 \frac{1}{\sqrt{3-s^2}}\,ds$

(ii) $\displaystyle\int_{-1}^1 \frac{1}{2^2+x^2}\,dx$

④ Differentiate the following functions with respect to x:

(i) $\arcsin(3x)$

(ii) $\arccos\left(\frac{1}{2}x\right)$

(iii) $\arctan(5x)$

(iv) $\arcsin\left(3x^2\right)$

(v) $\arctan\left(e^x\right)$

(vi) $3\arctan\left(1-x^2\right)$

⑤ Find the following indefinite integrals:

(i) $\displaystyle\int \frac{1}{4+x^2}\,dx$

(ii) $\displaystyle\int \frac{1}{1+4x^2}\,dx$

(iii) $\displaystyle\int \frac{1}{\sqrt{4-x^2}}\,dx$

(iv) $\displaystyle\int \frac{1}{\sqrt{1-4x^2}}\,dx$

⑥ Find the following indefinite integrals:

(i) $\displaystyle\int \frac{5}{x^2+36}\,dx$

(ii) $\displaystyle\int \frac{4}{25+4x^2}\,dx$

(iii) $\displaystyle\int \frac{1}{\sqrt{9-4x^2}}\,dx$

(iv) $\displaystyle\int \frac{7}{\sqrt{5-3x^2}}\,dx$

⑦ Evaluate the following definite integrals, leaving your answers in terms of π:

(i) $\displaystyle\int_0^3 \frac{1}{9+x^2}\,dx$

(ii) $\displaystyle\int_0^{\sqrt{2}} \frac{1}{\sqrt{4-x^2}}\,dx$

(iii) $\displaystyle\int_{-\frac{1}{\sqrt{3}}}^{\frac{1}{3}} \frac{1}{1+9x^2}\,dx$

(iv) $\displaystyle\int_0^{\frac{1}{4}} \frac{1}{\sqrt{1-4x^2}}\,dx$

⑧ The diagram below shows the curves $y = \dfrac{2}{1+x^2}$ and $y = \dfrac{1}{\sqrt{4-3x^2}}$.

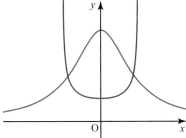

Figure 4.7

Show that the area between the curves is given by $\pi\left(1 - \dfrac{2}{3\sqrt{3}}\right)$.

⑨ Use implicit differentiation to prove:

(i) $\dfrac{d}{dx}\left(\arcsin\dfrac{x}{a}\right) = \dfrac{1}{\sqrt{a^2 - x^2}}$

(ii) $\dfrac{d}{dx}\left(\arctan\dfrac{x}{a}\right) = \dfrac{a}{a^2 + x^2}$

⑩ Differentiate the function $f(x) = x\arcsin\left(x^2\right)$.

⑪ The graph below shows the curve $y = \dfrac{1}{1+x^2}$.

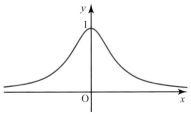

Figure 4.8

Find the total area under the curve.

⑫ Evaluate the following definite integral: $\displaystyle\int_{\sqrt{\frac{5}{6}}}^{\sqrt{\frac{5}{2}}} \dfrac{1}{5 + 2x^2}\,dx$

3 Partial fractions

Partial fractions can often be used in integration.

Prior knowledge

You need to know how to find partial fractions of the following types:

$\dfrac{qx + r}{(ax + b)(cx + d)}$ can be written in the form $\dfrac{A}{ax + b} + \dfrac{B}{cx + d}$

$\dfrac{px^2 + qx + r}{(ax + b)(cx + d)(ex + f)}$ can be written in the form

$\dfrac{A}{ax + b} + \dfrac{B}{cx + d} + \dfrac{C}{ex + f}$

$\dfrac{px^2 + qx + r}{(ax + b)(cx + d)^2}$ can be written in the form

$\dfrac{A}{ax + b} + \dfrac{B}{cx + d} + \dfrac{C}{(cx + d)^2}$

Example 4.8

Find $\int \dfrac{x - 17}{(x + 1)(x - 5)} \, \mathrm{d}x$.

Solution

$$\dfrac{x - 17}{(x + 1)(x - 5)} = \dfrac{A}{x + 1} + \dfrac{B}{x - 5}$$

$$x - 17 = A(x - 5) + B(x + 1)$$

Let $x = 5 \quad \Rightarrow -12 = 6B \quad \Rightarrow B = -2$

Let $x = -1 \Rightarrow -18 = -6A \Rightarrow A = 3$

$$\int \dfrac{x - 17}{(x + 1)(x - 5)} \, \mathrm{d}x = \int \dfrac{3}{x + 1} - \dfrac{2}{x - 5} \, \mathrm{d}x$$

$$= 3\ln|x + 1| - 2\ln|x - 5| + c$$

Example 4.9

Find $\int \dfrac{25(4x + 1)}{(3x - 1)(2x + 1)^2} \, \mathrm{d}x$.

Solution

$$\dfrac{25(4x + 1)}{(3x - 1)(2x + 1)^2} = \dfrac{A}{3x - 1} + \dfrac{B}{2x + 1} + \dfrac{C}{(2x + 1)^2}$$

$$25(4x + 1) = A(2x + 1)^2 + B(3x - 1)(2x + 1) + C(3x - 1)$$

Let $x = \frac{1}{3} \quad \Rightarrow 25 \times \frac{7}{3} = \frac{25}{9}A \quad \Rightarrow A = 21$

Let $x = -\frac{1}{2} \quad \Rightarrow -25 = -\frac{5}{2}C \quad \Rightarrow C = 10$

Let $x = 0 \quad \Rightarrow 25 = A - B - C \quad \Rightarrow B = -14$

$$\int \dfrac{25(4x + 1)}{(3x - 1)(2x + 1)^2} \, \mathrm{d}x = \int \dfrac{21}{3x - 1} - \dfrac{14}{2x + 1} + \dfrac{10}{(2x + 1)^2} \, \mathrm{d}x$$

$$= 7\ln|3x - 1| - 7\ln|2x + 1| - \dfrac{5}{2x + 1} + c$$

The next example shows you how to extend your knowledge of partial fractions to include a quadratic expression in the denominator that cannot be factorised, and to integrate them.

An expression of the form $\dfrac{px^2 + qx + r}{(ax + b)(cx^2 + d)}$ can be written in the form $\dfrac{A}{ax + b} + \dfrac{Bx + C}{cx^2 + d}$.

Example 4.10

Find $\int \dfrac{x-2}{(x+1)(x^2+2)}\,dx$.

Solution

$$\frac{x-2}{(x+1)(x^2+2)} = \frac{A}{x+1} + \frac{Bx+C}{x^2+2}$$

$$x-2 = A(x^2+2) + (Bx+C)(x+1)$$

Let $x=-1 \Rightarrow -3 = 3A \Rightarrow A = -1$

Let $x=0 \Rightarrow -2 = 2A+C \Rightarrow C = 0$

Equating coefficients of x^2: $0 = A+B \Rightarrow B = 1$

$$\int \frac{x-2}{(x+1)(x^2+2)}\,dx = \int \frac{-1}{x+1} + \frac{x}{x^2+2}\,dx$$

$$= -\ln|x+1| + \frac{1}{2}\ln|x^2+2| + c$$

$$= \ln\left|\frac{\sqrt{x^2+2}}{x+1}\right| + c$$

> You could substitute any value of x to find B, but equating coefficients is often easier.

> In the second term, the numerator, x, is half of the derivative of the denominator, so you can do this by inspection, or by using the substitution $u = x^2 + 2$.

In the example above, C turned out to be zero, which meant that each term could be integrated using methods you have met previously. In the next example, C is not zero.

Example 4.11

Find $\int \dfrac{9x-8}{(x+2)(x^2+9)}\,dx$.

Solution

> Write the function in partial fractions.

$$\frac{9x-8}{(x+2)(x^2+9)} = \frac{A}{x+2} + \frac{Bx+C}{x^2+9}$$

$$9x-8 = A(x^2+9) + (Bx+C)(x+2)$$

Let $x=-2 \Rightarrow -26 = 13A \Rightarrow A = -2$

Let $x=0 \Rightarrow -8 = 9A+2C \Rightarrow C = 5$

Equating coefficients of x^2: $0 = A+B \Rightarrow B = 2$

$$\int \frac{9x-8}{(x^2+9)(x+2)}\,dx = \int \left(\frac{2x+5}{x^2+9} - \frac{2}{x-2}\right)dx$$

$$= \int \left(\frac{2x}{x^2+9} + \frac{5}{x^2+9} - \frac{2}{x-2}\right)dx$$

$$= \ln(x^2+9) + \frac{5}{3}\arctan\frac{x}{3} - 2\ln|x-2| + c$$

$$= \ln\left(\frac{x^2+9}{(x-2)^2}\right) + \frac{5}{3}\arctan\frac{x}{3} + c$$

> The second term can be integrated using the standard arctan x result, with $a = 3$.

> The first term can be integrated by inspection, since the numerator is the derivative of the denominator.

Exercise 4.3

① (i) Show that $\dfrac{3(5x+1)}{(x+1)(5x-1)} \equiv \dfrac{2}{x+1} + \dfrac{5}{5x-1}$.

(ii) Use this result to find $\displaystyle\int \dfrac{3(5x+1)}{(x+1)(5x-1)} \, dx$.

② (i) Show that $\dfrac{3x+4}{(2x+3)(x+1)^2} \equiv -\dfrac{2}{2x+3} + \dfrac{1}{x+1} + \dfrac{1}{(x+1)^2}$.

(ii) Use this result to find $\displaystyle\int \dfrac{3x+4}{(2x+3)(x+1)^2} \, dx$.

③ (i) Show that $\dfrac{2x^2 - 3x + 5}{(x-3)(x^2+5)} \equiv \dfrac{1}{x-3} + \dfrac{x}{x^2+5}$.

(ii) Use this result to find $\displaystyle\int \dfrac{2x^2 - 3x + 5}{(x-3)(x^2+5)} \, dx$.

④ (i) Write $\dfrac{17-5x}{(x+7)(x^2+3)}$ in the form $\dfrac{A}{x+7} + \dfrac{Bx+C}{x^2+3}$.

(ii) Find $\displaystyle\int \dfrac{17-5x}{(x+7)(x^2+3)} \, dx$.

⑤ Find the following integrals:

(i) $\displaystyle\int \dfrac{4}{(x^2+1)(x+1)} \, dx$

(ii) $\displaystyle\int \dfrac{x+11}{(x^2+9)(x-2)} \, dx$

(iii) $\displaystyle\int \dfrac{5x^2 + 3x + 3}{(4x^2+1)(x+2)} \, dx$

⑥ Evaluate the following definite integrals:

(i) $\displaystyle\int_0^1 \dfrac{x+3}{(x+1)(x^2+1)} \, dx$

(ii) $\displaystyle\int_2^5 \dfrac{2x^2+3}{(x-1)(x^2+4)} \, dx$

(iii) $\displaystyle\int_0^2 \dfrac{x-6}{(x+1)(3x^2+4)} \, dx$

⑦ The graph shows part of the curve $y = \dfrac{4-3x}{(x^2+4)(x+3)}$.

Find the exact area of the shaded region.

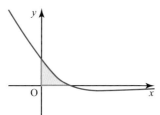

Figure 4.9

⑧ Express f(x) in a partial fraction form, where:

$$f(x) = \dfrac{6 + 2x + 2x^2 - 2x^3}{(x+1)(x-1)^2(x^2+1)},$$

Use this to show that: $\displaystyle\int f(x)\,dx = \ln \dfrac{(x^2+1)|x+1|}{|x-1|^3} - \dfrac{2}{x-1} + c$

⑨ Evaluate $\int_0^\infty \dfrac{1}{(x+1)^2(x+2)}\,\mathrm{d}x$.

⑩ Figure 4.10 shows the start of an infinite sequence of rectangles of width 1 and height $\left(\dfrac{1}{x^2-1}\right)$ where the values of x are restricted to the integers from 2 upwards.

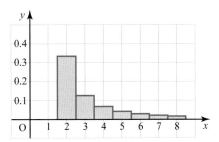

Figure 4.10

(i) Find the total area of the rectangles.

(ii) Show that the midpoints of the tops of the rectangles lie on the continuous curve $y = \dfrac{1}{x^2-1}$.

(iii) Find the exact area of the region bounded by the x-axis, the line $x = 4.5$ and the part of the curve $y = \dfrac{1}{x^2-1}$ for which $x \geqslant 1.5$.

(iv) Find the percentage error in using the area of the rectangles as an approximation for the area of the region described in part (iii).

4 Further integration

The previous section showed how differentiating the inverse trigonometric functions allows you to carry out new types of integration. These techniques can be extended to apply to less obvious integrals with a few algebraic tricks, as you will see in this section.

Example 4.12

Find $\int \dfrac{4}{x^2-2x+3}\,\mathrm{d}x$. ◀── | The quadratic in the denominator does not factorise – but it can be manipulated slightly to end up looking like one of the standard forms previously covered.

Solution

Complete the square on the denominator. ──▶

This now looks like the standard result for arctan x.

$$\int \dfrac{4}{x^2-2x+3}\,\mathrm{d}x = 4\int \dfrac{1}{(x-1)^2+2}\,\mathrm{d}x$$

$$= 4\int \dfrac{1}{u^2+2}\,\mathrm{d}u \text{ where } u = x-1 \text{ and } \mathrm{d}x = \mathrm{d}u$$

$$= 4 \times \dfrac{1}{\sqrt{2}}\arctan\left(\dfrac{u}{\sqrt{2}}\right) + c$$

$$= 2\sqrt{2}\arctan\left(\dfrac{x-1}{\sqrt{2}}\right) + c$$

Example 4.13

Find $\int \dfrac{5}{\sqrt{2 + 4x - 4x^2}}\, dx$.

Note

The clue as to how to proceed is that the denominator is the square root of a quadratic with negative coefficient of x^2, just as in the standard result:

$\int \dfrac{1}{\sqrt{a^2 - x^2}}\, dx$

$= \arcsin\left(\dfrac{x}{a}\right) + c.$

Solution

$\displaystyle\int \dfrac{5}{\sqrt{2 + 4x - 4x^2}}\, dx = \dfrac{5}{2}\int \dfrac{1}{\sqrt{\frac{1}{2} - \left(x^2 - x\right)}}\, dx$

> Take out a factor of 5 from the numerator and 4 from 'inside' the square root in the denominator.

> Complete the square, and adjust the constant (with care over the negative signs!).

$= \dfrac{5}{2}\int \dfrac{1}{\sqrt{\frac{3}{4} - \left(x - \frac{1}{2}\right)^2}}\, dx$

$= \dfrac{5}{2}\int \dfrac{1}{\sqrt{\frac{3}{4} - u^2}}\, du$ where $u = x - \frac{1}{2}$ and $du = dx$

$= \dfrac{5}{2}\arcsin\left(\dfrac{2u}{\sqrt{3}}\right) + c$

> These two lines might be omitted with practice.

$= \dfrac{5}{2}\arcsin\left(\dfrac{2x - 1}{\sqrt{3}}\right) + c$

ACTIVITY 4.4

Try to use the method in the example above to find the following integrals:

(i) $\displaystyle\int \dfrac{1}{x^2 + 2x - 2}\, dx$

(ii) $\displaystyle\int \dfrac{1}{\sqrt{3 - 2x + x^2}}\, dx$

Explain why the method does not work.
How can you predict which integrals of this form can be done using this method, and which cannot?

Trigonometric substitutions

You have seen that functions of the form $\dfrac{1}{\sqrt{a^2 - x^2}}$ can be integrated to give an arcsin function. This gives a clue to integrating other functions that involve $\sqrt{a^2 - x^2}$. This expression might remind you that $\cos u = \sqrt{1 - \sin^2 u}$, and so a substitution of the form $x = a\sin u$ may be useful.

Example 4.14

Find $\displaystyle\int \dfrac{1}{(1 - x^2)^{\frac{3}{2}}}\, dx$.

Solution

Let $x = \sin u \quad \Rightarrow \quad \dfrac{dx}{du} = \cos u$

$\displaystyle\int \dfrac{1}{(1 - x^2)^{\frac{3}{2}}}\, dx = \int \dfrac{1}{(1 - \sin^2 u)^{\frac{3}{2}}} \times \cos u\, du$

> Replace x with $\sin u$, and dx with $\cos u\, du$.

$= \displaystyle\int \dfrac{1}{(\cos^2 u)^{\frac{3}{2}}} \times \cos u\, du$

> $\cos^2 x = 1 - \sin^2 x$

$$= \int \frac{1}{\cos^3 u} \times \cos u \, du$$

$$= \int \frac{1}{\cos^2 u} \, du$$

$$= \int \sec^2 u \, du \longleftarrow \boxed{\sec^2 u \text{ is the derivative of } \tan u.}$$

$$= \tan u + c$$

$$= \frac{\sin u}{\cos u} + c$$

$$= \frac{x}{\sqrt{1 - x^2}} + c \longleftarrow \boxed{\text{Change back to the original variable of } x, \text{ using } \cos u = \sqrt{1 - \sin^2 u}.}$$

Example 4.15

Evaluate $\int_0^2 \sqrt{16 - x^2} \, dx$.

Solution

Let $x = 4 \sin u \Rightarrow \dfrac{dx}{du} = 4 \cos u$ $\longleftarrow$ $\boxed{\text{You need } x = 4 \cos u \text{ so that } x^2 = 16 \cos^2 u.}$

When $x = 0$, $\sin u = 0 \Rightarrow u = 0$ $\longleftarrow$

When $x = 2$, $\sin u = \dfrac{1}{2} \Rightarrow u = \dfrac{\pi}{6}$ $\boxed{\text{Change the limits of the integral from } x\text{-values to the equivalent } u\text{-values.}}$

$$\int_0^2 \sqrt{16 - x^2} \, dx = \int_0^{\frac{\pi}{6}} \sqrt{16 - 16 \sin^2 u} \times 4 \cos u \, du$$

$$= \int_0^{\frac{\pi}{6}} \sqrt{16 \cos^2 u} \times 4 \cos u \, du$$

$$= \int_0^{\frac{\pi}{6}} 4 \cos u \times 4 \cos u \, du$$

$$= 16 \int_0^{\frac{\pi}{6}} \cos^2 u \, du$$

$\boxed{\text{Using the double angle formula } \cos 2u = 2 \cos^2 u - 1.}$ $\longrightarrow$ $= 16 \int_0^{\frac{\pi}{6}} \dfrac{\cos 2u + 1}{2} \, du$

$$= 8 \left[\frac{1}{2} \sin 2u + u \right]_0^{\frac{\pi}{6}}$$

$$= 4 \sin \frac{\pi}{3} + 8 \times \frac{\pi}{6} - 0$$

$$= 2\sqrt{3} + \frac{4}{3}\pi$$

In a similar way, you have seen that functions of the form $\dfrac{1}{a^2 + x^2}$ can be integrated to give an arctan function. This gives a clue to finding other integrals that involve $a^2 + x^2$. This might remind you that $\sec^2 \theta = 1 + \tan^2 \theta$, and so a substitution of the form $x = a \tan u$ may be useful.

Example 4.16

Find $\int_0^1 \dfrac{1}{(1 + x^2)^{\frac{3}{2}}} \, dx$.

Solution

Let $x = \tan u \Rightarrow \dfrac{dx}{du} = \sec^2 u$ $\boxed{\text{Change the limits of the integral from } x\text{-values to the equivalent } u\text{-values.}}$

When $x = 0$, $\tan u = 0 \Rightarrow u = 0$ $\longleftarrow$

When $x = 1, \tan u = 1 \Rightarrow u = \frac{\pi}{4}$

$$\int_0^1 \frac{1}{(1 + x^2)^{\frac{3}{2}}}\,dx = \int_0^{\frac{\pi}{4}} \frac{1}{(1 + \tan^2 u)^{\frac{3}{2}}} \times \sec^2 u\,du$$

$$= \int_0^{\frac{\pi}{4}} \frac{1}{(\sec^2 u)^{\frac{3}{2}}} \times \sec^2 u\,du \quad \longleftarrow \boxed{\text{Using } \sec^2 u = 1 + \tan^2 u.}$$

$$= \int_0^{\frac{\pi}{4}} \frac{1}{\sec^3 u} \times \sec^2 u\,du$$

$$= \int_0^{\frac{\pi}{4}} \frac{1}{\sec u}\,du$$

$$= \int_0^{\frac{\pi}{4}} \cos u\,du$$

$$= \left[\sin u\right]_0^{\frac{\pi}{4}}$$

$$= \sin \frac{\pi}{4} - \sin 0$$

$$= \frac{1}{\sqrt{2}}$$

Exercise 4.4

① Using the substitution $x = 2\sin u$, find $\int_0^{\frac{1}{2}} \frac{1}{(4 - x^2)^{\frac{3}{2}}}\,dx$.

② Using the substitution $x = 3\sin u$, find $\int_0^3 \sqrt{9 - x^2}\,dx$.

③ Using the substitution $x = 2\tan u$, find $\int_0^2 \frac{1}{(4 + x^2)^{\frac{3}{2}}}\,dx$.

④ Find $\int \frac{3}{9x^2 + 6x + 5}\,dx$.

⑤ Find $\int \frac{1}{\sqrt{3 + 2x - x^2}}\,dx$.

⑥ (i) By writing $\arcsin x$ as $1 \times \arcsin x$, use integration by parts to find $\int \arcsin x\,dx$.

　 (ii) Use a similar method to find the following integrals:

　　 (a) $\int \arccos x\,dx$ 　　 (b) $\int \arctan x\,dx$ 　　 (c) $\int \arccot x\,dx$

⑦ Use a suitable substitution to evaluate:

　 (i) $\int_0^2 \frac{1}{(16 - x^2)^{\frac{3}{2}}}\,dx$ 　 (ii) $\int_{-\frac{1}{2}}^{\frac{1}{2}} \frac{1}{(1 + 4x^2)^{\frac{3}{2}}}\,dx$ 　 (iii) $\int_0^{\frac{2}{5}} \sqrt{4 - 25x^2}\,dx$

⑧ (i) Find $\int_0^b \sqrt{a^2 - x^2}\,dx$

　 where $a > b > 0$.

　 (ii) Draw a sketch to show the significance of the area you calculated in part (i), and explain both terms of your answer to part (i) geometrically.

⑨ (i) Use the substitution $x = a \sin u$ to prove the result

$$\int \frac{1}{\sqrt{a^2 - x^2}} \, dx = \arcsin \frac{x}{a} + c$$

(ii) Use the substitution $x = a \tan u$ to prove the result

$$\int \frac{1}{a^2 + x^2} \, dx = \frac{1}{a} \arctan \frac{x}{a} + c$$ ◄

> You have already proved these results in Question 9 in Exercise 4.2, by starting from differentiation of $\arcsin \frac{x}{a}$ and $\arctan \frac{x}{a}$.

⑩ Find the following integrals:

(i) $\displaystyle\int \frac{x + 1}{x^2 + 1} \, dx$

(ii) $\displaystyle\int \frac{x + 1}{x^2 + 2x + 3} \, dx$

(iii) $\displaystyle\int \frac{1}{x^2 + 2x + 3} \, dx$

⑪ Find the following integrals:

(i) $\displaystyle\int \frac{2 - x}{\sqrt{4 - x^2}} \, dx$

(ii) $\displaystyle\int \frac{1}{\sqrt{4x - x^2}} \, dx$

(iii) $\displaystyle\int \frac{2 - x}{\sqrt{4x - x^2}} \, dx$

⑫ Find $\dfrac{d}{dx}\left(\text{arcsec}\, x\right)$ and $\displaystyle\int \frac{1}{x\sqrt{x^2 - a^2}} \, dx$.

⑬ By considering the equation $y = \sqrt{r^2 - x^2}$ and integrating between 0 and r, prove that the area of a circle, radius r, is πr^2.

⑭ Given that

$$\int \sec^3 x \, dx = \frac{1}{2} \ln \left| \frac{\cos\left(\frac{x}{2}\right) + \sin\left(\frac{x}{2}\right)}{\cos\left(\frac{x}{2}\right) - \sin\left(\frac{x}{2}\right)} \right| + \frac{1}{2} \sec x \tan x$$

evaluate the following definite integral by performing an arctan x substitution, giving your answer to 4 d.p.

$$\int_0^1 \sqrt{1 + x^2} \, dx$$

(You will see an easier way to find the value of this integral in Chapter 7 on Hyperbolic functions.)

LEARNING OUTCOMES

When you have completed this chapter you should be able to:

➤ evaluate improper integrals where either the integrand is undefined at a value in the interval of integration or the interval of integration extends to infinity

➤ use the method of partial fractions in integration, including where the denominator has a quadratic factor of form $ax^2 + c$ and one linear term

➤ differentiate inverse trigonometric functions

➤ recognise integrals of functions of the form $\left(a^2 - x^2\right)^{-\frac{1}{2}}$ and $\left(a^2 + x^2\right)^{-1}$ and be able to integrate related functions by using trigonometric substitutions.

KEY POINTS

An improper integral is an integral in which either:

- at least one of the limits is infinity; or
- the function to be integrated approaches infinity at some point in the interval of integration.

1   Improper integrals involving a limit of infinity may be investigated by replacing the problem limit by the constant a and then finding, if possible, the limit of the value of the integral as $a \to \infty$. The value of the integral may be finite, in which case the integral is convergent, or it may be divergent, in which case it cannot be evaluated.

2   Improper integrals in which the functions to be integrated approach infinity at some point in the interval of integration, may be investigated by splitting the integral into two at the problem point (if the problem point is not one of the end points), and replacing the problem value with the constant a, and then finding, if possible, the limit of the value of the integral as $a \to \infty$.

3   $\displaystyle \int \frac{1}{\sqrt{a^2 - x^2}}\, \mathrm{d}x = \arcsin \frac{x}{a} + c$

4   $\displaystyle \int \frac{1}{a^2 + x^2}\, \mathrm{d}x = \frac{1}{a} \arctan \frac{x}{a} + c$

5   Integrals of the form $\displaystyle \int \frac{px^2 + qx + r}{(a + bx^2)(cx + d)}\, \mathrm{d}x$ can be found by first splitting the function into partial fractions of the form $\dfrac{Ax + B}{a + bx^2} + \dfrac{C}{cx + d}$.

6   Integrals that involve functions of the form $\sqrt{a^2 - x^2}$ may often be integrated using the substitution $x = a \sin u$.

7   Integrals that involve functions of the form $a^2 + x^2$ may often be integrated using the substitution $x = a \tan u$.

FUTURE USES

- You will meet some other integrals similar to those covered in this chapter, in the Hyperbolic functions chapter, and you will use similar techniques there.
- You will use many of the integration techniques covered in this chapter in the chapters Applications of integration, First order differential equations and Second order differential equations.

T **PS** ① A 2×2 matrix is given by $M = \begin{pmatrix} a & b \\ c & d \end{pmatrix}$.

Figure 1 was drawn using graphing software. It shows an arrowhead A and its images under M, M^2, M^3, M^4 and M^5.

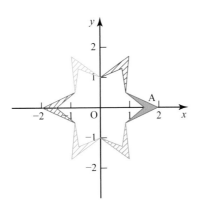

Figure 1

Find two sets of values for a, b, c and d.   [4 marks]

② (i) Express $\dfrac{1}{(2r + 3)(2r + 5)}$ in partial fractions.   [3 marks]

(ii) Using the method of differences, find $\displaystyle\sum_{r=1}^{n} \dfrac{1}{(2r + 3)(2r + 5)}$,

expressing your answer as a single fraction.   [4 marks]

(iii) Evaluate $\displaystyle\sum_{r=1}^{\infty} \dfrac{1}{(2r + 3)(2r + 5)}$.   [1 mark]

M ③ Prove by induction that $7^n + 2^{2n} + 1$ is divisible by 6 for all $n \in \mathbb{N}$.

[7 marks]

M ④ You are given the three related sequences a_1, a_2, a_3, …, b_1, b_2, b_3, … and c_1, c_2, c_3, … where

$$a_1 = 1 \qquad\qquad b_1 = 3 \qquad\qquad c_1 = \dfrac{a_1}{b_1}$$

$$a_2 = 1 + 3 \qquad\qquad b_2 = 5 + 7 \qquad\qquad c_2 = \dfrac{a_2}{b_2}$$

$$a_3 = 1 + 3 + 5 \qquad\qquad b_3 = 7 + 9 + 11 \qquad\qquad c_3 = \dfrac{a_3}{b_3}$$

$$a_4 = 1 + 3 + 5 + 7 \qquad b_4 = 9 + 11 + 13 + 15 \qquad c_4 = \dfrac{a_4}{b_4}$$

and this pattern continues.

Prove that c_n is independent of n and state its value.   [6 marks]

⑤ (i) Find the inverse of the matrix $\begin{pmatrix} k & 1 & 1 \\ 1 & k & 0 \\ 1 & 0 & k \end{pmatrix}$.

State any values of k for which the matrix has no inverse. **[6 marks]**

(ii) Describe how the following three planes intersect when $k = 0$.

$$kx + y + z = 5$$

$$x + ky = 0$$

$$x + kz = 3$$ **[2 marks]**

M

(iii) Prove that, for all values of k for which the following three planes intersect in one point, the coordinates of the point of intersection are independent of k.

$$kx + y + z = k$$

$$x + ky = 1$$

$$x + kz = 1$$ **[3 marks]**

⑥ (i) Show that the straight lines L and M are perpendicular but do not intersect.

$$L : \frac{x - 1}{1} = \frac{y + 3}{3} = \frac{z - 1}{-2}$$

$$M : \frac{x - 3}{5} = \frac{y - 2}{1} = \frac{z + 1}{4}$$ **[8 marks]**

PS

(ii) Find the coordinates of points A and B with A on L and B on M such that the distance AB is as small as possible. **[6 marks]**

⑦ Do not use your calculator in this question. Give your answer in exact form.

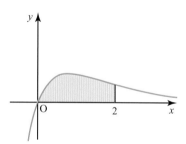

Figure 2

Calculate the area of the region of the plane shaded in the diagram. This region is bounded by the lines $y = 0$ and $x = 2$ and the curve

$$y = \frac{2x(6 - x)}{(3x + 2)(x^2 + 4)} \quad \text{where } 0 \leqslant x \leqslant 2.$$ **[10 marks]**

Polar coordinates

1 Polar coordinates

This nautilus shell forms a shape called an **equiangular spiral**. How could you describe this mathematically?

You will be familiar with using cartesian coordinates (x, y) to show the position of a point in a plane.

Figure 5.1 shows an alternative way to describe the position of a point P by giving:

■ its distance from a fixed point O, known as the **pole**;

■ the angle θ between OP and a line called the **initial line**.

The numbers (r, θ) are called the polar coordinates of P.

The length r is the distance of the point P from the origin.

The angle θ is usually measured in radians, in an anticlockwise direction from the initial line which is drawn horizontally to the right.

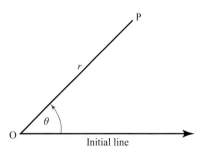

Discussion points

➜ Is it possible to provide more than one set of polar coordinates (r, θ) to define a given point P?

➜ If so, in how many ways can a point be defined?

Figure 5.1

At the point O, $r = 0$ and θ is undefined. Each pair of polar coordinates (r, θ) gives a unique point in the plane.

You may have noticed that adding or subtracting any integer multiple of 2π to the angle θ does not change the point P'.

For example, the point in Figure 5.2 below could be expressed as $\left(3, \dfrac{\pi}{3}\right), \left(3, \dfrac{7\pi}{3}\right), \left(3, \dfrac{13\pi}{3}\right), \left(3, -\dfrac{5\pi}{3}\right)$ and so on.

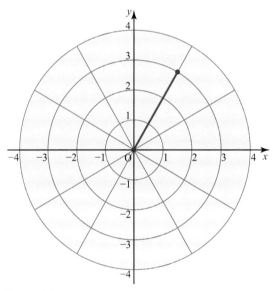

Figure 5.2

This means that each point P can be written in an infinite number of ways.

ACTIVITY 5.1

Check by drawing a diagram that the polar coordinates $\left(5, \dfrac{\pi}{6}\right), \left(5, \dfrac{13\pi}{6}\right) \left(5, -\dfrac{11\pi}{6}\right)$ and all describe the same point.

Give the polar coordinates for the point $\left(6, \dfrac{3\pi}{4}\right)$ in three other ways.

If you need to specify the polar coordinates of a point uniquely, you use the **principal polar coordinates**, where $r > 0$ and $-\pi < \theta \leq \pi$. This is similar to the convention used when writing a complex number in modulus argument form.

Converting between polar and cartesian coordinates

It is easy to convert between polar coordinates (r, θ) and cartesian coordinates (x, y).

From Figure 5.3 you can see:

$$x = r\cos\theta \qquad y = r\sin\theta \qquad r = \sqrt{x^2 + y^2} \qquad \tan\theta = \frac{y}{x}$$

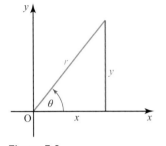

Figure 5.3

You need to be careful to choose the right quadrant when finding θ, as the equation $\tan\theta = \frac{y}{x}$ always gives two values of θ that differ by π. Always draw a sketch to make sure you know which angle is correct.

Example 5.1

(i) Find the cartesian coordinates of the following points:

(a) $\left(4, \frac{2\pi}{3}\right)$ (b) $\left(12, -\frac{\pi}{6}\right)$

(ii) Find the polar coordinates of the following points:

(a) $\left(-\sqrt{3}, 1\right)$ (b) $\left(4, -4\right)$

Solution

First draw a diagram to represent the coordinates of the point:

(i) (a) $4\cos\frac{\pi}{3} = 2$ so $x = -2$

$4\sin\frac{\pi}{3} = 2\sqrt{3}$ so $y = 2\sqrt{3}$

$\left(4, \frac{2\pi}{3}\right)$ has cartesian coordinates $\left(-2, 2\sqrt{3}\right)$.

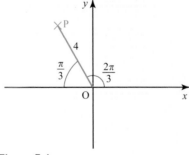

Figure 5.4

(i) (b) $12\cos\frac{\pi}{6} = 6\sqrt{3}$ so $x = 6\sqrt{3}$

$12\sin\frac{\pi}{6} = 6$ so $y = -6$

$\left(12, \frac{-\pi}{6}\right)$ has cartesian coordinates $\left(6\sqrt{3}, -6\right)$.

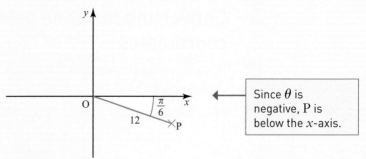

Since θ is negative, P is below the x-axis.

Figure 5.5

(ii) (a) $r = \sqrt{\left(\sqrt{3}\right)^2 + 1^2} = 2$

$\tan\alpha = \frac{1}{\sqrt{3}}$ so $\alpha = \frac{\pi}{6}$ and $\theta = \frac{5\pi}{6}$

$\left(-\sqrt{3}, 1\right)$ has polar coordinates $\left(2, \frac{5\pi}{6}\right)$.

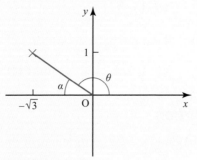

Figure 5.6

(ii) (b) $r = \sqrt{4^2 + 4^2} = 4\sqrt{2}$

$\tan\alpha = \frac{4}{4}$ so $\alpha = \frac{\pi}{4}$ so $\theta = -\frac{\pi}{4}$.

$(4, -4)$ has polar coordinates $\left(4\sqrt{2}, -\frac{\pi}{4}\right)$.

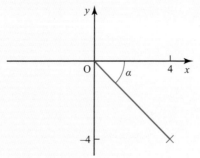

Figure 5.7

Exercise 5.1

① Find the cartesian coordinates of the following points:

(i) $\left(8, -\dfrac{\pi}{2}\right)$ (ii) $\left(8, -\dfrac{3\pi}{4}\right)$ (iii) $\left(8, \dfrac{\pi}{3}\right)$ (iv) $\left(8, \dfrac{5\pi}{6}\right)$

② Find the principal polar coordinates of the following points, giving answers as exact values or to three significant figures as appropriate:

(i) $(5, -12)$ (ii) $(-5, 0)$ (iii) $\left(-\sqrt{3}, -1\right)$ (iv) $(3, 4)$

③ Plot the points with polar coordinates $A\left(5, \dfrac{5\pi}{6}\right), B\left(3, -\dfrac{3\pi}{4}\right), C\left(5, -\dfrac{\pi}{6}\right)$ and $D\left(3, \dfrac{\pi}{4}\right)$.

Write down the name of the quadrilateral ABCD. Explain your answer.

④ Plot the points with polar coordinates $A\left(3, \dfrac{\pi}{5}\right), B\left(2, \dfrac{7\pi}{10}\right), C\left(3, -\dfrac{4\pi}{5}\right)$ and $D\left(4, -\dfrac{3\pi}{10}\right)$.

Write down the name of the quadrilateral ABCD. Explain your answer.

⑤

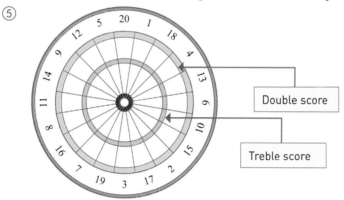

Figure 5.8

The diagram shows a dartboard made up of six concentric circles. The radii of the six circles are 6, 16, 99, 107, 162 and 170 mm respectively.

The smallest circle at the centre is called the inner bull and the next circle is called the outer bull. If a dart lands in either of these two regions it scores 50 or 25 points respectively.

The areas that get a double score or treble score are labelled. If a dart lands in one of these two rings it doubles or trebles the sector number.

The initial line passes through the middle of the sector labelled 6 and angles θ are measured in degrees from this line.

(i) Find the score in the region for which $16 < r < 99$ and $27° < \theta < 45°$.

(ii) Give conditions for r and θ that define the boundary between sectors 10 and 15.

(iii) Give conditions for r and θ for which the score is:

(a) treble 14

(b) 17

(c) double 18.

⑥ One vertex of an equilateral triangle has polar coordinates $A\left(4, \frac{\pi}{4}\right)$.

Find the polar coordinates of all the other possible vertices B and C of the triangle, when:

(i) the origin O is at the centre of the triangle

(ii) B is the origin

(iii) O is the midpoint of one of the sides of the triangle.

⑦ The diagram shows a regular pentagon OABCD in which A has cartesian coordinates $(5, 2)$.

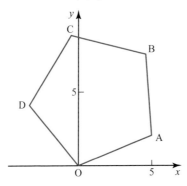

Figure 5.9

(i) Show that OB = 8.71, correct to 2 decimal places.

(ii) Find the polar coordinates of the vertices A, B, C and D, giving angles in radians.

(iii) Hence find the cartesian coordinates of the vertices B, C and D.

In parts (ii) and (iii) give your answers correct to two decimal places.

2 Sketching curves with polar equations

Prior knowledge

From the A Level Mathematics course you need to be familiar with using radians to represent an angle. You also need to know the definitions of the reciprocal trigonometric functions:

$$\sec\theta \equiv \frac{1}{\cos\theta}$$

$$\csc\theta \equiv \frac{1}{\sin\theta}$$

and the double angle formulae:

$$\sin 2\theta \equiv 2\sin\theta\cos\theta$$

$$\cos 2\theta \equiv \cos^2\theta - \sin^2\theta$$

You will be familiar with using cartesian equations such as $y = 2x^2 + 5$ to represent the relationship between the coordinates (x, y) of points on a curve. Curves can also be represented using the relationship between polar coordinates (r, θ) of points on the curve. The **polar equation** $r = f(\theta)$ is sometimes simpler than the cartesian equation, especially if the curve has rotational symmetry. Polar equations have many important applications, for example in the study of orbits.

Example 5.2 shows three ways to produce the curve with a specific polar equation $r = f(\theta)$.

Example 5.2

Investigate the curve with polar equation $r = 10\cos\theta$.

Note

Notice that some of the values of r have turned out to be negative. Although the distance from the origin cannot be negative, mathematicians sometimes interpret this by thinking about the distance being measured in the opposite direction. So the point for which $\theta = \dfrac{7\pi}{12}$ and $r = -2.6$ is equivalent to the point for which $\theta = -\dfrac{5\pi}{12}$ and $r = 2.6$

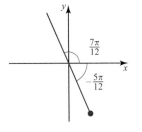

Solution

Method 1 – Plotting points

Start by making a table of values – this table has values of θ that increase in intervals of $\dfrac{\pi}{12}$, which gives a convenient number of points.

θ	0	$\dfrac{\pi}{12}$	$\dfrac{\pi}{6}$	$\dfrac{\pi}{4}$	$\dfrac{\pi}{3}$	$\dfrac{5\pi}{12}$	$\dfrac{\pi}{2}$	$\dfrac{7\pi}{12}$	$\dfrac{2\pi}{3}$	$\dfrac{3\pi}{4}$	$\dfrac{5\pi}{6}$	$\dfrac{11\pi}{12}$	π
r	10	9.7	8.7	7.1	5.0	2.6	0	−2.6	−5.0	−7.1	−8.7	−9.7	−10

Figure 5.10

If you take values of θ from 0 to $-\pi$ (instead of from 0 to π) you get the same points again; for example, $\theta = -\dfrac{\pi}{12} \Rightarrow r = 9.7$ which is the same point as $\left(-9.7, \dfrac{11\pi}{12}\right)$.

Plotting the points gives the curve shown in Figure 5.12. This curve looks like a circle. Methods 2 and 3 will both prove that it is in fact a circle.

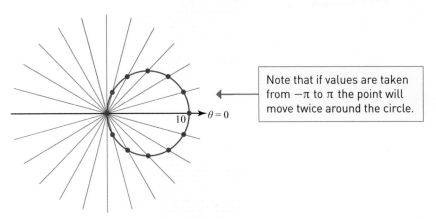

Note that if values are taken from $-\pi$ to π the point will move twice around the circle.

Figure 5.12

Method 2 – Convert to cartesian form

Multiply both sides by r.

If $r \neq 0$ then

$$r = 10\cos\theta$$

$$\Leftrightarrow r^2 = 10r\cos\theta$$

$$\Leftrightarrow x^2 + y^2 = 10x$$

You know from earlier that $x^2 + y^2 = r^2$ and $x = r\cos\theta$.

This can be rearranged to the cartesian form $(x-5)^2 + y^2 = 25$ which shows that the curve is a circle, centre $(5, 0)$, radius 5.

If $r = 0$ then $x = y = 0$ which also satisfies $x^2 + y^2 = 10x$.

Method 3 – Using geometrical reasoning

This method involves working backward from the answer.

The circle shown in Figure 5.13 has centre $(5, 0)$ and radius 5.

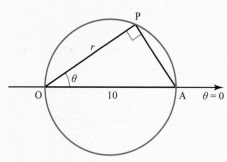

Figure 5.13

If P is the point with polar coordinates (r, θ) then triangle OPA is a right angle. Using trigonometry, $r = 10\cos\theta$ as required, and the same argument applies to points on the lower semicircle as the cosine function is even, i.e. $\cos\theta = \cos(-\theta)$.

The results from Activity 5.3 can be generalised:

- $r = a$ is a circle centre O, radius a.
- $\theta = k \, (-\pi < k \leqslant \pi)$ is a half line starting at the origin making an angle k with the initial line.

ACTIVITY 5.2

Find out how to use a graphical calculator or graphing software to draw a curve from its polar equation.

Check that you can adjust the scales on the axes so that, for the curve $r = 10\cos\theta$ from Example 5.2, you get a circle not an ellipse.

ACTIVITY 5.3

Sketch the curves with polar equations:
$$r = 7$$
$$\theta = \frac{\pi}{3}$$

Example 5.3

Describe the motion of a point along the curve $r = 1 + 2\cos\theta$ as θ increases from 0 to 2π.

Solution

The curve is shown in Figure 5.14.

Step 2
As θ increases to $\frac{2\pi}{3}$, r decreases to zero since $\cos\left(\frac{2\pi}{3}\right) = -\frac{1}{2}$.

Step 1
As θ increases from 0 to $\frac{\pi}{2}$, r decreases from 3 to 1.

Step 3
The curve touches the line $\theta = \frac{2\pi}{3}$ at the origin.

Step 4
For θ between $\frac{2\pi}{3}$ and $\frac{4\pi}{3}$, r is negative, with $r = -1$ at $\theta = \pi$.

Step 5
The curve touches the line $\theta = \frac{4\pi}{3}$ at the origin.

Step 6
r increases to 1 at $\theta = \frac{3\pi}{2}$ and then to 3 at $\theta = 2\pi$.

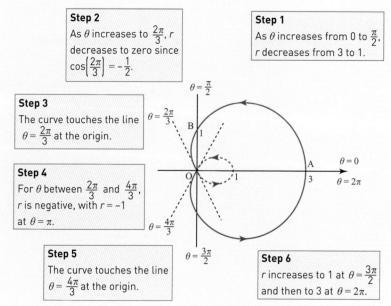

Figure 5.14

This double loop is one of a family of curves called limaçons (snail curves).

The diagram in Example 5.3 uses the convention that the parts of the curve for which $r < 0$ are shown by a broken line, whereas sections where $r > 0$ are shown by a solid line. In some applications it is physically impossible for r to be negative so it is worth distinguishing such portions in this way.

ACTIVITY 5.4

This is the curve $r = 2\cos 3\theta$.

■ How do the constants 2 and 3 in the equation relate to the shape of the curve?

■ Copy the sketch of the curve. Use arrows to show how a point would move around the curve as the value of θ varies from 0 to π.

Use a graphical calculator or graphing software to investigate the curve $r = k\sin n\theta$ where k and n are positive and n is an integer.

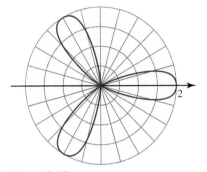

Figure 5.15

The type of curve shown in Activity 5.4 is called a **rhodonea** (rose curve).

Exercise 5.2

TECHNOLOGY

Use a graphical calculator or graphing software to check your graphs in this exercise. Remember that the scales used on the axes can affect how the shape of the graph appears.

① Sketch the curve given by the equations:

(i) $r = 5$ (ii) $\theta = -\dfrac{3\pi}{4}$

(iii) $r = 3\cos\theta$ (iv) $r = 2\sin\theta$

(v) $r = 3\theta$ for the interval $0 \le \theta \le 2\pi$

② Make a table of values of $r = 8\sin\theta$ for θ from 0 to π in intervals of $\dfrac{\pi}{12}$, giving answers to two decimal places where appropriate.

Explain what happens when $\pi \le \theta \le 2\pi$.

By plotting the points, confirm that the curve $r = 8\sin\theta$ represents a circle that is symmetrical about the y-axis.

Write down the cartesian equation of the circle.

③ Sketch the curves with equations $r = 3\cos 2\theta$ and $r = 3\cos 3\theta$ for the values of θ from 0 to 2π.

State the number of petals on each of the curves.

By considering the interval $0 \le \theta \le 2\pi$, explain why the curve $r = 3\cos n\theta$ has $2n$ petals when $n = 2$ but only n petals when $n = 3$.

④ The curve $r = \dfrac{4\theta}{\pi}$ for $-2\pi \le \theta \le 2\pi$ is called the **spiral of Archimedes**. Draw the curve.

⑤ A curve with polar equation $r = a(1 + \cos\theta)$ is called a **cardioid**.

(i) Draw the curve when $a = 8$. How do you think the curve got its name?

(ii) Sketch the curve with polar equation $r = a(1 - \cos\theta)$ when $a = 8$.

How does the shape of your graph compare to that in part (i)?

⑥ Prove that $r = a\sec\theta$ and $r = b\cosec\theta$, where a and b are non-zero constants, are the polar equations of two straight lines. Find their cartesian equations.

⑦ Example 5.3 introduced the family of curves of the form $r = a + b\cos\theta$ called **limaçons**.

   (i) Use a graphical calculator or a computer with graph plotting software to draw the curves $r = k + 3\cos\theta$ for $k = 2, 3, 4, 5, 6$.

   (ii) Investigate the shape of the curve $r = a + b\cos\theta$ for other values of a and b and use this to define the shape of the curve when:

     (a) $a = b$ (b) $a < b$

     (c) $a > b$ (d) $a \geqslant 2b$

   (iii) Investigate how the shape of the curve differs for polar curves of the form $r = a + b\sin\theta$.

⑧ A **lemniscate** has the equation $r^2 = a^2\cos 2\theta$ or $r^2 = a^2\sin 2\theta$.

Taking $r > 0$ initially, sketch these curves for $0 \leqslant \theta \leqslant 2\pi$ explaining clearly what happens in each interval of $\dfrac{\pi}{4}$ radians.

What happens if you consider values where $r < 0$?

⑨ The straight line L passes through the point A with polar coordinates (p, α) and is perpendicular to OA.

   (i) Prove that the polar equation of L is $r\cos(\theta - \alpha) = p$.

   (ii) Use the identity $\cos(\theta - \alpha) \equiv \cos\theta\cos\alpha + \sin\theta\sin\alpha$ to find the cartesian equation of L.

3 Finding the area enclosed by a polar curve

Prior knowledge

You need to be able to integrate polynomial functions and trigonometric functions of the form $a\sin bx$ and $a\cos bx$.

Look at the region in Figure 5.16 bounded by the lines OU and OV and the curve UV. To find the area of this region, start by dividing it up into smaller regions OPQ. Let OU and OV have angles $\theta = \alpha$ and $\theta = \beta$ respectively.

If the curve has equation $r = f(\theta)$, P and Q have coordinates (r, θ) and $(r + \delta r, \theta + \delta\theta)$.

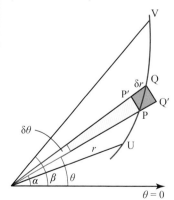

Figure 5.16

Let the area of OUV be A and the area of OPQ be δA.

The area δA lies between the circular sectors OPP' and OQQ', so:

$$\frac{1}{2}r^2\delta\theta < \delta A < \frac{1}{2}(r + \delta r)^2\,\delta\theta$$

> Remember that the area of a sector of a circle is given by $\frac{1}{2}r^2\theta$, where θ is in radians.

therefore

$$\frac{1}{2}r^2 < \frac{\delta A}{\delta\theta} < \frac{1}{2}(r + \delta r)^2$$

As $\delta\theta \to 0$, $\delta r \to 0$ and so $\frac{1}{2}(r + \delta r)^2 \to \frac{1}{2}r^2$. Therefore $\frac{\delta A}{\delta\theta}$ must also tend to $\frac{1}{2}r^2$ as $\delta\theta \to 0$.

But as $\delta\theta \to 0$, $\dfrac{\delta A}{\delta\theta} \to \dfrac{\mathrm{d}A}{\mathrm{d}\theta}$

Therefore $\dfrac{\mathrm{d}A}{\mathrm{d}\theta} = \frac{1}{2}r^2$.

Integrating both sides with respect to θ shows the result for the area of a region bounded by a polar curve and two straight lines $\theta = \alpha$ and $\theta = \beta$ is:

$$A = \int_{\alpha}^{\beta} \frac{1}{2}r^2\,\mathrm{d}\theta$$

Example 5.4

Figure 5.17 shows the curve $r = 1 + 2\cos\theta$ from Example 5.3.

Find the area of the inner loop of the limaçon $r = 1 + 2\cos\theta$.

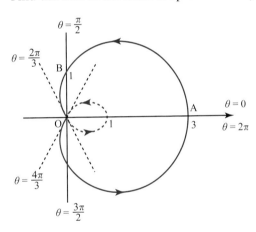

Figure 5.17

Solution

In Example 5.3 you saw that the inner loop is formed as θ varies from $\dfrac{2\pi}{3}$ to $\dfrac{4\pi}{3}$ so its area is given by

$$A = \int_{\frac{2\pi}{3}}^{\frac{4\pi}{3}} \frac{1}{2}r^2\mathrm{d}\theta = \int_{\frac{2\pi}{3}}^{\frac{4\pi}{3}} \frac{1}{2}(1 + 2\cos\theta)^2\,\mathrm{d}\theta$$

$$\Rightarrow A = \int_{\frac{2\pi}{3}}^{\frac{4\pi}{3}} \frac{1}{2}\left(1 + 4\cos\theta + 4\cos^2\theta\right)\mathrm{d}\theta$$

> Using $\cos^2\theta = \frac{1}{2}(1 + \cos 2\theta)$. Note that even though the value of r is negative between $\dfrac{2\pi}{3}$ and $\dfrac{4\pi}{3}$ the integrand $\frac{1}{2}r^2$ is always positive so there is no issue of 'negative areas' as there is when curves go below the x-axis when using cartesian coordinates.

$$= \left[\frac{3\theta}{2} + 2\sin\theta + \frac{1}{2}\sin 2\theta\right]_{\frac{2\pi}{3}}^{\frac{4\pi}{3}}$$

$$= \pi - \frac{3\sqrt{3}}{2}$$

Exercise 5.3

① (i) Check that the integral $\int \frac{1}{2} r^2 \, d\theta$ correctly gives the area of the circle $r = 10\cos\theta$ when it is evaluated from $-\frac{\pi}{2}$ to $\frac{\pi}{2}$.

   (ii) What happens when the integral is evaluated from 0 to 2π?

② A curve has equation $r = 5\cos 4\theta$.

   (i) Sketch the curve for the interval $0 \leqslant \theta \leqslant 2\pi$.

   (ii) Find the area of one loop of the curve.

③ A curve has equation $r = 3 + 3\sin\theta$.

   (i) Sketch the curve for the interval 0 to 2π.

   (ii) Find the area enclosed by the curve.

④ Find the area bounded by the spiral $r = \frac{4\theta}{\pi}$ from $\theta = 0$ to $\theta = 2\pi$ and the initial line.

⑤ For the limaçon $r = 1 + 2\cos\theta$ in Example 5.4, find:

   (i) the total area contained inside the outer loop

   (ii) the area between the two loops

⑥ Find the exact areas of the two portions into which the line $\theta = \frac{\pi}{2}$ divides the upper half of the cardioid $r = 8(1 + \cos\theta)$.

⑦ Sketch the lemniscate $r^2 = a^2 \cos 2\theta$ and find the area of one of its loops.

⑧ The diagram shows the **equiangular spiral** $r = ae^{k\theta}$ where a and k are positive constants and e is the exponential constant 2.71828…

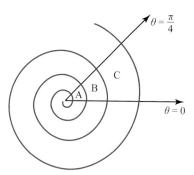

Figure 5.18

Prove that the areas A, B and C formed by the lines $\theta = 0$ and $\theta = \frac{\pi}{4}$ and the spiral form a geometric sequence and find its common ratio.

⑨ Find the area enclosed between the curves $r = 3 - 3\cos\theta$ and $r = 4\cos\theta$. Give your final answer to three significant figures.

LEARNING OUTCOMES

When you have completed this chapter you should be able to:

- ➤ understand and use polar coordinates
- ➤ convert from polar to cartesian coordinates and vice versa
- ➤ sketch curves with simple polar equations in the form $r = f(\theta)$
- ➤ find the area enclosed by a polar curve.

KEY POINTS

1 To convert from polar coordinates to cartesian coordinates $x = r\cos\theta$, $y = r\sin\theta$.

2 To convert from cartesian coordinates to polar coordinates $r = \sqrt{x^2 + y^2}$, $\theta = \arctan\dfrac{y}{x}$ ($\pm\pi$ if necessary).

3 The principal polar coordinates (r, θ) are those for which $r > 0$ and $-\pi < \theta \leqslant \pi$.

4 The area of a sector is $\displaystyle\int_{\alpha}^{\beta} \frac{1}{2} r^2 \, d\theta$.

FUTURE USES

■ If you study the *Further Pure with Technology* option you will use graphing software to plot, describe and generalise polar curves and find gradients and arc lengths of these curves.

6 Maclaurin series

'If I feel unhappy, I do mathematics to become happy. If I feel happy, I do mathematics to keep happy.'

Alfred Renyi, 1921–1970

The four photographs above were taken over a period of time. They build up to the final picture which shows the complete situation.

In this chapter you will meet a similar idea building up a polynomial series to represent a function. Instead of showing how the Olympic Stadium developed over time as new parts were added to it, you will be expressing a function as a series of ever increasing accuracy by adding on successive terms of a polynomial.

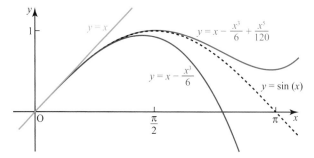

Figure 6.1

Discussion point

→ Look at the diagram above. Can $\sin x$ be represented as a polynomial for all values of x?

1 Polynomial approximations and Maclaurin series

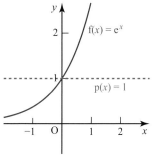

Figure 6.2

Since polynomial functions are easy to evaluate, differentiate and integrate (among other things), they are often useful approximations to more complicated functions. Here you will see one way of building these approximations, starting with the example of the exponential function: $f(x) = e^x$.

To build up a polynomial function $p(x)$ which approximates $f(x)$, start by making sure it has the correct value at $x = 0$ (where the graph cuts the y-axis).

Since $e^0 = 1$, the first term of the polynomial approximation is 1, and so $p(x) = 1 + \ldots$ (see Figure 6.2).

You can certainly find a better approximation than this. The next step is to consider the gradient of your approximation. The derivative of e^x is e^x, and so at $x = 0$ the gradient is 1.

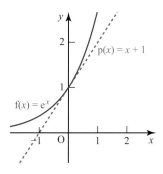

Figure 6.3

The next term of $p(x)$ will be a multiple of x and, since its derivative is 1, this will be x itself. So $p(x) = 1 + x + \ldots$ (Figure 6.3).

This is a better approximation, but, again, can be improved further. The next step is to ensure that the second derivatives of e^x and $p(x)$ have the same values when $x = 0$. The second derivative of $f(x) = e^x$ at $x = 0$ is also 1. The next term of $p(x)$ will be a multiple of x^2. Since the second derivative of x^2 is 2, the quadratic term must be $\frac{1}{2}x^2$ to give a second derivative of 1. So $p(x) = 1 + x + \frac{1}{2}x^2 + \ldots$, and the graph of this function is a much better approximation to $y = f(x)$ (Figure 6.4).

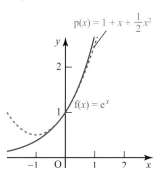

Figure 6.4

ACTIVITY 6.1

(i) Extend this method one more step and show that the cubic approximation to $f(x) = e^x$ is

$$p(x) = 1 + x + \frac{1}{2}x^2 + \frac{1}{6}x^3$$

(ii) Extend the method two further steps to find a degree 5 polynomial approximation for $f(x)$.

You will see from the graph below (Figure 6.5) that, for positive values of x around $x = 0$, the accuracy of the approximation improves as you add more terms. This is also true for negative x-values, although it is not so clear from the diagram.

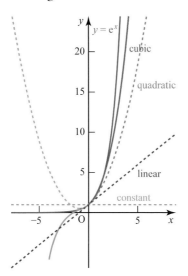

ACTIVITY 6.2

Write down the cubic approximation to e^x and substitute $(-x)$ in place of x to generate a cubic approximation to e^{-x}.

Multiply these two cubics together and comment on your answer.

Figure 6.5

Example 6.1

(i) In the polynomial approximation to e^x, find an expression for the term in x^r.

(ii) Hence express e^x as the sum of an infinite series.

(iii) Investigate whether the terms of the series converge for all values of x.

Solution

(i) Let the term in x^r be kx^r.

Differentiating this r times gives $kr \times (r-1) \times (r-2) \times \ldots \times 2 \times 1 = kr!$
The rth derivative of e^x at $x = 0$ is 1, so $kr! = 1$ and therefore $k = \dfrac{1}{r!}$.
So the term in x^r is $\dfrac{x^r}{r!}$.

(ii) $p(x) = \displaystyle\sum_{r=0}^{\infty} \frac{x^r}{r!}$

(iii) For $x = 2$, $p(x) = 1 + 2 + \dfrac{2^2}{2!} + \dfrac{2^3}{3!} + \dfrac{2^4}{4!} + \ldots$

In the 5th term, $4!$ is greater than 2^4 so this term is less than 1. After this, the terms continue to get smaller, so it appears that the terms converge.

If you take any value of x, at some point the value of $r!$ will be greater than x^r, so after that, the terms are less than 1 and will continue to get smaller. So it seems that the terms converge for all values of x.

Discussion point

→ Is it always the case that if the terms of a series converge, the sum of the series also converges?

In Example 6.1 you saw that the terms of the series for e^x converge, whatever the value of x. In fact, the sum of the terms also converges for all values of x. The formal proof of the convergence is beyond the scope of this book.

ACTIVITY 6.3

This spreadsheet shows the start of a method for approximating e^x.

	A	**B**
1	0.5	=SUM(B3:B10)
2		
3	0	1
4	=A3+1	=A1*B3/A4

Note

Notice the $ signs in the formula in cell B4: these indicate an **absolute reference**, meaning this part of the formula will always refer to cell A1, whereas the other parts of the formula are **relative references**, meaning they will change relative to where the formula is copied.

Copy these cells into your own spreadsheet. Copy and paste the formulae in Row 4 down to Row 10. Check the formula does indeed behave as explained above.

The number in cell A1 is the value of x, try changing it and watch the effect on the other cells. In particular, explain why the value in cell B1 is an approximation to e^x, and state the order of the polynomial approximation this example achieves.

Change your spreadsheet so that it gives a polynomial approximation up to the term in x^{10}. Use it to calculate the value, and the percentage error, of the Maclaurin approximation up to the term in x^{10}, for the calculation e^2.

Maclaurin approximations and series

In general, you can find a polynomial $p(x)$, of order n, for any function $f(x)$, for which its first n derivatives at $x = 0$ exist.

To ensure that $p(x)$ takes the same value as $f(x)$ for each derivative, at $x = 0$, then you need:

$$p(0) = f(0)$$
$$p'(0) = f'(0)$$
$$p''(0) = f''(0)$$
$$p^{(3)}(0) = f^{(3)}(0)$$

A third derivative $f'''(x)$ can be written as $f^{(3)}(x)$, and similarly for higher derivatives.

$$\dots$$
$$p^{(n)}(0) = f^{(n)}(0)$$

This is a list of $n + 1$ conditions, and you can see the general nth order polynomial has $n + 1$ constants to determine:

$$p(x) = a_0 + a_1 x + a_2 x^2 + a_3 x^3 + \dots + a_r x^r + \dots + a_n x^n$$

Substituting in $x = 0$ immediately gives:

$$a_0 = f(0)$$

Doing this with the derivatives of p gives the various values of a_r:

$$p'(x) = a_1 + 2a_2x + 3a_3x^2 + \ldots + ra_rx^{r-1} + \ldots + na_nx^{n-1}$$

so

$$a_1 = f'(0)$$

Then

$$p''(x) = 2a_2 + 6a_3x + \ldots + r(r-1)a_rx^{r-2} + \ldots + n(n-1)a_nx^{n-2}$$
$$2a_2 = f''(0)$$
$$a_2 = \frac{1}{2}f''(0)$$

and

$$p^{(3)}(0) = 6a_3 + \ldots + r(r-1)(r-2)a_rx^{r-3} + \ldots + n(n-1)(n-2)a_nx^{n-3}$$
$$6a_3 = f^{(3)}(0)$$
$$a_3 = \frac{1}{6}f^{(3)}(0)$$

In general, the nth derivative gives the condition

$$n!a_n = f^{(n)}(0)$$

so

$$a_n = \frac{1}{n!}f^{(n)}(0)$$

Putting all these back into the $p(x)$ definition to find the approximation for $f(x)$ gives:

$$f(x) \approx f(0) + f'(0)x + \frac{f''(0)}{2!}x^2 + \frac{f^{(3)}(0)}{3!}x^3 + \ldots + \frac{f^{(r)}(0)}{r!}x^r + \ldots + \frac{f^{(n)}(0)}{n!}x^n.$$

This is the **Maclaurin approximation** or **Maclaurin expansion** for $f(x)$ up to the term in x^n.

The accuracy of the approximation is usually defined as follows:

$$\text{error} = \text{approximate value} - \text{exact value}$$

$$\text{percentage error} = \frac{\text{approximate value} - \text{exact value}}{\text{exact value}} \times 100\%$$

Example 6.2

Find the Maclaurin expansion for $(1 - x)^{-1}$, as far as x^n.

Solution

Let $f(x) = (1 - x)^{-1}$.

$f(x) = (1 - x)^{-1}$	$f(0) = 1$
$f'(x) = (1 - x)^{-2}$	$f'(0) = 1$
$f''(x) = 2(1 - x)^{-3}$	$f''(0) = 2$
$f^{(3)}(x) = 6(1 - x)^{-4}$	$f^{(3)}(0) = 6$
$f^{(4)}(x) = 24(1 - x)^{-5}$	$f^{(4)}(0) = 24$
$\vdots$	$\vdots$
$f^{(n)}(x) = n!(1 - x)^{-(n+1)}$	$f^{(n)}(0) = n!$

It is useful to tabulate the function $f(x)$ and its derivatives, then evaluate them at $x = 0$.

Table 6.1

So

$$(1 - x)^{-1} \approx 1 + x + \frac{2}{2!}x^2 + \frac{6}{3!}x^3 + \frac{24}{4!}x^4 + \ldots + \frac{n!}{n!}x^n$$

$$= 1 + x + x^2 + x^3 + x^4 + \ldots + x^n$$

ACTIVITY 6.4

Compare the result from Example 6.2 with the binomial expansion for $(1 - x)^{-1}$ (see Chapter 7 in A Level Year 2 Mathematics) and the sum to infinity of the geometric series $1 + x + x^2 + x^3 + \ldots$

Note

A Maclaurin *expansion* for a function involves a finite number of terms and is an approximation to the function. A Maclaurin *series* is the sum of an infinite number of terms.

If the function and *all* its derivatives exist at $x = 0$ then, of course, this expansion could be continued indefinitely, in which case you would get an infinite order polynomial:

$$f(x) = f(0) + f'(0)x + \frac{f''(0)}{2!}x^2 + \frac{f^{(3)}(0)}{3!}x^3 + \ldots + \frac{f^{(r)}(0)}{r!}x^r + \ldots$$

This is known as the **Maclaurin series** for $f(x)$. Care must be taken with infinite polynomials like this, but if the sum of the series up to and including the term in x^n tends to a limit as n tends to infinity, and this limit is $f(x)$, then you can say that the expansion **converges** to $f(x)$.

Validity of Maclaurin series

The example above showed that the nth Maclaurin expansion for $(1 - x)^{-1}$ is the geometric series $1 + x + x^2 + x^3 + \ldots + x^n$. If you let n tend to infinity this would become the **sum to infinity**, but you will already know that this sum only converges if $|x| < 1$. This is an example of a Maclaurin series which only converges for a limited range of x values – these are described as the values for which the series is **valid**.

Other Maclaurin series are valid for different ranges of x. As you saw in Example 6.1, the Maclaurin series for e^x

$$e^x \equiv 1 + x + \frac{1}{2!}x^2 + \frac{1}{3!}x^3 + \ldots + \frac{1}{r!}x^r + \ldots$$

is valid for *all* values of x.

A power series may be regarded as incomplete without a statement of the values of x for which it is valid. However, if there is no such statement it may be taken that it is valid for all x.

Example 6.3	Find the first three non-zero terms and the general term of the Maclaurin series for $\sin x$.

Solution

Let $f(x) = \sin x$.

$f(x) = \sin x$	$f(0) = 0$
$f'(x) = \cos x$	$f'(0) = 1$
$f''(x) = -\sin x$	$f''(0) = 0$
$f^{(3)}(x) = -\cos x$	$f^{(3)}(x) = -1$
$f^{(4)}(x) = \sin x$	$f^{(4)}(0) = 0$
$f^{(5)}(x) = \cos x$	$f^{(5)}(0) = 1$
...	...
$f^{(2r)}(x) = (-1)^r \sin x$	$f^{(2r)}(0) = 0$
$f^{(2r+1)}(x) = (-1)^r \cos x$	$f^{(2r+1)}(0) = (-1)^r$

Note

All the odd derivatives are zero at $x = 0$. The even derivatives alternate between 1 and -1 at $x = 0$.

Table 6.2

Using $f(x) = f(0) + f'(0)x + \frac{f''(0)}{2!}x^2 + \frac{f^{(3)}(0)}{3!}x^3 + \ldots + \frac{f^{(r)}(0)}{r!}x^r + \ldots$

gives

$$\sin x = x - \frac{1}{3!}x^3 + \frac{1}{5!}x^5 - \ldots + \frac{(-1)^r x^{2r+1}}{(2r+1)!} + \ldots$$

Note

Note that for this series, x is measured in radians, as the rules for differentiating $\sin x$ and $\cos x$ require x to be measured in radians.

Exercise 6.1

Note

Make sure that you are working in radians.

① (i) Show that the Maclaurin series for $\cos x$ is:

$$\cos x = 1 - \frac{1}{2!}x^2 + \frac{1}{4!}x^4 - \frac{1}{6!}x^6 + \ldots + \frac{(-1)^r x^{2r}}{(2r)!} + \ldots$$

(ii) Use the first three terms of the series to calculate an approximate value for $\cos(0.1)$.

(iii) Use your calculator to find $\cos(0.1)$ and find the percentage error in your answer to (ii).

② The Maclaurin series for e^x is $1 + x + \frac{1}{2!}x^2 + \frac{1}{3!}x^3 + \ldots + \frac{1}{r!}x^r + \ldots$

(i) Use this to calculate $\frac{1}{\sqrt{e}}$ correct to five decimal places, stating how many terms you need to use to be sure.

(ii) Use the e^x function on your calculator to calculate the percentage error in this approximation.

③ (i) Write down the cubic approximation to $\sin x$.

(ii) Use this approximation to rewrite the equation $\sin x = x^2$ as a polynomial equation.

(iii) By solving this polynomial equation, show that an approximate root of the original equation is

$$x = \sqrt{15} - 3$$

④ (i) Explain why it is not possible to find Maclaurin expansions for $f(x) = \ln x$.

(ii) (a) Show that the Maclaurin series for $\ln(1 + x)$ is

$$x - \frac{x^2}{2} + \frac{x^3}{3} - \frac{x^4}{4} + \ldots + \frac{(-1)^{n+1} x^n}{n} + \ldots$$

(b) This series is valid for $-1 < x \leqslant 1$ only; by drawing graphs of $y = \ln(1 + x)$ and several successive approximations show that this is plausible.

⑤ The third Maclaurin approximation to a function is $f(x) = 1 - \frac{3}{2}x^2 + \frac{5}{2}x^3$.

Write down the values of $f'(0)$, $f''(0)$ and $f^{(3)}(0)$.

Sketch the graph of $y = f(x)$ near $x = 0$.

⑥ Find the Maclaurin expansion for $f(x) = \tan x$ up to the term in x^3.

⑦ An approximate rule used by builders to find the length, c, of a circular arc ABC is

$$c = \frac{8b - a}{3},$$

where a and b are as shown in Figure 6.6.

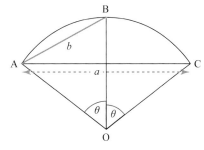

Figure 6.6

(i)   If O is the centre of the circle, show that $b = 2r\sin\left(\dfrac{\theta}{2}\right)$ and $a = 2r\sin(\theta)$.

(ii)   Using the cubic approximation to $\sin x$, show that $8b - a = 6r\theta$. Hence verify the rule.

(iii)   Find the percentage error caused by using this rule when $\theta = \dfrac{\pi}{3}$.

⑧ A curve passes through the point $(0, 2)$; its gradient is given by the differential equation $\dfrac{\mathrm{d}y}{\mathrm{d}x} = 1 - xy$. Assume that the equation of this curve can be expressed as the Maclaurin series

$$y = a_0 + a_1 x + a_2 x^2 + a_3 x^3 + a_4 x^4 + \dots$$

(i)   Find a_0.

(ii)   Show that
$$a_1 + 2a_2 x + 3a_3 x^2 + 4a_4 x^3 + \dots = 1 - 2x - a_1 x^2 - a_2 x^3 - a_3 x^4 - \dots$$

(iii)   Equate coefficients to find the first seven terms of the Maclaurin series.

(iv)   Draw graphs to compare the solution given by these seven terms with a solution generated (step by step) on a computer.

⑨ (i)   Write down the Maclaurin expansions of

    (a) $\cos\theta$ (b) $\sin\theta$ (c) $\cos\theta + \mathrm{i}\sin\theta$

    giving your answers in ascending powers of θ.

(ii)   Substitute $x = \mathrm{i}\theta$ in the Maclaurin series for e^x and simplify the terms.

(iii)   Show that your answers to (i)(c) and (ii) are the same, and hence
$$\mathrm{e}^{\mathrm{i}\theta} = \cos\theta + \mathrm{i}\sin\theta$$

(See page 227 in Chapter 10 Complex numbers.)

⑩ In this question y_n and a_n are used to denote $\mathrm{f}^{(n)}(x)$ and $\mathrm{f}^{(n)}(0)$ respectively.

(i)   Let $\mathrm{f}(x) = \arcsin x$.
    Show that $\left(1 - x^2\right)y_1^2 = 1$ and $\left(1 - x^2\right)y_2 - xy_1 = 0$.

(ii)   Find a_1 and a_2.

(iii)   Prove by induction that $\left(1 - x^2\right)y_{n+2} - (2n + 1)xy_{n+1} - n^2 y_n = 0$, and deduce that $a_{n+2} = n^2 a_n$.

(iv)   Find the Maclaurin expansion of $\arcsin x$, giving the first three non-zero terms and the general term.

2 Using Maclaurin series for standard functions

You will often need to use the Maclaurin series for some common functions, which are listed below, together with the values of x for which the expansion is valid.

Note

It is always good practice to state the values of x for which it is valid when you write down a Maclaurin series, particularly in cases where this is not for all x.

$e^x = 1 + x + \dfrac{x^2}{2!} + \ldots + \dfrac{x^r}{r!} + \ldots$	Valid for all x
$\ln(1 + x) = x - \dfrac{x^2}{2} + \dfrac{x^3}{3} - \ldots + (-1)^{r+1} \dfrac{x^r}{r} + \ldots$	Valid for $-1 < x \leqslant 1$
$\sin x = x - \dfrac{x^3}{3!} + \dfrac{x^5}{5!} - \ldots + (-1)^r \dfrac{x^{2r+1}}{(2r+1)!} + \ldots$	Valid for all x
$\cos x = 1 - \dfrac{x^2}{2!} + \dfrac{x^4}{4!} - \ldots + (-1)^r \dfrac{x^{2r}}{(2r)!} + \ldots$	Valid for all x
$(1 + x)^n = 1 + nx + \dfrac{n(n-1)}{2!}x^2 + \ldots$ $+ \dfrac{n(n-1)\ldots(n-r+1)}{r!}x^r + \ldots$	Valid for $\lvert x \rvert < 1, \ n \in \mathbb{R}$

Table 6.3

ACTIVITY 6.5

(i) What happens when you differentiate these series, term by term?

(ii) What do you notice about the powers in the series for $\sin x$ and $\cos x$? How does this relate to the symmetry of the graphs of $\sin x$ and $\cos x$?

These standard series can be used to find Maclaurin series for related functions.

Example 6.4

(i) Find the Maclaurin expansion for e^{-2x} up to the term in x^4.

(ii) Find the general term for the expansion.

(iii) For what values of x is the expansion valid?

Solution

(i) $e^x = 1 + x + \dfrac{x^2}{2!} + \dfrac{x^3}{3!} + \dfrac{x^4}{4!} + \ldots$

Substituting $-2x$ for x:

$e^{-2x} = 1 + (-2x) + \dfrac{(-2x)^2}{2!} + \dfrac{(-2x)^3}{3!} + \dfrac{(-2x)^4}{4!} + \ldots$

$= 1 - 2x + \dfrac{4x^2}{2} - \dfrac{8x^3}{6} + \dfrac{16x^4}{24} + \ldots$

$= 1 - 2x + 2x^2 - \dfrac{4x^3}{3} + \dfrac{2x^4}{3} + \ldots$

(ii) General term $= \dfrac{(-2x)^r}{r!} = \dfrac{(-2)^r}{r!}x^r$

(iii) Since the expansion for e^x is valid for all values of x, this expansion is also valid for all values of x.

Sometimes a Maclaurin series can be found by adapting one or more known Maclaurin series. An example you will have already verified is that of differentiating the series for $\sin x$ to obtain the series for $\cos x$. It is important to question whether this is a justifiable method:

- Is it valid to integrate or differentiate an infinite series term by term?

- Can you form the product of two infinite series by multiplying terms?

- Is the series obtained identical to the series that would have been obtained by evaluating the derivatives?

Answering these questions in detail is beyond the scope of this book, but you may take it that the answer to all of them is, 'Yes, subject to certain conditions.' In the work in this book you may safely assume the conditions are met, but strange things can happen with infinite series in other situations.

ACTIVITY 6.6

Try these suggested methods of deriving new series, and explain why they work.

(i) The Maclaurin series for $\ln(1 + x)$ can be found by integrating the terms of the binomial series for $\dfrac{1}{1 + x}$. Why is the constant of integration zero?

(ii) The start of the Maclaurin series for $\dfrac{e^x}{1 + x}$ can be found by multiplying together the first four terms of the series for e^x and $(1 + x)^{-1}$, then discarding all terms in x^4 and higher powers.

(iii) The first few terms of the Maclaurin series for $\sec x$ can be found from the first three terms of the series for $(1 + y)^{-1}$ when $y = -\dfrac{x^2}{2!} + \dfrac{x^4}{4!}$.

Exercise 6.2

① (i) Write down the first four non-zero terms in Maclaurin series for $\cos u$.

(ii) Substitute $u = 2x$ in this series to obtain the first four non-zero terms in the Maclaurin series for $\cos 2x$.

② Use known Maclaurin series to find the Maclaurin series for the following functions as far as the term in x^4:

(i) $\sin 3x$

(ii) $\ln(1 + 2x)$

(iii) $e^{\frac{1}{2}x}$

③ Use known Maclaurin series to find the Maclaurin series for the following functions as far as the term in x^4:

(i) $\sin^2 x$ (ii) $\ln(1 + \sin x)$

(iii) $e^{-x}\sin x$ (iv) $e^{\sin x}$

④ (i) Find $\displaystyle\int \frac{1}{\sqrt{1 - 4x^2}}\,dx$.

(ii) By expanding $\left(1 - 4x^2\right)^{-\frac{1}{2}}$ and integrating term by term, or otherwise, find the series expansion for $\arcsin 2x$, when $|x| < \frac{1}{2}$ as far as the term in x^7.

(iii) Use your expansion to find an approximate value for $\arcsin 0.5$ and find the percentage error for this approximate value.

⑤ In this questions give all numerical answers to four decimal places.

(i) Put $x = 1$ in the expansion
$$\ln(1 + x) \approx x - \frac{x^2}{2} + \frac{x^3}{3} - \ldots - \frac{x^{10}}{10}$$
and calculate an estimate of $\ln 2$.

> Approximately 1000 terms would be needed to obtain an estimate of $\ln 2$ accurate to 3 d.p. by this method.

(ii) Show that $\ln 2 = -\ln\left(1 - \frac{1}{2}\right)$ and hence estimate $\ln 2$ by summing six terms.

(iii) Write down the series for $\ln(1 + x) - \ln(1 - x)$ as far as the first three non-zero terms and estimate $\ln 2$ by summing these terms using a suitable value of x.

⑥ (i) By integrating $\dfrac{1}{1 + x^2}$ and its Maclaurin expansion, show that the Maclaurin series for $\arctan x$ is
$$x - \frac{x^3}{3} + \frac{x^5}{5} - \frac{x^7}{7} + \ldots$$

> This is known as Gregory's series, after Scottish mathematician James Gregory. It is valid for $|x| \leqslant 1$.

(ii) By putting $x = 1$ show that
$$\frac{\pi}{4} = 1 - \frac{1}{3} + \frac{1}{5} - \frac{1}{7} + \ldots$$

> This is known as Leibniz's series. It converges very slowly.

(iii) Show that:

(a) $\dfrac{\pi}{4} = \arctan\left(\dfrac{1}{2}\right) + \arctan\left(\dfrac{1}{3}\right)$

> Known as Euler's formula for π.

(b) $\dfrac{\pi}{4} = 4\arctan\left(\dfrac{1}{5}\right) - \arctan\left(\dfrac{1}{239}\right)$

> Known as Machin's formula.

(iv) Use Machin's formula with Gregory's series to find the value of π to five decimal places.

⑦ (i) Write down the first four terms of the series for $\dfrac{1}{1 - x}$.

(ii) By comparing with part (i), find an expression for the sum of the series $1 + 2x + 3x^2 + 4x^3 + \ldots$

⑧ (i) Prove by induction that:
$$f(x) = e^x \sin x \implies f^{(n)}(x) = 2^{\frac{\pi}{2}} e^x \sin\left(x + \frac{n\pi}{4}\right)$$

Use this result to obtain the Maclaurin series for $e^x \sin x$ as far as x^6.

(ii) Multiply the cubic Maclaurin approximation for e^x by the cubic approximation for $\sin x$ and comment on your answer.

(iii) Find a Maclaurin approximation for $e^x \cos x$ by multiplying the cubic Maclaurin approximation for e^x by the quartic Maclaurin approximation for $\cos x$, giving as many terms in your answer as you think justifiable.

⑨ (i) Given that $f(x) = e^{2x} \sin 3x$, show that $f''(x) = 4f'(x) - 13f(x)$.

(ii) Differentiate this result twice to find expressions for $f^{(3)}(x)$ and $f^{(4)}(x)$ in terms of lower derivatives.

(iii) Hence find the Maclaurin expansion for $f(x)$ up to the term in x^4.

⑩ A projectile is launched from O with initial velocity $\begin{pmatrix} u \\ v \end{pmatrix}$ relative to horizontal and vertical axes through O. The path of the projectile may be modelled in various ways. Table 6.4 (below) shows the position (x, y) of the projectile at time t after launch, as given by two different models. Both models assume that g (gravitational acceleration) is constant.

	Assumptions	Position at time t
Model 1	No air resistance	$x = ut$ $y = vt - \dfrac{1}{2} gt^2$
Model 2	Air resistance is proportional to the velocity (with proportionality constant k).	$x = \dfrac{u}{k}\left(1 - e^{-kt}\right)$ $y = \dfrac{g + kv}{k^2}\left(1 - e^{-kt}\right) - \dfrac{gt}{k}$

Table 6.4

Use the Maclaurin expansion for e^{-kt}, where k is constant, to show that the results given by Model 1 are a special case of the results from Model 2, with $k = 0$.

⑪ Figure 6.7 illustrates a Maclaurin expansion. Find it.
(You may assume that $0 < r < 1$.)

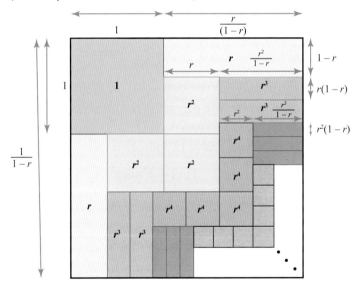

Figure 6.7

LEARNING OUTCOMES

When you have completed this chapter you should be able to:

➤ find the Maclaurin series of a function, including the general term

➤ know that a Maclaurin series may converge only for a restricted set of values of x

➤ recognise and use the Maclaurin series of standard functions: e^x, $\ln(1 + x)$, $\sin x$, $\cos x$ and $(1 + x)^n$.

KEY POINTS

1 The general form of the Maclaurin series for $f(x)$ is:

$$f(x) = f(0) + f'(0)x + \frac{f''(0)}{2!}x^2 + \frac{f^{(3)}(0)}{3!}x^3 + \ldots + \frac{f^{(r)}(0)}{r!}x^r + \ldots$$

2 Series which are valid for all x:

$$e^x = 1 + x + \frac{x^2}{2!} + \frac{x^3}{3!} + \ldots + \frac{x^r}{r!} + \ldots$$

$$\sin x = x - \frac{x^3}{3!} + \frac{x^5}{5!} - \frac{x^7}{7!} + \ldots + \frac{(-1)^r x^{2r+1}}{(2r+1)!} + \ldots$$

$$\cos x = 1 - \frac{x^2}{2!} + \frac{x^4}{4!} - \frac{x^6}{6!} + \ldots + \frac{(-1)^r x^{2r}}{(2r)!} + \ldots$$

3 Series valid for $|x| \leqslant 1$:

$$\arctan x = x - \frac{x^3}{3} + \frac{x^5}{5} - \frac{x^7}{7} + \ldots + \frac{(-1)^{r-1} x^{2r-1}}{2r-1} + \ldots$$

4 Series valid for $-1 < x \leqslant 1$:

$$\ln(1 + x) = x - \frac{x^2}{2} + \frac{x^3}{3} - \ldots + \frac{(-1)^{r-1} x^r}{r} + \ldots$$

5 Series where validity depends on n:

$$(1 + x)^n = 1 + nx + \frac{n(n-1)}{2!}x^2 + \ldots + \frac{n(n-1)\ldots(n-r+1)}{r!}x^r + \ldots$$

If n is a positive integer: the series terminates after $n + 1$ terms, and is valid for all x.

If n is not a positive integer: the series is valid for $|x| < 1$; also for $|x| = 1$ if $n \geqslant 1$; and for $x = -1$ if $n > 0$.

Review: Complex numbers

1 Working with complex numbers

The different types of numbers in the number system can be represented in a diagram as in Figure R.1.

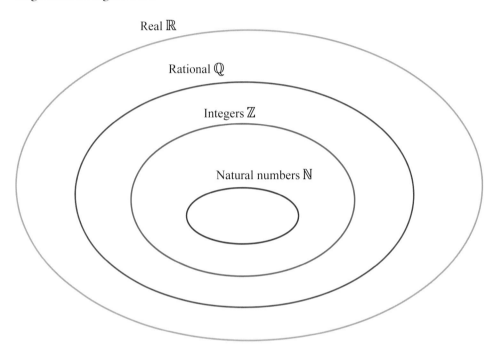

Figure R.1

The number system can be extended to include complex numbers, denoted $\mathbb{C}$, by introducing a new number $i = \sqrt{-1}$. Any number z of the form $x + yi$, where x and y are real, is called a **complex number**. x is called the **real part** of the complex number, denoted by $\text{Re}(z)$, and y is called the **imaginary part**, denoted by $\text{Im}(z)$. If $y = 0$ the number $z = x + yi$ is wholly real, so the set $\mathbb{R}$ is a subset of the set $\mathbb{C}$.

$z^* = x - yi$ is the **complex conjugate** of $z = x + yi$.

Example R.1

Use the quadratic formula to solve the quadratic equation $z^2 - 2z + 10 = 0$, simplifying your answer as far as possible.

Solution

$$z^2 - 2z + 10 = 0$$

$$z = \frac{2 \pm \sqrt{4 - (4 \times 1 \times 10)}}{2 \times 1}$$

> Using the quadratic formula with $a = 1$, $b = -2$ and $c = 10$.

$$= \frac{2 \pm \sqrt{-36}}{2}$$

> $\sqrt{-36} = \sqrt{-1}\sqrt{36} = 6i.$

$$= \frac{2 \pm 6i}{2}$$

$$= 1 \pm 3i$$

The roots have a real part and an imaginary part. The roots form a **conjugate pair** as they have the same real part and the imaginary parts have the opposite signs.

> 1 is the real part of the complex numbers, denoted $\mathrm{Re}(z)$

$$1 \pm 3i$$

> 3 is the imaginary part of the complex numbers, denoted $\mathrm{Im}(z)$.

Example R.2

Calculate:

(i) $(2 - 3i) + (1 + 5i)$

(ii) $(2 - 3i) - (1 + 5i)$

(iii) $(2 - 3i)(1 + 5i)$

(iv) $\dfrac{2 - 3i}{1 + 5i}$

Solution

(i) $(2 - 3i) + (1 + 5i) = (2 + 1) + (-3 + 5)i$

> Add the real parts and add the imaginary parts.

$$= 3 + 2i$$

(ii) $(2 - 3i) - (1 + 5i) = (2 - 1) + (-3 - 5)i$

> Subtract the real parts and subtract the imaginary parts.

$$= 1 - 8i$$

(iii) $(2 - 3i)(1 + 5i) = 2 + 10i - 3i - 15i^2$

> Multiply out the brackets in the usual way and simplify.

> When simplifying it is important to remember that $i^2 = -1$.

$$= 2 + 7i - 15(-1)$$

$$= 17 + 7i$$

(iv) $\dfrac{2 - 3i}{1 + 5i} \times \dfrac{1 - 5i}{1 - 5i}$

> Multiply numerator and denominator by the conjugate of the denominator, then simplify.

$$= \frac{2 - 10i - 3i + 15i^2}{1 - 5i + 5i - 25i^2}$$

$$= \frac{-13 - 13i}{26} = \frac{-1 - i}{2} \text{ or } -\frac{1}{2} - \frac{1}{2}i$$

Do not use a calculator in this exercise.

① (i) Write down the values of:

(a) i^2 (b) i^3 (c) i^4 (d) i^5 (e) i^6

(ii) Explain how you would quickly work out the value of i^n for any positive integer value n.

② Find the following:

(i) $2i(3 - 4i) + i(6 - i)$ (ii) $(2 - 5i)^2$ (iii) $(3 + i)(6 - i)(2 - 5i)$

③ Find the following:

(i) $\dfrac{5i}{2 - i}$ (ii) $\dfrac{5 + i}{2 - i}$ (iii) $\dfrac{5 - i}{2 + i}$

④ Solve the equation $(7 + 3i)z = (5 + i)(2 - 9i)$, giving your answer in the form $a + bi$.

⑤ Given that the complex numbers

$$z_1 = (2 - a) + 5bi$$

$$z_2 = a^2 + \left(b^2 + 6\right)i$$

are equal, find the possible values of a and b.

Hence list the possible complex numbers z_1 and z_2.

⑥ For the complex number $z = 2 - 5i$ find $\dfrac{1}{z} + \dfrac{1}{z^*}$ in its simplest form.

Write down the value of $\dfrac{1}{z} + \dfrac{1}{z^*}$ for $z = 2 + 5i$.

⑦ For all complex numbers $z = x + yi$ show that $z + z^*$ and zz^* are both real.

⑧ Find the values of the real numbers a and b which satisfy

$$\frac{a}{2 - i} + \frac{2b}{1 + i} = \frac{5i}{2 - i}$$

2 Representing complex numbers geometrically

A complex number $x + yi$ can be represented by the point with cartesian coordinates (x, y).

For example, in Figure R.2

$5 + 3i$ is represented by $(5, 3)$

$2 - 3i$ is represented by $(2, -3)$

$-6i$ is represented by $(0, -6)$.

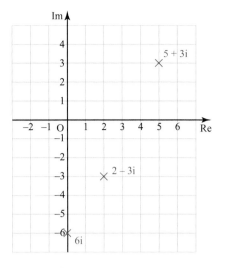

Figure R.2

All real numbers are represented by points on the x-axis, which is therefore called the **real axis**. Purely imaginary numbers which have no real component (of the form $0 + y\text{i}$) give points on the y-axis, which is called the **imaginary axis**.

These axis are labelled as Re and Im.

This geometrical illustration of complex numbers is called the **complex plane** or the **Argand diagram**.

Representing the sum and difference of complex numbers geometrically

Example R.3

Given two complex numbers z_1 and z_2 draw separate Argand diagrams to show geometrically:

(i) $z_1 + z_2$

(ii) $z_1 - z_2$

Solution

(i)

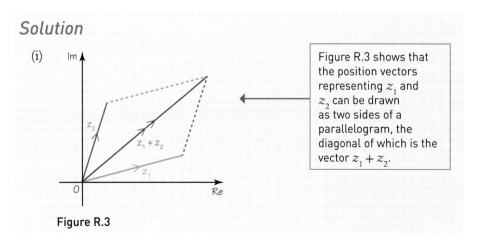

Figure R.3 shows that the position vectors representing z_1 and z_2 can be drawn as two sides of a parallelogram, the diagonal of which is the vector $z_1 + z_2$.

Figure R.3

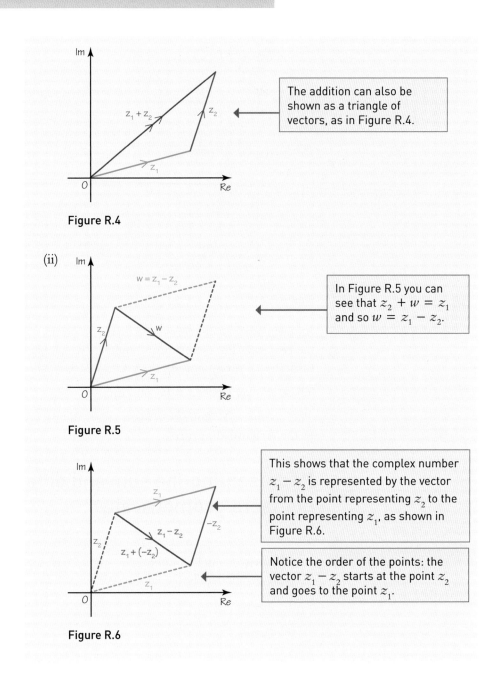

Figure R.4

The addition can also be shown as a triangle of vectors, as in Figure R.4.

(ii)

Figure R.5

In Figure R.5 you can see that $z_2 + w = z_1$ and so $w = z_1 - z_2$.

Figure R.6

This shows that the complex number $z_1 - z_2$ is represented by the vector from the point representing z_2 to the point representing z_1, as shown in Figure R.6.

Notice the order of the points: the vector $z_1 - z_2$ starts at the point z_2 and goes to the point z_1.

① (i) Plot the points $z_1 = 3 + 5i$ and $z_2 = -4 + i$ on an Argand diagram.

(ii) Plot the points $-z_1$ and $-z_2$ and describe the geometrical connection between these points and the original points.

(iii) Plot the points z_1^* and z_2^* and describe the geometrical connection between these points and the original points.

② Given that $z = 3 - i$, represent the following by points on a single Argand diagram.

(i) z (ii) $-z$ (iii) z^* (iv) $-z^*$

(v) iz (vi) $-iz$ (vii) iz^* (viii) $(iz)^*$

③ Given that $z = 2 + 3i$ and $w = 1 - 2i$, represent the following complex numbers on an Argand diagram.

(i) z (ii) w (iii) $z + w$

(iv) $z - w$ (v) $w - z$

④ Given that $z = -\dfrac{1}{2} - \dfrac{\sqrt{3}}{2}i$:

(i) (a) Calculate z^0, z^1 and z^2.

(b) Plot the points A, B and C representing z^0, z^1 and z^2 on an Argand diagram.

(ii) By finding the lengths AB, AC and BC, show that triangle ABC is equilateral.

⑤ (i) Simplify the complex number $z_1 = \dfrac{2i}{3 - 7i}$ and find the complex number $z_2 = iz_1$ where $z_1 = \dfrac{2i}{3 - 7i}$.

(ii) Plot the points A and B representing the complex numbers $z_1 = \dfrac{2i}{3 - 7i}$ and $z_2 = iz_1$ on an Argand diagram.

(iii) Describe the geometrical relationship between z_1 and z_2.

(iv) Show that the points O, A and B form an isosceles triangle.

KEY POINTS

1   Complex numbers are of the form $z = x + yi$ with $i^2 = -1$.
    x is called the real part, $\mathrm{Re}(z)$, and y is called the imaginary part, $\mathrm{Im}(z)$.

2   The conjugate of $z = x + yi$ is $z^* = x - yi$.

3   To add or subtract complex   numbers, add or subtract the real and imaginary parts separately.
$$(x_1 + y_1i) \pm (x_2 + y_2i) = (x_1 \pm x_2) + (y_1 \pm y_2)i$$

4   Multiplication: Expand the brackets then simplify using the fact that $i^2 = -1$

5   Division: Write as a fraction, then multiply top and bottom by the conjugate of the bottom and simplify the answer.

6   Two complex numbers $z_1 = x_1 + y_1i$ and $z_2 = x_2 + y_2i$ are equal only if $x_1 = x_2$ and $y_1 = y_2$.

7   The complex number $z = x + yi$ can be represented geometrically as the point (x, y). This is known as an Argand diagram.

Hyperbolic functions

As is well-known, Physics became a science only after the invention of differential calculus.

Riemann

Discussion point

→ How would you describe the curved shape in this suspension bridge?

1 Hyperbolic functions

The curved shape formed by the wires holding up a suspension bridge (as in the photo above) is a catenary. A catenary is also formed when a chain is hung between two posts.

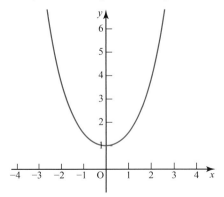

Figure 7.1

Note

The **sinh** function is pronounced in many different ways by different people, but most commonly as 'shine', or 'sine-aitch', or 'cinch' (with a soft 'c').

Figure 7.1 shows the same curve drawn with a different horizontal scale. If you think it looks like a quadratic, you are in good company. Galileo made the same observation but on investigation found that it was close to, but not exactly the same as, a quadratic curve.

The equation of this curve in its simplest form is actually $y = \dfrac{e^x + e^{-x}}{2}$ and this function is called $y = \cosh x$.

A closely related curve is $y = \sinh x$. Its equation is $y = \dfrac{e^x - e^{-x}}{2}$.

Hyperbolic functions and circular functions

Example 7.1

Simplify $\cosh^2 x - \sinh^2 x$.

Solution

$$\cosh^2 x - \sinh^2 x = \left(\frac{e^x + e^{-x}}{2} \right)^2 - \left(\frac{e^x - e^{-x}}{2} \right)^2$$

$$= \frac{e^{2x} + 2 + e^{-2x}}{4} - \frac{e^{2x} - 2 + e^{-2x}}{4}$$

$$= \frac{e^{2x} + 2 + e^{-2x} - e^{2x} + 2 - e^{-2x}}{4}$$

$$= \frac{4}{4}$$

$$= 1$$

Discussion point

→ Compare the result in the example above to a similar result involving the more familiar functions of sine and cosine.

You will have worked with the **circular functions** (often referred to as the **trigonometric functions**, because of their uses in measuring triangles) throughout your A-level Mathematics studies. They are properly called the circular functions because they describe the coordinates of a point moving in a circle – they **parameterise** the circle, since

$$x = \cos \theta \text{ and } y = \sin \theta$$

give the circle $x^2 + y^2 = 1$.

Note

This is, of course, where the identity
$\cos^2 x + \sin^2 x \equiv 1$
comes from.

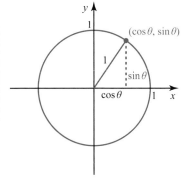

Figure 7.2

If you use parametric equations $x = \cosh t$, $y = \sinh t$, you get one branch of the rectangular hyperbola $x^2 - y^2 = 1$, rather than a circle. For this reason the sinh and cosh functions and other related functions are known as the **hyperbolic functions**.

Discussion point

➜ Why do the parametric equations $x = \cosh t$, $y = \sinh t$ only give one of the branches of the hyperbola? Which branch do they give?

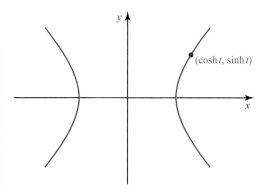

Figure 7.3

Graphs and properties

The hyperbolic functions have similar properties to the circular functions, but their graphs are *not* periodic.

Note

■ Since $\cosh x = \frac{1}{2}(e^x + e^{-x})$ the graph of $y = \cosh x$ lies midway between the graphs of $y = e^x$ and $y = e^{-x}$.

■ Notice that the function has a minimum point at $(0, 1)$.

■ The domain of the cosh function is $x \in \mathbb{R}$ and its range is $y \geqslant 1$.

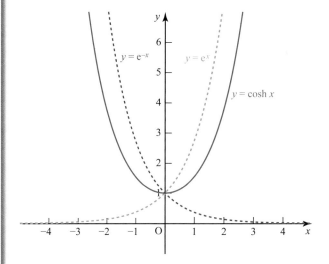

Figure 7.4

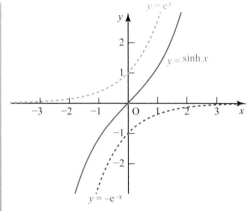

Figure 7.5

The similarities with the circular functions are more obvious when it comes to identities relating $\sinh x$ and $\cosh x$. The reasons for the similarities between hyperbolic and circular functions will become more apparent in Chapter 10 on complex numbers.

The comparison with the circular functions also motivates a definition of the ratio of $\sinh x$ to $\cosh x$, to give an equivalent to the circular $\tan \theta$ function: $\tanh x$.

$$\tanh x = \frac{\sinh x}{\cosh x} = \frac{e^x - e^{-x}}{e^x + e^{-x}}$$

or (by dividing top and bottom by e^{2x})

$$\tanh x = \frac{\sinh x}{\cosh x} = \frac{1 - e^{-2x}}{1 + e^{-2x}}$$

This helps to visualise the function, since you can see that $y \to 1$ as $x \to \infty$ and $y \to -1$ as $x \to -\infty$.

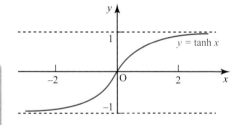

Figure 7.6

As you saw in Example 7.1 on page 135.

$$\cosh^2 x - \sinh^2 x = 1$$

In fact, most of the circular trigonometric identities you already use have an equivalent hyperbolic identity.

Note

You will notice that this table includes three further hyperbolic functions: sech, cosech and coth. These are related to cosh, sinh and tanh in the same way that the reciprocal trigonometric functions sec, cosec and cot are related to cos, sin and tan.

$\cos^2\theta + \sin^2\theta \equiv 1$	$\cosh^2 x - \sinh^2 x \equiv 1$
$\cos 2\theta \equiv \cos^2\theta - \sin^2\theta$	$\cosh 2x \equiv \cosh^2 x + \sinh^2 x$
$\sin 2\theta \equiv 2\sin\theta\cos\theta$	$\sinh 2x \equiv 2\sinh x \cosh x$
$\tan\theta \equiv \dfrac{\sin\theta}{\cos\theta}$	$\tanh x \equiv \dfrac{\sinh x}{\cosh x}$
$\sec\theta \equiv \dfrac{1}{\cos\theta}$	$\operatorname{sech} x \equiv \dfrac{1}{\cosh x}$
$\operatorname{cosec}\theta \equiv \dfrac{1}{\sin\theta}$	$\operatorname{cosech} x \equiv \dfrac{1}{\sinh x}$
$\cot\theta \equiv \dfrac{1}{\tan\theta}$	$\coth x \equiv \dfrac{1}{\tanh x}$

Table 7.1

Example 7.2

Solve the equation:

$$\cosh x = 2\sinh x - 1$$

Solution

It is often easiest to convert the hyperbolic functions into their definitions in terms of e^x.

$$\cosh x = 2\sinh x - 1$$

$$\frac{e^x + e^{-x}}{2} = e^x - e^{-x} - 1$$

$$0 = e^x - 3e^{-x} - 2$$

$$\left(e^x\right)^2 - 3 - 2e^x = 0 \quad \longleftarrow \quad \boxed{\text{Multiply by } e^x, \text{ to get a quadratic in } e^x.}$$

$$\left(e^x\right)^2 - 2e^x - 3 = 0$$

$$\left(e^x - 3\right)\left(e^x + 1\right) = 0$$

$$e^x = 3 \quad \longleftarrow \quad \boxed{\text{Since } e^x \text{ can't be negative.}}$$

$$x = \ln 3$$

The derivatives of the hyperbolic functions

Differentiating shows that

$$\frac{d}{dx}(\cosh x) = \frac{d}{dx}\left(\frac{e^x + e^{-x}}{2}\right) = \frac{e^x - e^{-x}}{2} = \sinh x$$

and

$$\frac{d}{dx}(\sinh x) = \frac{d}{dx}\left(\frac{e^x - e^{-x}}{2}\right) = \frac{e^x + e^{-x}}{2} = \cosh x$$

Exercise 7.1

1. Using the definitions of the hyperbolic functions, prove that $\sinh 2x = 2\cosh x \sinh x$.

2. Given that $\sinh x = 2$, find the exact values of $\cosh x$ and $\tanh x$.

3. (i) Rewrite the equation
   $$\cosh x + \sinh x = -1$$
   in terms of e^x, showing that it simplifies to:
   $$3e^x - e^{-x} + 2 = 0$$

   (ii) Multiply by e^x to create a quadratic in e^x.

   (iii) Solve the quadratic equation to show that the only real root of this equation is $x = -\ln 3$.

4. Use the chain rule or product rule (as appropriate) to differentiate the following functions, with respect to x:
   (i) $\sinh 4x$ (ii) $\cosh x^2$
   (iii) $\cosh^2 x$ (iv) $\cosh x \sinh x$

5. Find the following integrals, using the suggested methods:

   (i) $\displaystyle\int \sinh 3x \, dx$ (use a substitution of $u = 3x$, if necessary)

   (ii) $\displaystyle\int x \sinh x \, dx$ (use integration by parts)

   (iii) $\displaystyle\int x \cosh\left(1 + x^2\right) dx$ (use a substitution of $u = 1 + x^2$, if necessary)

6. Find all the real roots of each of the following equations:
   (i) $10\cosh x - 2\sinh x = 11$ (ii) $\cosh x - 5\sinh x = 5$
   (iii) $7\cosh x + 4\sinh x = 3$

7. Given that
   $$\sinh x + \sinh y = \frac{25}{12}$$
   $$\cosh x - \cosh y = \frac{5}{12}$$
   show that
   $$2e^x = 5 + 2e^{-y}$$
   and
   $$3e^{-x} = -5 + 3e^y$$
   Hence find the real values of x and y.

⑧ The diagram represents a cable hanging between two points A and B, where AB is horizontal. The lowest point of the cable, O, is taken as the origin of the coordinate system.

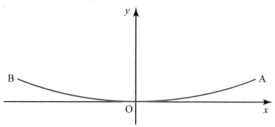

Figure 7.7

If the cable is flexible and has uniform density then the curve formed is a **catenary** with equation:

$$y = c\left(\cosh\left(\frac{x}{c}\right) - 1\right)$$

where c is a constant.

For a particular cable $c = 20\,$m and AB $= 16\,$m. Find the sag of the cable, i.e. the distance of O below AB, and the angle that the tangent at A makes with the horizontal.

⑨ (i) Using the definition of $\cosh x$ in terms of the exponential function, prove:

$$\cosh^2 x = \tfrac{1}{2}(\cosh 2x + 1)$$

(ii) Deduce that $\sinh^2 x = \tfrac{1}{2}(\cosh 2x - 1)$.

(iii) Hence find $\int \cosh^2 x\,dx$ and $\int \sinh^2 x\,dx$.

⑩ Find the exact coordinate of the turning point of the graph $y = e^{2x}\sinh 5x$.

⑪ Find the exact area between the curve of $y = 5 - 4\cosh x$ and the x-axis.

⑫ Find the Maclaurin series for $\cosh x$ including the general term:

(i) by finding the values of successive derivatives at $x = 0$

(ii) by using the definition $\cosh x = \tfrac{1}{2}\left(e^x + e^{-x}\right)$.

(The Maclaurin series for $\cosh x$ and $\sinh x$ are valid for all values of x, like the ones for $\cos x$ and $\sin x$.)

⑬ (i) Sketch the curve $y = \cosh x$ and the line $y = x$ on the same axes. Prove that $\cosh x > x$ for all x.

(ii) Prove that the point on the curve $y = \cosh x$ which is closest to the line $y = x$ has coordinates $\left(\ln\left(1 + \sqrt{2}\right), \sqrt{2}\right)$ and mark this point on your graph.

⑭ Find conditions on a, b and c which are necessary and sufficient to ensure that the equation $a\cosh x + b\sinh x = c$ has:

(i) two distinct real roots   (ii) exactly one real root

(iii) no real roots.

⑮ A function is defined as:
$$f(x) = 2\tanh x - 2\tanh^3 x$$

(i) Calculate the value of x for which $f(x) = 0$ (and show there is only one).

(ii) Show there are exactly two turning points and calculate their coordinates.

(iii) Show that $\int_0^\infty f(x)\,dx = 1$.

(iv) Sketch the graph for both positive and negative x, and indicate the area calculated in part (iii).

2 Inverse hyperbolic functions

Example 7.3

Solve the equation $\sinh x = 2$.

Solution

$$\sinh x = 2$$

$$\frac{e^x - e^{-x}}{2} = 2$$

$$e^x - e^{-x} = 4 \quad \longleftarrow \boxed{\text{Multiply through by } e^x.}$$

$$e^{2x} - 1 = 4e^x \quad \longleftarrow$$

$$e^{2x} - 4e^x - 1 = 0 \quad \longleftarrow \boxed{\text{This is a quadratic in } e^x.}$$

$$e^x = \frac{4 \pm \sqrt{20}}{2} \quad \longleftarrow \boxed{\text{Using the quadratic formula.}}$$

$$= 2 \pm \sqrt{5}$$

$$x = \ln(2 \pm \sqrt{5}) \quad \longleftarrow \boxed{2 - \sqrt{5} \text{ is negative so } \ln(2 - \sqrt{5}) \text{ is not real.}}$$

The solution of the equation is $x = \ln(2 + \sqrt{5})$.

The inverse function of the sinh function is denoted by arsinh, or sometimes $\sinh^{-1}$.

So the example above shows that $\operatorname{arsinh} 2 = \ln(2 + \sqrt{5})$.

The graph below shows the curves $y = \sinh x$ and $y = \operatorname{arsinh} x$.

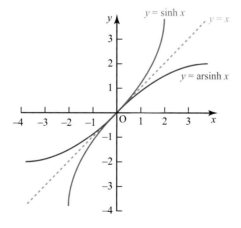

Figure 7.8

As for any function and its inverse, the curves are reflections of each other in the line $y = x$.

This allows you to find an expression for arsinh x, using a similar method to Example 7.3.

$$y = \text{arsinh}\, x \Rightarrow x = \sinh y$$

$$x = \frac{e^y - e^{-y}}{2}$$

$$2x = e^y - e^{-y}$$

$$e^{2y} - 2xe^y - 1 = 0$$

$$e^y = \frac{2x \pm \sqrt{4x^2 + 4}}{2} = x + \sqrt{x^2 + 1}$$

$$y = \ln\left(x + \sqrt{x^2 + 1}\right)$$

> $\sqrt{x^2 + 1}$ is greater than x, so $x - \sqrt{x^2 + 1}$ is negative. As e^y cannot be negative for any real y, the negative square root is discarded.

So arsinh $x = \ln\left(x + \sqrt{x^2 + 1}\right)$

This result gives an easy method to solve an equation like $\sinh x = 2$.

$$\sinh x = 2 \Rightarrow x = \text{arsinh}\, 2 = \ln\left(2 + \sqrt{2^2 + 1}\right) = \ln(2 + \sqrt{5})$$

You can find the inverse of the tanh function in a similar way.

$$y = \text{artanh}\, x \Rightarrow x = \tanh y$$

$$x = \frac{e^{2y} - 1}{e^{2y} + 1}$$

$$xe^{2y} + x = e^{2y} - 1$$

$$e^{2y}(1 - x) = 1 + x$$

$$e^{2y} = \frac{1 + x}{1 - x}$$

$$2y = \ln\left(\frac{1 + x}{1 - x}\right)$$

$$y = \frac{1}{2}\ln\left(\frac{1 + x}{1 - x}\right)$$

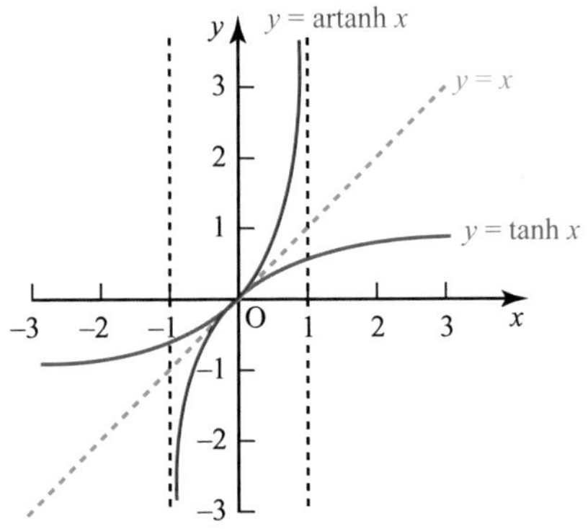

Figure 7.9

So artanh $x = \frac{1}{2}\ln\left(\frac{1 + x}{1 - x}\right)$.

Notice that both the sinh function and the tanh function are one-to-one, and so the inverse functions are defined over the whole domain.

Example 7.4

Solve the equation $\cosh x = 2$.

Solution

$$\cosh x = 2$$

$$\frac{e^x + e^{-x}}{2} = 2$$

$$e^x + e^{-x} = 4$$

$$e^{2x} + 1 = 4e^x$$

$$e^{2x} - 4e^x + 1 = 0$$

$$e^x = \frac{4 \pm \sqrt{12}}{2}$$

$$= 2 \pm \sqrt{3}$$

$$x = \ln(2 \pm \sqrt{3})$$

> Multiply through by e^x.

> This is a quadratic in e^x.

> Using the quadratic formula.

> $2 - \sqrt{3}$ is positive so the equation has two real roots.

The two roots of the equation in Example 7.4 are shown on the graph below.

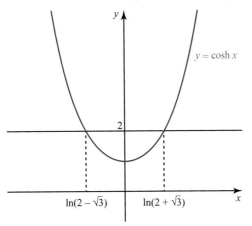

Figure 7.10

You can see from the symmetry of the graph that one of the roots is the negative of the other.

> **Note**
>
> It is probably not immediately obvious to you that $\ln(2 - \sqrt{3})$ is the negative of $\ln(2 + \sqrt{3})$. However, it is quite easy to prove that this is the case by adding them together.
>
> $$\ln(2 + \sqrt{3}) + \ln(2 - \sqrt{3}) = \ln\big((2 + \sqrt{3})(2 - \sqrt{3})\big)$$
> $$= \ln(4 - 3)$$
> $$= \ln 1$$
> $$= 0$$
>
> Since $\ln(2 + \sqrt{3}) + \ln(2 - \sqrt{3}) = 0$, $\ln(2 - \sqrt{3}) = -\ln(2 + \sqrt{3})$.

Example 7.4 shows that the equation $\cosh x = k$ has two real roots if $k > 1$.

The inverse of the cosh function is denoted by arcosh (or sometimes $\cosh^{-1}$). For this to be a function, arcosh k must have only one value.

The graph of $x = \cosh y$ is shown in Figure 7.11.

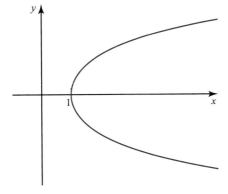

Figure 7.11

Notice that this graph goes both above and below the x-axis, so the negative part has to be excluded to give the function $y = \text{arcosh}\, x$.

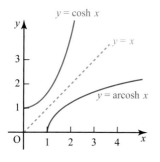

Figure 7.12

If the domain of the cosh function is restricted to the non-negative real numbers, i.e. to $x \geqslant 0$, then the function is one-to-one, with the graph shown by the red line in Figure 7.12. This restricted cosh function has an inverse function – the arcosh function.

Partly because of the restriction in the domain of arcosh x the derivation is more complicated:

$$y = \text{arcosh}\, x$$

$$x = \cosh y$$

$$2x = e^y + e^{-y}$$

$$e^{2y} - 2xe^y + 1 = 0$$

$$e^y = \frac{2x \pm \sqrt{4x^2 - 4}}{2}$$

$$= x \pm \sqrt{x^2 - 1}$$

$$y = \ln\!\left(x + \sqrt{x^2 - 1}\right) \text{ or } \ln\!\left(x - \sqrt{x^2 - 1}\right)$$

As in Example 7.4, the second root is the negative of the first (as you can see in Figure 7.13 showing the $y = \cosh x$ curve).

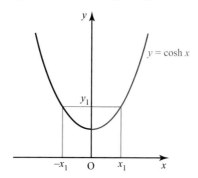

Figure 7.13

Since arcosh $x > 0$ by definition, the positive root is the required one.

Therefore:

$$\text{arcosh}\, x = \ln\!\left(x + \sqrt{x^2 - 1}\right)$$

It is important to understand that there is a difference between solving the equation $\cosh x = k$ and finding arcosh k. Compare this with solving an equation like $\sin x = 0.5$. When you use the inverse sine button on your calculator, it gives you just one answer – the principal value. You then need to work out any other roots in the required range. Similarly, if you use the square root button on your calculator to solve the equation $x^2 = 2$, you get the positive square root only, but the full solution of the equation is $x = \pm\sqrt{2}$.

Differentiating the inverse hyperbolic functions

It is possible to differentiate the logarithmic versions of arcosh x and arsinh x, but it is easier to work with the hyperbolic functions themselves and their identities:

$$y = \operatorname{arcosh} x$$

$$\cosh y = x$$

$$\sinh y \frac{dy}{dx} = 1 \quad \longleftarrow \quad \boxed{\text{Differentiating implicitly.}}$$

$$\frac{dy}{dx} = \frac{1}{\sinh y}$$

$$= \frac{1}{\pm\sqrt{\cosh^2 y - 1}} \quad \longleftarrow \quad \boxed{\begin{array}{l}\text{Using the identity}\\ \cosh^2 y - \sinh^2 y \equiv 1.\end{array}}$$

$$= \frac{1}{\pm\sqrt{x^2 - 1}}$$

Since the gradient of $y = \operatorname{arcosh} x$ is always positive you must take the positive square root, and so:

$$\frac{d}{dx}\left(\operatorname{arcosh} x\right) = \frac{1}{\sqrt{x^2 - 1}}$$

Exercise 7.2

① Find the exact value of each of the following, giving your answers as logarithms:

    (i)   arcosh 3   (ii) arsinh 1   (iii) artanh 0.5

    (iv) arsinh (-2) (v) arcosh $\dfrac{5}{4}$ (vi) artanh$\left(-\dfrac{2}{3}\right)$

② Solve the following equations:

    (i) $\sinh x = 3$ (ii) $\cosh x = 3$ (iii) $\tanh x = 0.2$

③ Differentiate the following:

    (i)   arsinh $3x$ (ii) arcosh $4x$

    (iii) arsinh $(2x^2)$ (iv) arcosh $(2x + 1)$

④ (i)   Use the chain rule to differentiate

        $\ln\left(x + \sqrt{x^2 - 1}\right)$.

   (ii) Show, by multiplying the numerator and denominator by $\left(x^2 - 1\right)^{\frac{1}{2}}$, that your answer simplifies to

        $\dfrac{1}{\sqrt{x^2 - 1}}$.

   (iii) Differentiate $\ln\left(x + \sqrt{x^2 + 1}\right)$ to show that the derivative of arsinh x is

        $\dfrac{1}{\sqrt{x^2 + 1}}$.

⑤ Use implicit differentiation to show that:

(i) $\dfrac{d}{dx}\left(\operatorname{arsinh}\dfrac{x}{a}\right) = \dfrac{1}{\sqrt{x^2 + a^2}}$

(ii) $\dfrac{d}{dx}\left(\operatorname{arcosh}\dfrac{x}{a}\right) = \dfrac{1}{\sqrt{x^2 - a^2}}$

⑥ (i) Prove that $\dfrac{d}{dx}\left(\operatorname{artanh}x\right) = \dfrac{1}{1 - x^2}$.

(ii) By using partial fractions and integrating, deduce the logarithmic form of artanh x.

⑦ Integrate the following with respect to x:

(i) arcosh x

(ii) arsinh x

(iii) artanh x

> Hint: write
> $\operatorname{arcosh}x = 1 \times \operatorname{arcosh}x$
> and integrate by parts.

⑧ (i) Find $\displaystyle\int x\sinh\left(x^2\right)dx$.

(ii) By writing $x^3 \sinh x^2$ as $x^2(x \sinh x^2)$, or otherwise, find $\displaystyle\int x^3 \sinh x^2\, dx$.

⑨ Prove that the curves $y = \operatorname{arsinh}x$ and $y = \operatorname{arcosh}2x$ intersect where $x = \dfrac{1}{\sqrt{3}}$.

Sketch the curves on the same axes and shade the region bounded by the x-axis and the curves.

Find the area of the shaded region.

3 Integration using inverse hyperbolic functions

From the derivatives of arcosh x and arsinh x, given on page 144, it is clear by integrating that

$$\int \frac{1}{\sqrt{x^2 - 1}}dx = \operatorname{arcosh}x + c$$

and

$$\int \frac{1}{\sqrt{x^2 + 1}}dx = \operatorname{arsinh}x + c.$$

ACTIVITY 7.3

Use the chain rule and the derivatives for arsinh x and arcosh x given above, to show that

■ $\dfrac{d}{dx}\left(\operatorname{arsinh}\dfrac{x}{a}\right) = \dfrac{1}{\sqrt{x^2 + a^2}}$

■ $\dfrac{d}{dx}\left(\operatorname{arcosh}\dfrac{x}{a}\right) = \dfrac{1}{\sqrt{x^2 - a^2}}$

Note

When evaluating a definite integral, it is usually easier to work with the logarithmic form.

The results in Activity 7.3 lead to the following results:

$$\int \frac{1}{\sqrt{x^2 + a^2}} \, dx = \operatorname{arsinh} \frac{x}{a} + c \text{ or } \ln\left(x + \sqrt{x^2 + a^2}\right) + c$$

$$\int \frac{1}{\sqrt{x^2 - a^2}} \, dx = \operatorname{arcosh} \frac{x}{a} + c \text{ or } \ln\left(x - \sqrt{x^2 + a^2}\right) + c$$

By comparing these results with those from Chapter 4 you should see that the range of functions that you can integrate has been extended. Just as before, more complicated examples use techniques such as taking out constant factors or completing the square. You can often choose an appropriate substitution, even if the integral is not quite a standard one.

Example 7.5

Find $\int \frac{1}{\sqrt{9x^2 - 25}} \, dx$.

To use the standard integral, the coefficient of x^2 must be 1, so you need to take out a factor of 9, which becomes 3 when it leaves the square root.

Solution

$$\int \frac{1}{\sqrt{9x^2 - 25}} \, dx = \frac{1}{3} \int \frac{1}{\sqrt{x^2 - \frac{25}{9}}} \, dx$$

This is now a standard integral with $a = \frac{5}{3}$.

$$= \frac{1}{3} \operatorname{arcosh} \left(\frac{x}{\left(\frac{5}{3}\right)} \right) + c$$

$$= \frac{1}{3} \operatorname{arcosh} \left(\frac{3x}{5} \right) + c$$

Exercise 7.3

① Use the standard results to find the following indefinite integrals:

(i) $\int \frac{1}{\sqrt{x^2 - 4}} \, dx$ (ii) $\int \frac{1}{\sqrt{x^2 + 4}} \, dx$

② Use the standard results to find the following definite integrals, giving your answers in terms of logarithms.

(i) $\int_3^6 \frac{1}{\sqrt{x^2 - 9}} \, dx$ (ii) $\int_3^6 \frac{1}{\sqrt{x^2 + 9}} \, dx$

③ Find the following indefinite integrals:

(i) $\int \frac{1}{\sqrt{9x^2 - 1}} \, dx$ (ii) $\int \frac{1}{\sqrt{9x^2 + 1}} \, dx$

(iii) $\int \frac{1}{\sqrt{4x^2 + 9}} \, dx$ (iv) $\int \frac{1}{\sqrt{4x^2 - 9}} \, dx$

④ Evaluate the following definite integrals, giving your answers as logarithms:

(i) $\int_1^2 \frac{1}{\sqrt{25x^2 - 16}} \, dx$ (ii) $\int_0^2 \frac{1}{\sqrt{9x^2 - 4}} \, dx$

⑤ (i) Prove that $\cosh^2 x = \frac{1}{2}(\cosh 2x + 1)$.

(ii) Use the substitution $x = 2 \sinh u$ and the result from (i) to find $\int \sqrt{x^2 + 4} \, dx$.

⑥ (i) Prove that $\sinh^2 x = \frac{1}{2}(\cosh 2x - 1)$.

(ii) Use a suitable substitution and the result from (ii) to find $\int \sqrt{x^2 - 9} \, dx$.

⑦ (i) Use the substitution $x = a \sinh u$ to prove that

$$\int \frac{1}{\sqrt{x^2 + a^2}} \, dx = \operatorname{arsinh} \frac{x}{a} + c.$$

(ii) Use the substitution $x = a \cosh u$ to prove that

$$\int \frac{1}{\sqrt{x^2 - a^2}} \, dx = \operatorname{arcosh} \frac{x}{a} + c.$$

⑧ Show that $\displaystyle\int_0^4 \frac{4x + 1}{\sqrt{x^2 + 9}} \, dx = 8 + \ln 3$.

⑨ (i) Write $4x^2 + 12x - 40$ in completed square form.

(ii) Evaluate the definite integral

$$\int_{10}^{20} \frac{1}{\sqrt{4x^2 + 12x - 40}} \, dx.$$

⑩ The points $P_1\,(a\cos\theta, a\sin\theta)$ and $P_2\,(a\cosh\phi, a\sinh\phi)$ lie on the circle $x^2 + y^2 = a^2$ and the rectangular hyperbola $x^2 - y^2 = a^2$ respectively (see Figure 7.14).

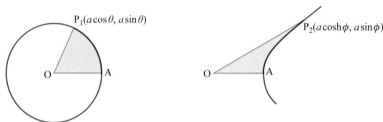

Figure 7.14

In both diagrams A = (1, 0).

Prove that area OAP_1 is proportional to θ and that area OAP_2 is proportional to ϕ, with the same constant of proportionality.

LEARNING OUTCOMES

When you have completed this chapter you should be able to:

➤ know the definitions of the hyperbolic functions and their domains and ranges, and be able to sketch their graphs

➤ understand and use the identity $\cosh^2 x - \sinh^2 x \equiv 1$

➤ differentiate and integrate hyperbolic functions

➤ understand and use the definitions of the inverse hyperbolic functions and know their domains and ranges

➤ derive and use the logarithmic forms of the inverse hyperbolic functions

➤ recognise integrals of functions of the form $\dfrac{1}{\sqrt{x^2 + a^2}}$ and $\dfrac{1}{\sqrt{x^2 - a^2}}$ and be able to integrate related functions by using substitutions.

KEY POINTS

1 $\cosh x = \dfrac{e^x + e^{-x}}{2}$ $\sinh x = \dfrac{e^x - e^{-x}}{2}$

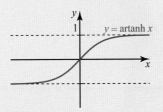

Figure 7.15

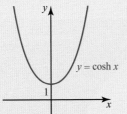

Figure 7.16

$\tanh x = \dfrac{e^x - e^{-x}}{e^x + e^{-x}} = \dfrac{e^{2x} - 1}{e^{2x} + 1}$

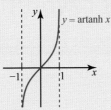

Figure 7.17

2 $\cosh^2 x - \sinh^2 x \equiv 1$

3 $\dfrac{d}{dx}(\cosh x) = \sinh x$ $\displaystyle\int \cosh x \, dx = \sinh x + c$

   $\dfrac{d}{dx}(\sinh x) = \cosh x$ $\displaystyle\int \sinh x \, dx = \cosh x + c$

4 $\operatorname{arcosh} x = \ln\left(x + \sqrt{x^2 - 1}\right)$ $\operatorname{arsinh} x = \ln\left(x + \sqrt{x^2 + 1}\right)$

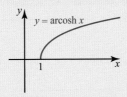

Figure 7.18

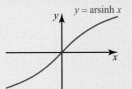

Figure 7.19

$\operatorname{artanh} x = \frac{1}{2} \ln\left(\frac{1+x}{1-x}\right)$

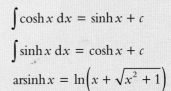

Figure 7.20

5 $\displaystyle\int \dfrac{1}{\sqrt{x^2 - a^2}} \, dx = \operatorname{arcosh}\dfrac{x}{a} + c$ or $\ln\left(x + \sqrt{x^2 - a^2}\right) + c$

   $\displaystyle\int \dfrac{1}{\sqrt{x^2 + a^2}} \, dx = \operatorname{arsinh}\dfrac{x}{a} + c$ or $\ln\left(x + \sqrt{x^2 + a^2}\right) + c$

8 Applications of integration

Mathematics is not only real, but it is the only reality.

Martin Gardner

Discussion point

→ What plane shape would need to be rotated through 360° to produce a solid in the shape of an egg?

1 Volumes of revolution

When the shaded region in Figure 8.1 is rotated by 360° about the *x*-axis the solid obtained, illustrated in Figure 8.2, is called a solid of revolution.

Discussion points

→ How could you use the formula for the volume of a cone to work out the volume of the solid shown in Figure 8.2?

→ Do you know formulae that will allow you to find the volumes of all other solids of revolution?

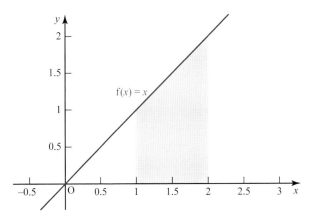

Figure 8.1

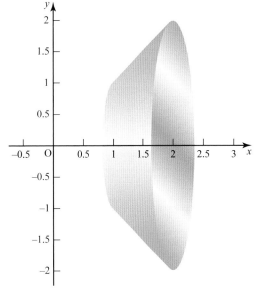

Figure 8.2

If the line $y = x$ in Figure 8.1 is replaced with a curve, then the calculation is less obvious.

You already know that you can use integration to calculate the area under a curve. In Figure 8.3, each of the rectangles has a height of y (which depends on the value of x) and a width of δx, where δx is small.

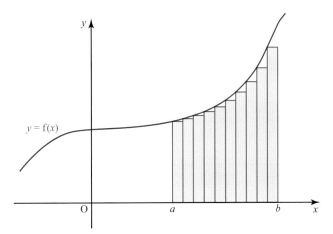

Figure 8.3

The total area of the rectangles is given by $\displaystyle\sum_{x=a}^{x=b} y\,\delta x$.

As the rectangles become thinner, the approximation for the area becomes more accurate. In the limiting case, as $\delta x \to 0$, the sum becomes an integral and the expression for the area is exact.

$$A = \int_a^b y\,\mathrm{d}x$$

The same ideas can be used to find the volume of a solid of revolution.

Look at the shaded region in Figure 8.4 and the solid of revolution it would form when rotated $360°$ about the x-axis in Figure 8.5. When the green strip is rotated, it forms a disc.

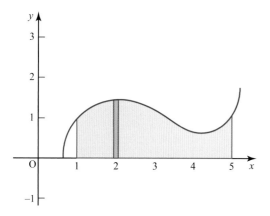

Figure 8.4

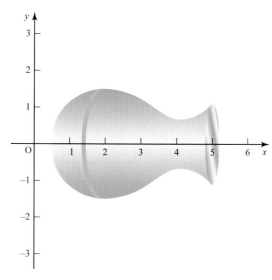

Figure 8.5

The green disc shown in Figure 8.5 is approximately cylindrical, with radius y and thickness δx, so its volume is given by:

$$\delta V = \pi y^2 \delta x$$

The volume of the complete solid is then approximately the sum of all of these discs, and this approximation will be better as the thickness of the discs gets smaller.

$$V \approx \sum \delta V$$

$$V \approx \sum_{x=a}^{x=b} \pi y^2 \delta x$$

The limit of this sum, as $\delta x \to 0$, becomes an integral, and you then have an accurate formula for the volume of revolution:

$$V = \int_a^b \pi y^2 \, \mathrm{d}x$$

Example 8.1

The region between the curve $y = x^2$, the x-axis and the lines $x = 1$ and $x = 3$ is rotated through $360°$ about the x-axis.

Find the volume of revolution which is formed.

Solution

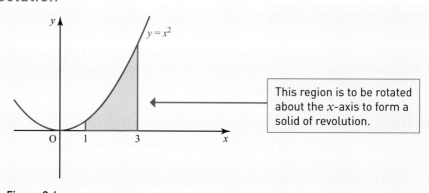

This region is to be rotated about the x-axis to form a solid of revolution.

Figure 8.6

Using the formula $V = \int_a^b \pi y^2 \, dx$

$$V = \int_1^3 \pi \left(x^2\right)^2 dx \quad \longleftarrow \boxed{\text{Replacing } y \text{ with } x^2.}$$

$$= \pi \int_1^3 x^4 \, dx$$

$$= \pi \left[\frac{x^5}{5}\right]_1^3$$

$$= \frac{\pi}{5}(243 - 1)$$

$$= \frac{242\pi}{5} \quad \longleftarrow$$

Depending on the circumstances, you may want to write this as 152 cubic units (to 3 s.f.). Unless a decimal answer is required, it is common to leave π in the answer and so keep the answer exact.

Rotation around the y-axis

You can also form a solid of revolution by rotating a region about the y-axis. The diagram in Figure 8.8 shows the solid which is obtained when a region between part of the curve from Figure 8.7 and the y-axis is rotated through 360° about the y-axis.

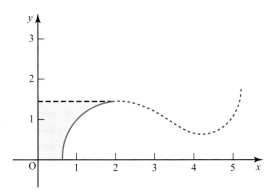

Figure 8.7

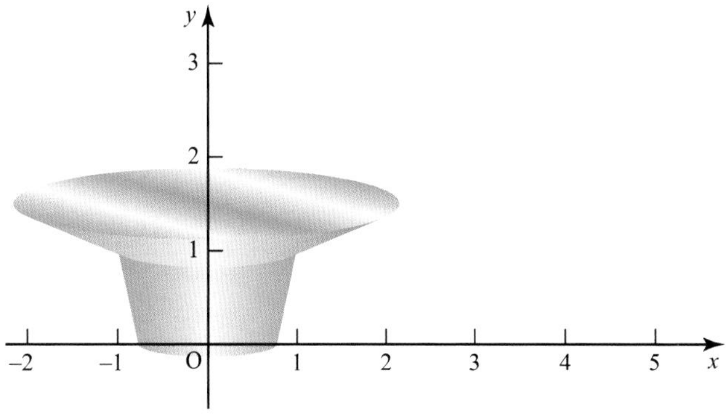

Figure 8.8

The formula for the volume can be obtained in a similar way, so that

$$V_y = \int \pi x^2 \, dy$$

The limits in this case are y-values rather than x-values.

$$V_y = \int_{y=q}^{y=p} \pi x^2 \, dy$$

Since the integration is with respect to y, the limits can just be written as p and q.

$$V_y = \int_q^p \pi x^2 \, dy \quad \longleftarrow \boxed{\begin{array}{l}\text{You would need to write } x \text{ in terms of } y \text{ so} \\ \text{that you can integrate with respect to } y.\end{array}}$$

Example 8.2

The region between the curve $y = x^2$, the y-axis and the lines $y = 2$ and $y = 5$ is rotated $360°$ around the y-axis. Find the volume of revolution obtained.

Solution

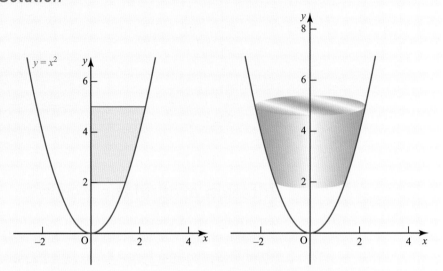

Figure 8.9   Figure 8.10

Using $V = \displaystyle\int_q^p \pi x^2 \, \mathrm{d}y$

$V = \displaystyle\int_2^5 \pi y \, \mathrm{d}y$ ⟵ Notice that $x^2 = y$.

$= \pi \left[\dfrac{y^2}{2} \right]_2^5$

$= \pi \left(\dfrac{25}{2} - \dfrac{4}{2} \right)$

$= \dfrac{21\pi}{2}$ cubic units

Exercise 8.1

① Name six common objects which could be generated as solids of revolution.

② Figure 8.11 shows the line $y = 3x$.

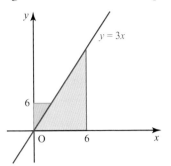

Figure 8.11

(i) Describe the solid obtained by rotating the purple region by 360° around the x-axis.

(ii) Use the formula $V = \pi \int_{x_1}^{x_2} y^2 \, dx$ to calculate the volume of the solid.

(iii) Describe the solid obtained by rotating the red region by 360° around the y-axis.

(iv) Use the formula $V = \pi \int_{y_1}^{y_2} x^2 \, dy$ to calculate the volume of the solid.

(v) Use the formula for the volume of a cone to show that both your answers are correct.

③ Figure 8.12 shows the region under the curve $y = \dfrac{1}{x}$ between $x = 1$ and $x = 2$.

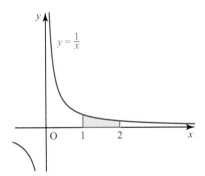

Figure 8.12

Find the volume of the solid of revolution formed by rotating this region through 360° about the x-axis.

④ A region is bounded by the lines $y = x + 2$, the x-axis, the y-axis and the line $x = 2$.

(i) Draw a sketch to show this region.

(ii) Find the volume of the solid obtained by rotating the region through 360° about the x-axis.

⑤ A region is bounded by the curve $y = x^2 - 1$ and the x-axis.

(i) Draw a sketch to show this region.

(ii) Find the volume of the solid obtained by rotating the region through 360° about the x-axis.

⑥ A region is bounded by the curve $y = \sqrt{x}$, the y-axis and the line $y = 2$.

(i) Draw a sketch to show this region.

(ii) Find the volume of the solid obtained by rotating the region through $360°$ about the y-axis.

⑦ (i) Sketch the graph of $y = (x - 3)^2$ for values of x between $x = -1$ and $x = 5$.

Shade in the region under your curve, between $x = 0$ and $x = 2$.

(ii) Calculate the area of the shaded region.

(iii) The shaded region is rotated about the x-axis to form a volume of revolution. Calculate this volume.

⑧ A mathematical model for a large plant pot is obtained by rotating the part of the curve $y = 0.1x^2$ which is between $x = 10$ and $x = 25$ through $360°$ about the y-axis and then adding a flat base. Units are in centimetres.

(i) Draw a sketch of the curve and shade in the area of the cross-section of the pot, indicating which line will form its base.

(ii) Garden compost is sold in litres. How many litres will be required to fill the pot to a depth of $45\,\text{cm}$? (Ignore the thickness of the pot.)

⑨ Figure 8.13 shows the curve $y = x^2 - 4$.

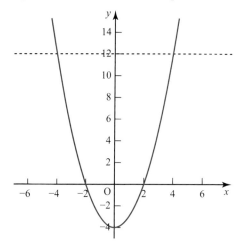

Figure 8.13

The region R is formed by the line $y = 12$, the y-axis, the x-axis and the curve $y = x^2 - 4$, for positive values of x.

(i) Make a sketch copy of the graph and shade the region R.

The inside of a vase is formed by rotating the region R through $360°$ about the y-axis. Each unit of x and y represents $2\,\text{cm}$.

(ii) Write down an expression for the volume of revolution of the region R about the y-axis.

(iii) Find the capacity of the vase in litres.

(iv) Show that when the vase is filled to $\dfrac{5}{6}$ of its internal height it is three–quarter full.

⑩ Figure 8.14 shows the circle $x^2 + y^2 = 25$ and the line $y = 4$.

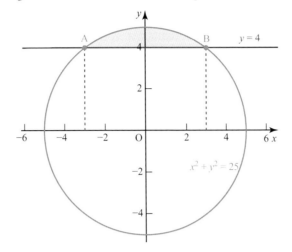

Figure 8.14

(i) Find the coordinates of the points A and B where the circle and line intersect.

(ii) A napkin ring is formed by rotating the shaded region through 360° about the x-axis. Find the volume of the napkin ring.

⑪ When $y = \dfrac{1}{x}$ is rotated around the x-axis, for $1 \leqslant x \leqslant a$, the resulting solid is a 'trumpet' shape which is sometimes called 'Gabriel's Horn' or 'Torricelli's Trumpet'.

(i) Find, in terms of a, the volume of the shape bounded by $y = \dfrac{1}{x}$, $x = 1$, $x = a$ and the x-axis, when rotated around the x-axis.

(ii) Show whether the volume of the 'trumpet' is defined as $a \to \infty$, and, if defined, find its value.

⑫ The function f(x) describes a semicircle, radius r, at a distance R above the origin:

$$f(x) = \sqrt{r^2 - x^2} + R$$

Use this function, and a similar function describing the other half of the circle, and your knowledge of volumes of revolution, to prove that the volume of a torus, as shown in Figure 8.16, is $2\pi^2 r^2 R$ or equivalently, $(2\pi R) \times (\pi r^2)$.

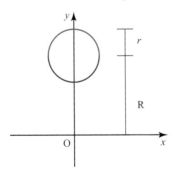

Figure 8.15

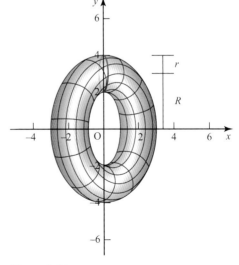

Figure 8.16

2 The mean value of a function

Sometimes you may need to find out the mean (average) value of a function over a particular domain.

Figure 8.17 shows a curve $y = f(x)$. The area under the curve between $x = a$ and $x = b$ is shaded blue.

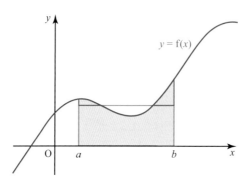

Figure 8.17

The green rectangle has been drawn so that it has the same area as the shaded region.

The height of this rectangle is the mean value of $f(x)$ for $a \leqslant x \leqslant b$.

The area of the shaded region is given by $\int_a^b f(x)\,dx$. ← | This is the same as the area of the rectangle.

The width of the rectangle is $b - a$.

> 'On the interval $[a, b]$' is another way of saying 'for $a \leqslant x \leqslant b$'.

So the mean value of the function $f(x)$ on the interval $[a, b]$ is given by $\dfrac{1}{b-a}\int_a^b f(x)\,dx$.

ACTIVITY 8.1

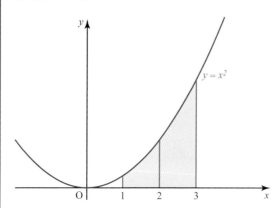

Figure 8.18

Figure 8.18 shows the curve $y = f(x)$, where $f(x) = x^2$.

(i)   Find the area under the curve between $x = 1$ and $x = 3$. Use your answer to find the mean value of the curve over this interval.

(ii) Find the mean of $f(1)$, $f(2)$ and $f(3)$.

(iii) Find the mean of $f(1)$, $f(1.5)$, $f(2)$, $f(2.5)$, $f(3)$.

(iv) Use a spreadsheet to find the mean of larger numbers of equally spaced points on the curve.

(v)   Compare your results with your answer to (i).

Discussion point

→ How is finding the mean of a function similar to finding the mean value of a set of numbers? How is it different?

Example 8.3

Calculate the mean value of the function $f(x) = x^2$ on the interval $[3, 10]$.

Solution

$$\text{Mean value} = \frac{1}{10 - 3} \int_3^{10} x^2 \, dx$$

$$= \frac{1}{7} \left[\frac{x^3}{3} \right]_3^{10}$$

$$= \frac{1}{7} \left(\frac{1000}{3} - \frac{27}{3} \right)$$

$$= \frac{139}{3}$$

Sometimes the mean value is not very informative. For example, a periodic, oscillating function like $f(x) = \sin x$ will have a mean value of 0, over the period 0 to 2π. If instead you use the **root mean square** value, defined as

$$\sqrt{\frac{1}{b - a} \int_a^b \left(f(x) \right)^2} \,, \text{ you get a measure of how far away the function is from}$$

zero on average. This might remind you of the root mean square deviation in statistics (it has a very similar structure). Notice that the function is squared before it is integrated, and the square root occurs at the very end of the process. When the function is periodic this root mean square value is usually calculated over an interval of a whole number of periods and is useful in many applications in physics and statistics.

ACTIVITY 8.2

This involves finding the integral of $\sin^2 t$. You can use the double angle formula for $\cos 2t$ to help you do this.

▶ (i) Find the root mean square value of $a \sin t$ over a single period.

(ii) In the UK, mains electricity is usually supplied as alternating current at a nominal 240 volts. This value is the root mean square of the supply voltage, which is the function $a \sin t$. Show that the supply peaks at about 339 V.

Exercise 8.2

① Find the mean value of $f(x) = \sqrt{x}$ between $x = 0$ and $x = 5$.

② Find the mean value of $f(x) = \dfrac{1}{x^2}$ for $1 \leqslant x \leqslant 4$.

③ (i) Find the mean value of $f(x) = (x - 1)^3$ on the interval $[0, 2]$.

(ii) Explain your answer to (i) with the aid of a sketch graph.

④ Find the mean value of the following functions over the given sets of values for x. In each case, draw the graph of the function and indicate the mean value of the function on your graph.

(i) $y = 4 - x^2$ on the interval $[-2, 2]$

(ii) $y = 2e^{-x}$ on the interval $[0, 1]$

(iii) $y = \dfrac{1}{x}$ on the interval $[1, 3]$

(iv) $y = \cos x$ on the interval $\left[0, \dfrac{\pi}{2} \right]$

⑤ (i) Find the mean value of f$(x) = (x-1)^4$ on the interval $[0,2]$.

(ii) Find the mean value of f$(x) = (x-1)^4$ on the interval $[0,1]$.

(iii) Explain with the aid of a sketch graph why your answers to (i) and (ii) are the same.

⑥ Figure 8.19 shows the curve $y = x^2 - 2x + 2$. The area under the curve between $x = 0$ and $x = 2$ is approximated by two rectangles.

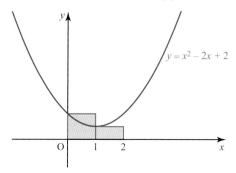

$y = x^2 - 2x + 2$

Figure 8.19

(i) Find the total area of the two rectangles.

(ii) Find the mean of the heights of the two rectangles.

(iii) Repeat parts (i) and (ii) using four rectangles.

(iv) Repeat parts (i) and (ii) using eight rectangles.

(v) Find the exact value of the area under the curve between $x = 0$ and $x = 2$, and the mean value of the function f$(x) = x^2 - 2x + 2$ for $0 \leqslant x \leqslant 2$ and comment on your results.

⑦ (i) Use the identity $\sin^2 x = \frac{1}{2}(1 - \cos 2x)$ to find the integral of $\sin^2 x$.

(ii) Show that the mean value of the function f$(x) = \sin(x)$ on the interval $[0, 2\pi]$ is 0.

(iii) Find the mean value of the function g$(x) = \sin^2 x$ on the interval $[0, 2\pi]$.

(iv) Find the root mean square of the function f(x) (i.e. the square root of your answer to (iii)).

(v) An electricity supply provides alternating current with peak voltage 169.7 volts. What is the root mean square of the voltage?

3 General integration

Hyperbolic and circular substitutions

In Chapter 4 you used trigonometric substitutions (circular functions) to integrate certain functions, and in Chapter 7 you used hyperbolic substitutions to integrate similar looking functions. You should now be in a position to distinguish which method is appropriate, alongside all the other techniques of integration you have learned.

The standard integrals and their connections to the identities are clues to which technique to use:

Function	Underlying relationships
$\displaystyle\int \frac{1}{\sqrt{a^2 - x^2}}\,\mathrm{d}x = \arcsin\left(\frac{x}{a}\right) + c$	This works because $\dfrac{\mathrm{d}}{\mathrm{d}x}\big(\sin(x)\big) = \cos(x)$ and $\cos(x) = \sqrt{1 - \sin^2(x)}$.
$\displaystyle\int \frac{1}{a^2 + x^2}\,\mathrm{d}x = \frac{1}{a}\arctan\left(\frac{x}{a}\right) + c$	This works because $\dfrac{\mathrm{d}}{\mathrm{d}x}\big(\tan(x)\big) = \sec^2(x)$ and $\sec^2(x) = 1 + \tan^2(x)$.
$\displaystyle\int \frac{1}{\sqrt{x^2 - a^2}}\,\mathrm{d}x = \operatorname{arcosh}\left(\frac{x}{a}\right) + c$	This works because $\dfrac{\mathrm{d}}{\mathrm{d}x}\big(\cosh(x)\big) = \sinh(x)$ and $\sinh(x) = \sqrt{\cosh^2(x) - 1}$.
$\displaystyle\int \frac{1}{\sqrt{a^2 + x^2}}\,\mathrm{d}x = \operatorname{arsinh}\left(\frac{x}{a}\right) + c$	This works because $\dfrac{\mathrm{d}}{\mathrm{d}x}\big(\sinh(x)\big) = \cosh(x)$ and $\cosh(x) = \sqrt{1 + \sinh^2(x)}$.

Table 8.1

It is helpful to recognise how the trigonometric and hyperbolic identities suggest a useful substitution. While the above results are standard and can be quoted, you may often find integrals which have a slightly different form, but similar terms, where making an appropriate trigonometric or hyperbolic substitution will simplify the calculations.

Example 8.4

Find the following integrals:

(i) $\displaystyle\int \frac{1}{x^2 + 4}\,\mathrm{d}x$

(ii) $\displaystyle\int \frac{1}{x^2 - 4}\,\mathrm{d}x$

(iii) $\displaystyle\int \frac{1}{\sqrt{x^2 - 4}}\,\mathrm{d}x$

(iv) $\displaystyle\int \sqrt{x^2 - 4}\,\mathrm{d}x$

(v) $\displaystyle\int \frac{1}{\sqrt{x^2 + 4x + 13}}\,\mathrm{d}x$

(vi) $\displaystyle\int \frac{13}{(x - 3)(x^2 + 4)}\,\mathrm{d}x$

Solution

(i) Use the standard result (or a $x = 2\tan u$ substitution) to get
$$\int \frac{1}{x^2 + 4}\,\mathrm{d}x = \frac{1}{2}\arctan\left(\frac{x}{2}\right) + c$$

(ii) Although it looks like the standard examples in the table above, the denominator is the difference of two squares, and so should be factorised and then partial fractions used:
$$\frac{1}{4}\int \frac{1}{x - 2} - \frac{1}{x + 2}\,\mathrm{d}x = \frac{1}{4}\big(\ln|x - 2| - \ln|x + 2|\big) + c$$
$$= \frac{1}{4}\ln\left|\frac{x - 2}{x + 2}\right| + c$$

(iii) Use the standard result (or a $x = 2\cosh u$ substitution) to get

$$\int \frac{1}{\sqrt{x^2 - 4}}\,dx = \operatorname{arcosh}\left(\frac{x}{2}\right) + c$$

(iv) This is not a standard result, but it looks like the form of a $\cosh u$ identity. Try the substitution $x = 2\cosh u$ then $\dfrac{dx}{du} = 2\sinh u$ and so

$$\int \sqrt{x^2 - 4}\,dx = \int \sqrt{4\cosh^2 u - 4} \times 2\sinh u\,du$$

$$= 4\int \sinh^2 u\,du$$

$$= \int (e^{2u} - 2 + e^{-2u})\,du$$

$$= \frac{1}{2}e^{2u} - 2u - \frac{1}{2}e^{-2u} + c$$

$$= \sinh 2u - 2\operatorname{arcosh}\left(\frac{x}{2}\right) + c$$

$$= 2\sinh u \cosh u - 2\operatorname{arcosh}\left(\frac{x}{2}\right) + c$$

$$= \frac{1}{2}x\sqrt{x^2 - 4} - 2\operatorname{arcosh}\left(\frac{x}{2}\right) + c$$

$$= \frac{1}{2}x\sqrt{x^2 - 4} - 2\ln\left|\frac{x}{2} + \frac{1}{2}\sqrt{x^2 - 4}\right| + c$$

(v) The quadratic in the denominator has no real roots since $b^2 - 4ac = 16 - 4 \times 13 < 0$.

Complete the square:

$$\int \frac{1}{\sqrt{(x^2 + 4x + 13)}}\,dx = \int \frac{1}{\sqrt{(x + 2)^2 + 9}}\,dx$$

It looks like a $\sinh u$ substitution will help: so let $x + 2 = 3\sinh u$, then $dx = 3\cosh u\,du$

$$= \int \frac{1}{\sqrt{9\sinh^2 u + 9}} \times 3\cosh u\,du$$

$$= \int 1\,du$$

$$= u + c$$

$$= \operatorname{arsinh}\left(\frac{x + 2}{3}\right) + c$$

(vi) Using partial fractions gives

$$\int \frac{13}{(x - 3)(x^2 + 4)}\,dx = \int \frac{1}{x - 3} - \frac{x + 3}{x^2 + 4}\,dx$$

$$= \ln|x - 3| - \int \frac{x}{x^2 + 4}\,dx - \int \frac{3}{x^2 + 4}\,dx + c$$

$$= \ln|x - 3| - \frac{1}{2}\ln|x^2 + 4| - \frac{3}{2}\arctan\left(\frac{x}{2}\right) + c$$

Exercise 8.3

① Find the following integrals:

(i) $\displaystyle\int \frac{1}{\sqrt{9-x^2}}\,dx$ 　(ii) $\displaystyle\int \frac{1}{\sqrt{x^2-9}}\,dx$ 　(iii) $\displaystyle\int \frac{1}{\sqrt{9+x^2}}\,dx$

(iv) $\displaystyle\int \frac{1}{9+x^2}\,dx$ 　(v) $\displaystyle\int \frac{1}{9-x^2}\,dx$

② Evaluate the following, giving your answers in exact form:

(i) $\displaystyle\int_0^5 \frac{1}{x^2+25}\,dx$ 　(ii) $\displaystyle\int_0^4 \frac{1}{25-x^2}\,dx$ 　(iii) $\displaystyle\int_0^5 \frac{1}{\sqrt{25+x^2}}\,dx$

(iv) $\displaystyle\int_5^{10} \frac{1}{\sqrt{x^2-25}}\,dx$ 　(v) $\displaystyle\int_0^5 \frac{1}{\sqrt{25-x^2}}\,dx$

③ Find:

(i) $\displaystyle\int \frac{1}{\sqrt{4-9x^2}}\,dx$ 　(ii) $\displaystyle\int \frac{1}{\sqrt{4+9x^2}}\,dx$ 　(iii) $\displaystyle\int \frac{1}{\sqrt{9x^2-4}}\,dx$

④ Evaluate the following, giving your answers in exact form:

(i) $\displaystyle\int_0^1 \frac{1}{3x^2+1}\,dx$ 　(ii) $\displaystyle\int_{\frac{2}{\sqrt{3}}}^{\frac{3}{\sqrt{3}}} \frac{1}{\sqrt{3x^2-4}}\,dx$ 　(iii) $\displaystyle\int_0^1 \frac{1}{\sqrt{4-3x^2}}\,dx$

⑤ Show that the volume of the solid of revolution formed by rotating the function $f(x) = \dfrac{1}{\sqrt{4+x^2}}$ between $x = 0$ and $x = 5$ about the x-axis, is 1.87 units^3 to 3 d.p.

⑥ Calculate the mean value of the function $f(x) = \dfrac{3x}{\sqrt{x^2-9}}$ on the interval $[5, 6]$.

⑦ Use a suitable trigonometric or hyperbolic substitution to find $\displaystyle\int_0^6 \frac{1}{\left(x^2+64\right)^{\frac{3}{2}}}\,dx$, giving your answer in exact form.

⑧ Use partial fractions to find $\displaystyle\int \frac{4x^2-6x+5}{(x-1)^2\left(x^2+2\right)}\,dx$.

⑨ Find $\displaystyle\int_0^{\sqrt{2}} \frac{3x-1}{x^2+2}\,dx$, giving your answer in exact form.

⑩ Figure 8.20 shows the curve $y = f(x)$, where $f(x) = \dfrac{10}{(x+1)(x+2)\left(x^2+1\right)}$.

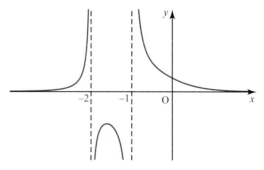

Figure 8.20

(i) Find the area under the graph between $x = 0$ and $x = 1$.

(ii) Why is it not possible to find the area under the graph for all values of x?

(iii) For which of the following intervals does the area under the graph converge? Give the area in any cases which do converge.

(a)　$[-3, -2]$　　(b)　$[-2, -1]$　　(c)　$[-1, \infty]$　　(d)　$[0, \infty]$

⑪　Figure 8.21 shows the graph of $y = f(x)$, where $f(x) = x^2(2 - x)$.

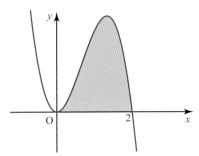

Figure 8.21

(i) Find the area of the shaded region.

(ii) The line $y = a$ is drawn so that the area under the line between $x = 0$ and $x = 2$ is the same as the area in (i). Find the value of a.

(iii) Find the volume of the solid formed by rotating the shaded region through $360°$ about the x-axis.

(iv) The line $y = b$ is drawn so that when the region under the line between $x = 0$ and $x = 2$ is rotated through $360°$ about the x-axis, the cylinder formed has the same volume as the solid in (iii). Find the value of b.

⑫　Evaluate $\displaystyle\int_0^1 \frac{2}{\sqrt{x^2 + 6x + 10}}\, dx$, giving your answer in exact form.

LEARNING OUTCOMES

When you have completed this chapter you should be able to:

➤ derive formulae for and calculate the volumes of solids generated by rotating a plane region about the x-axis

➤ derive formulae for and calculate the volumes of solids generated by rotating a plane region about the y-axis

➤ evaluate the mean value of a function on a given interval

➤ recognise integrals of functions of the form $\left(a^2 - x^2\right)^{-\frac{1}{2}}, \left(a^2 + x^2\right)^{-1},$ $\left(x^2 - a^2\right)^{-\frac{1}{2}}$ and $\left(x^2 + a^2\right)^{-\frac{1}{2}},$ and be able to integrate related functions by using trigonometric or hyperbolic substitutions.

KEY POINTS

1 Volume of revolution about the x-axis: $V = \int_a^b \pi y^2 \mathrm{d}x$

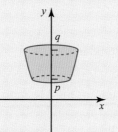

Figure 8.22

2 Volume of revolution about the y-axis: $V = \int_p^q \pi x^2 \mathrm{d}y$

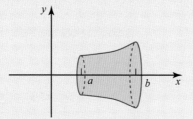

Figure 8.23

3 The mean value of a function $\mathrm{f}(x)$ on the interval $[a, b]$ is given by $\dfrac{1}{b-a} \int \mathrm{f}(x) \mathrm{d}x$.

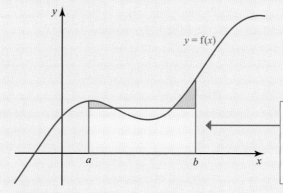

> The mean value is the height of the rectangle which has the same area as the area under the curve.

Figure 8.24

4 $\displaystyle\int \frac{1}{\sqrt{a^2 - x^2}} \mathrm{d}x = \arcsin\frac{x}{a} + c$

$\displaystyle\int \frac{1}{a^2 + x^2} \mathrm{d}x = \frac{1}{a}\arctan\frac{x}{a} + c$

$\displaystyle\int \frac{1}{\sqrt{x^2 - a^2}} \mathrm{d}x = \operatorname{arcosh}\frac{x}{a} + c$

$\displaystyle\int \frac{1}{\sqrt{x^2 + a^2}} \mathrm{d}x = \operatorname{arsinh}\frac{x}{a} + c$

FUTURE USES

■ The volume of revolution is required in mechanics in order to find the centre of mass of a solid of revolution.

M **T** ① Figure 1 shows the distance of the Moon from the Earth on each day in January 2016. Day 0 is 1 January, Day 1 is 2 January and so on. A spreadsheet has been used to fit a polynomial trend line to the data.

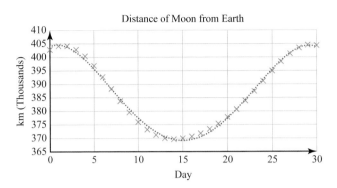

Figure 1

The equation of the trend line is

$$y = -0.884x^4 + 53x^3 - 840x^2 + 1332x + 403\,991$$

where y is the distance from the Moon to the Earth in km and x is the day number. This is used as a model for the distance of the Moon from the Earth.

(i) Explain why $y = -0.884x^4 + 53x^3 - 840x^2 + 1332x + 403\,991$ is not a suitable model for the distance of the Moon from the Earth in the long term. **[1 mark]**

(ii) Use the model to find the mean distance of the Moon from the Earth in the time period Day 0 to Day 30. **[3 marks]**

(iii) A website gives the average distance between the Moon and the Earth as 384 400 km. Compare this value with your answer to (ii) and comment. **[1 mark]**

PS ② The following equation has $1 + i$ as one of its roots. m is a real constant.

$$(m + 2)x^3 + (m^2 - 8)x^2 + (m + 3)x - 2 = 0$$

Find the other roots of the equation. **[6 marks]**

MP ③ Figure 2 shows the curve $y = \dfrac{1}{x}$. The area below the curve between $x = 1$ and $x = a$ is shaded.

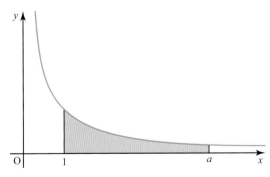

Figure 2

(a) (i) Find the shaded area in terms of a. [2 marks]

(ii) Show that the area becomes infinite as $a \to \infty$. [1 mark]

(b) Show that the volume of revolution about the x-axis of $y = \dfrac{1}{x}$ for $x \geqslant 1$ is π. [3 marks]

MP (4) (i) Prove that $\cosh 2x = 2\sinh^2 x + 1$. [3 marks]

Figure 3 shows the curve $y = \mathrm{f}(x)$, where $\mathrm{f}(x) = 2\sinh^2 x - 3\cosh x$.

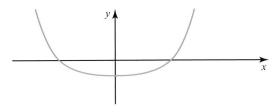

Figure 3

(ii) Show that the curve $y = \mathrm{f}(x)$ crosses the x-axis at $x = \pm\ln(2 + \sqrt{3})$. [3 marks]

(iii) Find the exact area bounded by the curve $y = \mathrm{f}(x)$ and the x-axis. [8 marks]

T PS (5) In this question, the notation y_n denotes the nth derivative of y with respect to x.

(i) Given $y = \ln(\cos x)$ where $0 \leqslant x < \dfrac{\pi}{2}$, show that $y_3 + 2y_2 y_1 = 0$. [3 marks]

(ii) Hence show that $y_4 = -4y_2 y_1^2 - 2y_2^2$. Using this result, or otherwise, obtain the Maclaurin expansion of y up to the term in x^4. [5 marks]

(iii) Taking $x = \dfrac{\pi}{3}$, deduce the approximation $\ln(2) \approx \dfrac{\pi^2}{18}\left(1 + \dfrac{\pi^2}{54}\right)$. [3 marks]

(iv) Taking $x = \dfrac{\pi}{4}$, deduce another approximation for $\ln(2)$.

Which of these approximations is better and why? [4 marks]

PS T (6) Figure 4 shows a circle with centre C and radius a. O is a point on a diameter so that OC has length $\frac{1}{2}a$. Q is a point on the circle such that the angle QCO is ϕ. The tangent to the circle at Q intersects the line CO at A. P is the foot of the perpendicular from O to this tangent. The distance OP is r and the angle POA is θ.

Referred to the system which has its origin at O and its initial line in the direction OA, the point P has polar coordinates (r, θ).

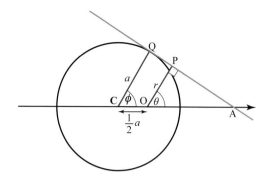

Figure 4

(i) Show that $\theta = \phi$ for $0 \leqslant \theta \leqslant \dfrac{\pi}{2}$. [1 mark]

(ii) Show that, as θ, varies with $0 \leqslant \theta \leqslant \dfrac{\pi}{2}$, the path of P has polar equation

$$r = a\left(1 - \frac{1}{2}\cos\theta\right).$$ [4 marks]

(iii) Draw a diagram showing a case where $\dfrac{\pi}{2} \leqslant \theta \leqslant \pi$ and explain why

the polar equation given in (ii) is still valid for $\dfrac{\pi}{2} \leqslant \theta \leqslant \pi$.

Explain further why the polar equation given in (ii) is valid for all $0 \leqslant \theta \leqslant 2\pi$. [4 marks]

(iv) Sketch the path of P. (You may use your calculator.) [2 marks]

(v) Show that the cartesian equation of the path of P is given by
$$x^2 + y^2 = a\left(\sqrt{x^2 + y^2} - \frac{1}{2}x\right),$$

where O is the origin of the cartesian coordinates, the x-axis is in the direction OA and the y-axis is perpendicular to OA. [2 marks]

Review: Roots of polynomials

1 Roots and coefficients

The relationships between the roots and coefficients of polynomial equations form patterns. These patterns hold for complex roots as well as for real roots.

Roots and coefficients of quadratic equations

For the general quadratic equation $az^2 + bz + c = 0$, with roots α and β:

- $\alpha + \beta = -\dfrac{b}{a}$
- $\alpha\beta = \dfrac{c}{a}$

> **Note**
>
> You can derive these results by comparing the factorised form
> $(z - \alpha)(z - \beta) = 0$ with the equation $z^2 + \dfrac{b}{a}z + \dfrac{c}{a} = 0$.

For example: the roots α and β of the quadratic equation $3z^2 + 2z - 7 = 0$ satisfy $\alpha + \beta = -\dfrac{2}{3}$ and $\alpha\beta = -\dfrac{7}{3}$.

Roots and coefficients of cubic equations

For the general cubic equation $az^3 + bz^2 + cz + d = 0$, with roots α, β and γ:

- $\alpha + \beta + \gamma = -\dfrac{b}{a}$
- $\alpha\beta + \beta\gamma + \gamma\alpha = \dfrac{c}{a}$
- $\alpha\beta\gamma = -\dfrac{d}{a}$

> **Note**
>
> As for quadratic equations, you can derive these results by comparing the factorised form $(z - \alpha)(z - \beta)(z - \gamma) = 0$ with the equation
> $z^3 + \dfrac{b}{a}z^2 + \dfrac{c}{a}z + \dfrac{d}{a} = 0$.

Roots and coefficients of quartic equations

For the general quartic equation $az^4 + bz^3 + cz^2 + dz + e = 0$, with roots α, β, γ and δ:

- $\sum \alpha = \alpha + \beta + \gamma + \delta = -\dfrac{b}{a}$
- $\sum \alpha\beta = \alpha\beta + \alpha\gamma + \alpha\delta + \beta\gamma + \beta\delta + \gamma\delta = \dfrac{c}{a}$
- $\sum \alpha\beta\gamma = \alpha\beta\gamma + \beta\gamma\delta + \gamma\delta\alpha + \delta\alpha\beta = -\dfrac{d}{a}$
- $\alpha\beta\gamma\delta = \dfrac{e}{a}$

> **Note**
>
> Shorthand notation can be used to save writing out all the combinations.
> So $\sum \alpha$ means the sum of the individual roots, $\sum \alpha\beta$ means the sum of all possible products of pairs of roots, and so on.

Forming new equations

The relationships between roots and coefficients can often be used to form a new equation with roots that are related to the roots of the original equation. Two different methods are shown in the following example.

Example R.1

The quadratic equation $2z^2 - 5z + 8 = 0$ has roots α and β. Find a quadratic equation with roots $\alpha + 2$ and $\beta + 2$.

Solution 1

From the original equation.

$$\alpha + \beta = \frac{5}{2}$$

$$\alpha\beta = \frac{8}{2} = 4$$

For the new equation: $-\dfrac{b}{a} = (\alpha + 2) + (\beta + 2)$

$$= \alpha + \beta + 4$$

$$= \frac{5}{2} + 4$$

$$= \frac{13}{2}$$

$$\frac{c}{a} = (\alpha + 2)(\beta + 2)$$

$$= \alpha\beta + 2(\alpha + \beta) + 4$$

$$= 4 + 2 \times \frac{5}{2} + 4$$

$$= 13$$

Taking $a = 2$ makes the coefficients integers. Any other value for a would give a multiple of the same equation.

Taking $a = 2$ gives $b = -13$ and $c = 26$.

So the new equation is $2z^2 - 13z + 26 = 0$.

Solution 2 (substitution method)

Think of the graph $y = f(x)$. The new graph is found by translating the original graph 2 units to the right, so the new graph has equation $y = f(x - 2)$.

In the new equation, $w = z + 2$ so $z = w - 2$.

Replacing z in the original equation:

$$2(w - 2)^2 - 5(w - 2) + 8 = 0$$

$$2w^2 - 8w + 8 - 5w + 10 + 8 = 0$$

$$2w^2 - 13w + 26 = 0$$

The substitution method shown above is often more efficient when dealing with higher order equations.

Example R.2

The cubic equation $z^3 - 3z^2 + z + 4 = 0$ has roots α, β and γ. Find a cubic equation with roots $2\alpha - 1, 2\beta - 1$ and $2\gamma - 1$.

Solution

In the new equation, $w = 2z - 1$ so $z = \dfrac{w+1}{2}$.

Replacing z in the original equation:

$$\left(\frac{w+1}{2}\right)^3 - 3\left(\frac{w+1}{2}\right)^2 + \left(\frac{w+1}{2}\right) + 4 = 0$$

$$\frac{(w+1)^3}{8} - \frac{3(w+1)^2}{4} + \frac{w+1}{2} + 4 = 0$$

$$(w+1)^3 - 6(w+1)^2 + 4(w+1) + 32 = 0$$

$$w^3 + 3w^2 + 3w + 1 - 6w^2 - 12w - 6 + 4w + 4 + 32 = 0$$

$$w^3 - 3w^2 - 5w + 29 = 0$$

> Multiply through by 8. It is usually easier to clear the fractions before expanding the brackets.

Exercise R.1

① The quadratic equation $2z^2 - 3z - 5 = 0$ has roots α and β.

Write down the values of:

(i) $\alpha + \beta$ (ii) $\alpha\beta$

② The cubic equation $3z^3 + z^2 + 4z - 7 = 0$ has roots α, β and γ.

Write down the values of:

(i) $\alpha + \beta + \gamma$ (ii) $\alpha\beta + \beta\gamma + \gamma\alpha$ (iii) $\alpha\beta\gamma$

③ The quartic equation $z^4 - 2z^3 - 5z^2 - 3 = 0$ has roots α, β and γ.

Write down the values of:

(i) $\alpha + \beta + \gamma + \delta$ (ii) $\alpha\beta + \alpha\gamma + \alpha\delta + \beta\gamma + \beta\delta + \gamma\delta$

(iii) $\alpha\beta\gamma + \beta\gamma\delta + \gamma\delta\alpha + \delta\alpha\beta$ (iv) $\alpha\beta\gamma\delta$

④ Find a cubic equation with roots $2, -4$ and 5.

⑤ The quadratic equation $3z^2 + 2z + 4 = 0$ has roots α and β.

Find quadratic equations with roots:

(i) $\alpha - 1$ and $\beta - 1$ (ii) 2α and 2β (iii) $2\alpha + 3$ and $2\beta + 3$

⑥ The cubic equation $z^3 - 3z^2 - z + 1 = 0$ has roots α, β and γ.

Find cubic equations with roots:

(i) $\alpha + 2, \beta + 2$ and $\gamma + 2$

(ii) $\frac{1}{2}\alpha, \frac{1}{2}\beta$ and $\frac{1}{2}\gamma$

(iii) $1 - 3\alpha, 1 - 3\beta$ and $1 - 3\gamma$

⑦ The cubic equation $f(x) = 0$, where $f(x) = x^3 - 4x^2 - 3x + k$, has a pair of equal roots.

(i) Find the two possible values of the roots of the equation and the corresponding values of k.

(ii) Sketch the graph of $y = f(x)$ for each of the possible values of k.

⑧ One of the roots of the equation $x^2 + px + q = 0$ is three times the other. Prove that $3p^2 = 16q$.

⑨ One of the roots of the equation $z^3 - 7z^2 + k = 0$ is twice one of the other roots.

Solve the equation and find the value of k.

⑩ The equation $x^4 + px^3 - qx - 4 = 0$ has roots $\alpha, 2\alpha, \beta$ and $-\beta$.

Find the two possible sets of values for p and q, and give the roots of the equation in each case.

2 Complex roots of polynomial equations

All polynomial equations of degree n have exactly n roots, some of which may be complex and some of which may be repeated.

You often need to use the factor theorem to help you solve polynomial equations.

Example R.3

(i) Show that $z = 2$ is a root of the equation $z^3 + 2z^2 - 3z - 10 = 0$.

(ii) Solve the equation.

Solution

$$f(z) = z^3 + 2z^2 - 3z - 10$$

$$f(2) = 2^3 + 2 \times 2^2 - 3 \times 2 - 10$$

$$= 8 + 8 - 6 - 10$$

$$= 0$$

> The factor theorem states $f(a) = 0 \Leftrightarrow (x - a)$ is a factor of $f(x)$.

$f(2) = 0$ so $(z - 2)$ is a factor of $f(z)$.

$$z^3 + 2z^2 - 3z - 10 = 0$$

$$(z - 2)(z^2 + 4z + 5) = 0$$

> Factorise using inspection or polynomial division.

$$z = 2 \text{ or } z^2 + 4z + 5 = 0$$

$$z = \frac{-4 \pm \sqrt{16 - 4 \times 1 \times 5}}{2} = \frac{-4 \pm \sqrt{-4}}{2} = -2 \pm i$$

So the solution of the equation is $z = 2$, $z = 2 + i$, $z = 2 - i$.

Notice that in the example above, the two complex roots are conjugates of each other.

If a polynomial equation has real coefficients, any complex roots always occur in conjugate pairs.

Sometimes the relationships between roots and coefficients of polynomial equations can be useful in solving problems involving polynomial equations with complex roots.

Example R.4

Find a cubic equation with roots $1, 2 + i$ and $2 - i$.

> **Note**
>
> It would be possible to solve this problem by writing the equation in factorised form and multiplying out. However, this would get quite messy and the method shown here is more efficient.

Solution

$$\sum \alpha = 1 + 2 + i + 2 - i = 5$$
$$\sum \alpha\beta = 1(2 + i) + 1(2 - i) + (2 + i)(2 - i)$$
$$= 2 + i + 2 - i + 4 + 1 = 9$$
$$\alpha\beta\gamma = 1(2 + i)(2 - i)$$
$$= 4 + 1 = 5$$
So $-\dfrac{b}{a} = 5, \dfrac{c}{a} = 9, -\dfrac{d}{a} = 5$

Taking $a = 1$ gives $b = -5, c = 9, d = -5$. ◄── If you use a different value for a, you will get a multiple of the same equation.
The equation is $z^3 - 5z^2 + 9z - 5 = 0$.

The next example shows two methods of solving the same problem.

Example R.5

One of the roots of the equation $2z^3 - 5z^2 + 12z - 5 = 0$ is $1 + 2i$.

Solve the equation and show the roots on an Argand diagram.

Solution 1

Since one of the roots is $1 + 2i$, another root is $1 - 2i$.
So a quadratic factor is $(z - 1 - 2i)(z - 1 + 2i)$

> Writing the expression in this form shows that it is the difference of two squares.

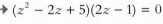
$$= \big((z - 1) - 2i\big)\big((z - 1) + 2i\big)$$
$$= (z - 1)^2 + 4$$
$$= z^2 - 2z + 5$$
$$2z^3 - 5z^2 + 12z - 5 = 0$$

> Find the linear factor by inspection.

$(z^2 - 2z + 5)(2z - 1) = 0$

The roots are $1 + 2i$, $1 - 2i$ and $\dfrac{1}{2}$.

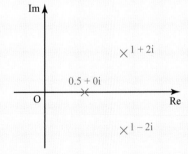

Figure R.1

Solution 2

Since one of the roots is $1 + 2i$, another root is $1 - 2i$.

The sum of the three roots is $\dfrac{5}{2}$. ◄──

So $1 + 2i + 1 - 2i + y = \dfrac{5}{2} \Rightarrow y = \dfrac{1}{2}$.

Using $\alpha + \beta + \gamma = \dfrac{b}{a}$

The roots are $1 + 2i$, $1 - 2i$ and $\dfrac{1}{2}$.

① One of the roots of the quadratic equation $x^2 + px + q = 0$ is $2 - 5i$.

    (i)   Write down the other root of the equation.

    (ii)   Find the values of p and q.

② (i)   Show that $x = -1$ is a root of the equation $z^3 - 5z^2 + 19z + 25 = 0$.

    (ii)   Find the other two roots.

    (iii) Show the three roots on an Argand diagram.

③ Find a quartic equation with roots 2, 5, $3 + 2i$ and $3 - 2i$.

④ The cubic equation $x^3 + px^2 - 2x + 4 = 0$ has a root of -2.
Find the value of p and solve the equation.

⑤ One of the roots of the cubic equation $3z^3 - 14z^2 + 47z - 26 = 0$
is $2 + 3i$. Solve the equation.

⑥ The four roots of a quartic equation are shown on the Argand diagram in Figure R.2.

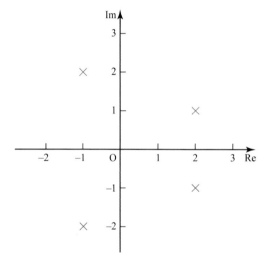

Figure R.2

Find the equation.

⑦ The cubic equation $z^3 + 4z^2 + 14z + 20 = 0$ has an integer root. Solve the equation and show the roots on an Argand diagram.

⑧ (i)   Given that $2 - i$ is a root of the equation $2z^3 + az^2 + 14z + b = 0$, find the values of a and b.

    (ii)   Solve the equation $2z^3 + az^2 + 14z + b = 0$

⑨ One of the roots of the equation $4z^4 - 8z^3 + 9z^2 - 2z + 2 = 0$ is $1 - i$.
Solve the equation.

⑩ One of the roots of the equation $z^4 - 4z^3 + pz^2 + qz + 50 = 0$ is $-1 + 2i$.
Find the values of p and q and solve the equation.

KEY POINTS

1 If α and β are the roots of the quadratic equation $az^2 + bz + c = 0$, then
$$\alpha + \beta = -\frac{b}{a},$$
$$\alpha\beta = \frac{c}{a}.$$

2 If α, β and γ are the roots of the cubic equation $az^3 + bz^2 + cz + d = 0$, then
$$\sum \alpha = \alpha + \beta + \gamma = -\frac{b}{a},$$
$$\sum \alpha\beta = \alpha\beta + \beta\gamma + \gamma\alpha = \frac{c}{a},$$
$$\alpha\beta\gamma = -\frac{d}{a}.$$

3 If α, β, γ and δ are the roots of the quartic equation
$az^4 + bz^3 + cz^2 + dz + e = 0$, then
$$\sum \alpha = \alpha + \beta + \gamma + \delta = -\frac{b}{a},$$
$$\sum \alpha\beta = \alpha\beta + \alpha\gamma + \alpha\delta + \beta\gamma + \beta\delta + \gamma\delta = \frac{c}{a},$$
$$\sum \alpha\beta\gamma = \alpha\beta\gamma + \beta\gamma\delta + \gamma\delta\alpha + \delta\alpha\beta = -\frac{d}{a},$$
$$\alpha\beta\gamma\delta = \frac{e}{a}.$$

4 All of these formulae may be summarised using the shorthand sigma notation for elementary symmetric functions as follows:
$$\sum \alpha = -\frac{b}{a}$$
$$\sum \alpha\beta = \frac{c}{a}$$
$$\sum \alpha\beta\gamma = -\frac{d}{a}$$
$$\sum \alpha\beta\gamma\delta = \frac{e}{a}$$

etc. (using the convention that polynomials of degree n are labelled $az^n + bz^{n-1} + \ldots = 0$ and have roots α, β, γ, ...)

5 A polynomial equation of degree n has n roots, taking into account complex roots and repeated roots.

6 In the case of polynomial equations with real coefficients, complex roots always occur in conjugate pairs.

9 First order differential equations

> It isn't that they can't see the solution.
> It's that they can't see the problem.
>
> G.K. Chesterton

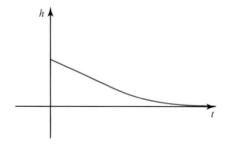

Figure 9.1

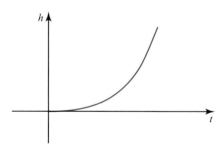

Figure 9.2

Discussion point

One of these graphs applies to an aeroplane landing and the other to one taking off. Which is which? What do these graphs tell you about $\frac{\mathrm{d}h}{\mathrm{d}t}$ in each of these circumstances?

1 Modelling rates of change

Modelling

Modelling is the process of representing situations in the real world in mathematical form. It is an important skill. Sometimes rates of change are involved and in those cases the models involve differential equations.

Forming differential equations

If you are given sufficient information about the rate of change of a quantity, such as the caffeine level in the body or the height of water in a harbour, you can form a differential equation to model the situation.

It is important to look carefully at the problem before writing down an equivalent mathematical statement. You have to decide whether you need a model for a rate of change with respect to time or with respect to another variable such as distance or height. You need to be familiar with the language used in these different cases.

If the altitude, h, of an aircraft is being considered, the phrase **rate of change of altitude** might be used. This actually means the **rate of change of altitude with respect to time**. You could write it as $\dfrac{dh}{dt}$ where t stands for time.

However, you might be more interested in how the altitude of the aircraft changes with the horizontal distance it has travelled. In this case you would talk about the **rate of change of altitude with respect to horizontal distance**, and you could write it as $\dfrac{dh}{dx}$ where x stands for the horizontal distance travelled.

Notation

- Any equation which involves a derivative such as $\dfrac{dq}{dt}, \dfrac{dy}{dt}$, or $\dfrac{d^2x}{dt^2}$, is called a differential equation.

- A differential equation which involves a first derivative such as $\dfrac{dq}{dt}$ is called a **first order differential equation**.

> This is just like the convention of naming polynomials e.g. cubics which contain x^3 terms, but might also contain x^2 and x terms too.

- One which involves a second order derivative such as $\dfrac{d^2x}{dt^2}$ is called a **second order differential equation**. A second order differential equation may also involve first derivatives as well – it is the *highest* derivative that matters.

- A third order differential equation involves a third order derivative (e.g. $\dfrac{d^3y}{dx^3}$), and so on.

You should be aware of two shorthand notations.

- Differentiation with respect to time is often indicated by writing a dot above the variable.

  For example $\dfrac{dx}{dt}$ may be written as $\dot{x}$ | You would say this as 'x dot'.

  and $\dfrac{d^2y}{dt^2}$ may be written as $\ddot{y}$. | You would say this as 'y double dot'.

- Differentiation with respect to x may be denoted by the use of the symbol $'$.

For example y' means $\dfrac{dy}{dx}$ | You would say this as 'y dash'. |

and f'' means $\dfrac{d^2f}{dx^2}$. | You would say this as 'f double dash'. |

The following examples show how differential equations can be formed from descriptions of a wide variety of situations.

Example 9.1

> **Note**
>
> ρ is the Greek letter 'Rho'; be careful as it looks very similar to the letter p.

A simple model of the atmosphere above the Earth's surface is as a perfect gas at constant temperature. This means that:

- the pressure, p, at any point is proportional to the density, ρ;

- the rate at which the pressure decreases with respect to height, z, is proportional to the density.

Use this information to find an expression for $\dfrac{dp}{dz}$ in terms of p.

Solution

Since the pressure, p, at any point is proportional to the density at that point, you can write:

$$p = c_1\rho$$

where c_1 is a positive constant.

Since the rate of change of pressure, p, with height, z, is proportional to the density, ρ, you can write:

$$\frac{dp}{dz} = -c_2\rho$$

| Negative sign needed since pressure *decreases* with height (assume $c_2 > 0$). |

where c_2 is a positive constant.

Substituting for ρ from the first equation into the second gives

$$\frac{dp}{dz} = -\frac{c_2}{c_1}p$$

| c is a constant formed from c_1 and c_2. |

$$= -cp$$

Example 9.2

The volume, V, of a spherical raindrop of radius r is decreasing (due to evaporation) at a rate proportional to its surface area, S. Find an expression for $\dfrac{dr}{dt}$.

Solution

Note that the problem description contains four variables, V, S, r and t, but that V and S are themselves functions of the radius r:

$$V = \frac{4}{3}\pi r^3 \quad \text{and} \quad S = 4\pi r^2.$$

You can use this to write the problem more simply in terms of the two variables, r and t.

The wording of the problem gives:

$$\frac{dV}{dt} = -kS$$

> The next step is to write this equation in terms of r.

$$\frac{dV}{dt} = -k\left(4\pi r^2\right)$$

where k is a positive constant.

The left-hand side of the differential equation may be rewritten using the chain rule in the form:

$$\frac{dV}{dt} = \frac{dV}{dr} \times \frac{dr}{dt}$$

Since $V = \frac{4}{3}\pi r^3$ so $\frac{dV}{dr} = 4\pi r^2$, and then $\frac{dV}{dt} = 4\pi r^2 \times \frac{dr}{dt}$.

The differential equation is now written:

$$4\pi r^2 \times \frac{dr}{dt} = -4k\pi r^2$$

$$\frac{dr}{dt} = -k$$

> This very simple differential equation tells you that the radius decreases at a constant rate. You can solve this equation directly, by integration.

For a model to be useful you need to not only set up a differential equation but also solve it.

You are familiar with the idea that integration is the inverse of differentiation. In taking this step you often solve a differential equation.

So, for example, if

$$\frac{dy}{dx} = 2x - 1$$

> This is a differential equation.

then

$$y = x^2 - x + c$$

> This is the solution (c is an arbitrary constant).

Note that the solution contains a constant c. This is because for this **first order** differential equation you carried out *one* integration (which always introduces the possibility of an arbitrary constant). If the original equation was a **second order** differential equation, you would have introduced *two* constants. This is true for solutions of all differential equations. Whatever method is used to solve them, the solution will contain a number of constants equal to the order of the differential equation.

A solution which contains constants like this is called the **general solution** of the differential equation. A general solution forms a family of solutions and can be thought of, and represented as, a set of curves on a graph.

For example, the differential equation:

$$\frac{dy}{dx} = 2x - 1,$$

with general solution:

$$y(x) = x^2 - x + c,$$

will produce particular curves for different values of c. e.g. $y = x^2 - x + 3$ or $y = x^2 - x - 1$.

Figure 9.3 shows some of these possibilities.

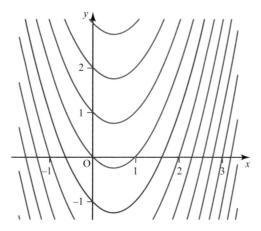

Figure 9.3

If you are investigating the outputs from graphing software you may come across a diagram like Figure 9.3. It is called a tangent field and consists of many small line segments that show the gradients of potential solutions at a large number of points. It can help you to see what the family of solution curves is likely to look like.

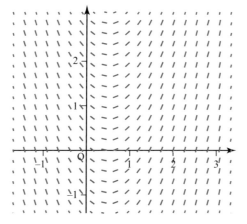

Figure 9.4

Suppose that you are now given one extra piece of information, such as $y = -1$ when $x = 1$. You can now use this information to find the particular value of c, and therefore the particular curve that fits your information.

Substituting $y = -1$ and $x = 1$ into the general solution gives

$$-1 = 1^2 - 1 + c$$

$$c = -1$$

The solution which fits your extra information is $y = x^2 - x - 1$. This is the **particular solution** of the differential equation. It is highlighted in blue in Figure 9.3 and passes through $(1, -1)$. Once a particular solution is found you can use it to find any value for y given a value of x.

The extra piece of information that allows you to identify the particular solution for your problem is called a **condition**. It effectively gives you the coordinates of one of the points through which the required solution curve passes. In the case of a **second order differential equation**, the general solution contains *two constants of integration*, and so *two conditions* are needed to uniquely specify a particular solution. For a **third order differential equation**, *three conditions* are needed, and so on.

Verification of solutions

You will meet situations in which you think you know the solution of a differential equation without actually having to solve it. It may be that the situation, or one like it, is familiar, or you may have experimental data that suggest a solution. In such situations you need to check or **verify** that your solution does indeed satisfy the differential equation. You do this by substituting the solution into the differential equation, a process called **verification**. You would also have to verify that the proposed solution satisfies any conditions that are given.

Verification is also helpful in cases where you think you have found a solution, but want to check that it fits the original problem.

Example 9.3

Verify that $y = Ae^{-3x}$ is the general solution of the differential equation $\dfrac{dy}{dx} = -3y$.

Solution

Differentiating $y = Ae^{-3x}$ gives

$$\frac{dy}{dx} = -3Ae^{-3x}$$

$$= -3\left(Ae^{-3x}\right)$$

$$= -3y,$$

so $y = Ae^{-3x}$ does satisfy the original differential equation, and has one arbitrary constant, so it is the general solution.

Some possible solution curves (red) are shown over the tangent field (blue) for this differential equation in Figure 9.5.

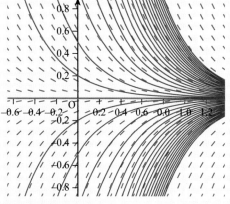

Figure 9.5

Example 9.4

Verify that:

$$v = 20e^{-2t} + 5$$

is the particular solution of the differential equation:

$$\frac{dv}{dt} = 10 - 2v$$

that satisfies the condition $v = 25$ when $t = 0$.

Solution

Differentiating $v = 20e^{-2t} + 5$ gives:

$$\frac{dv}{dt} = -40e^{-2t}$$

Start with the right-hand side of the differential equation:

$$10 - 2v = 10 - 2\left(20e^{-2t} + 5\right)$$

$$= 10 - 40e^{-2t} - 10$$

$$= -40e^{-2t}$$

$$= \frac{dv}{dt}$$

So the solution $v = 20e^{-2t} + 5$ does satisfy the differential equation $\frac{dv}{dt} = 10 - 2v$.

Checking the condition, substitute $t = 0$ into the solution:

$$v = 20e^{-2\times0} + 5$$

$$= 20 + 5$$

$$= 25$$

So the solution satisfies the condition too, and is therefore the particular solution required.

Exercise 9.1

① During the decay of a radioactive substance, the rate at which mass is lost is proportional to the mass present at that instant. Write down a differential equation to describe this relationship.

② The rate at which the population of a particular country increases is proportional to its population. Currently the population is 68 million, and it is increasing at a rate of 2 million per year. Form a differential equation for P, the population of the country in millions.

③ The rate at which water leaves a tank is modelled as being proportional to the square root of the height of water in the tank. Initially the height of water is $100\,\text{cm}$ and water is leaving at the rate of $20\,\text{cm}^3$ per second. Form a differential equation that describes this model.

④ Verify that $P = 40e^{0.2t}$ is a solution of $\frac{dP}{dt} = \frac{P}{5}$.

⑤ The temperature, T, of an object changes such that

$$\frac{dT}{dt} = 2 - 0.1T$$

(i) Verify that $T = 20 + 60e^{-0.1t}$ is a solution of this differential equation.

(ii) Verify that $T = 20 - 18e^{-0.1t}$ is also a solution.

(iii) One of these solutions models the temperature change of a cup of coffee placed in a room, and the other models that of a cold carton of juice just taken from a fridge. Which solution corresponds to which object?

⑥ A radioactive substance decays so that its mass, m mg, decreases according to the equation

$$\frac{dm}{dt} = -\frac{m}{100}$$

where time (t) is measured in hours.

(i) Verify that $m = 20e^{-\frac{t}{100}}$ is a solution of this differential equation for a particular sample of this substance.

(ii) Find the half-life (the time taken for the mass to halve) for this substance.

(iii) What is the initial mass of this sample?

(iv) Write down the solution that would apply if the initial mass were 50 mg.

⑦ Air is escaping from a spherical balloon at a rate proportional to its surface area. Given that the air is escaping at $4\,\text{cm}^3\,\text{s}^{-1}$ when the radius is 10 cm, find an expression for the rate of change of the radius, $\frac{dr}{dt}$.

⑧ The water pressure in the sea increases with depth. At depth h below the surface, the rate of pressure increase with respect to depth is proportional to the density of sea water. The constant of proportionality is the acceleration due to gravity, g ($9.8\,\text{m}\,\text{s}^{-2}$).

The density ρ (in kg m^{-3}) of sea water in part of an ocean is modelled by:

$$\rho = 1000(1 + 0.001h) \qquad 0 \leqslant h \leqslant 100$$

$$\rho = 1100 \qquad\qquad h > 100$$

Find a differential equation to model this situation.

⑨ A volume V of water is held in a tank that has a square base, side 2 m, and height 4 m. Initially the tank is full. Water leaves the tank through a hole at the bottom at a rate of $\sqrt{20h}$ litres per second, where h is the depth of water in the tank. Find expressions for $\frac{dV}{dt}$ and $\frac{dh}{dt}$ in terms of h.

⑩ Verify that $y = -\dfrac{2}{x^2 + 2}$ is the solution of the differential equation

$$\frac{dy}{dx} = xy^2$$

that satisfies the initial condition $y = -1$ when $x = 0$.

⑪ By integrating both sides find the general solution of the differential equation

$$\frac{dy}{dx} = 3x^2 - x + 1$$

Find the particular solution that satisfies the condition $y = 4$ when $x = 1$. Sketch a number of curves to illustrate the family of solutions including the particular solution that passes through the point $(1, 4)$.

⑫ Which of the following are possible solutions of the differential equation $\dfrac{dy}{dx} = -8y$?

(i) $y = 4e^{-8x}$ (ii) $y = 8e^{-4x}$ (iii) $y = 4e^{-8x} + 2$

(iv) $y = 4e^{-8x} + 8$ (v) $y = 8e^{-8x}$

⑬ (i) Verify that $T = \alpha + Ae^{-kt}$ is a solution of the differential equation

$$\dfrac{dT}{dt} = -k(T - \alpha), \quad (k > 0)$$

   (which models Newton's law of cooling).

(ii) Find, in terms of A and α, the initial and final temperatures of the object.

(iii) An object at a temperature of $90°C$ is placed in a room at $25°C$. State the values of α and A in this case.

(iv) Sketch this family of curves including cases where A is both positive and negative.

⑭ Heat-seeking missiles are designed to follow any object that emits heat. For example, an anti-aircraft missile can be 'locked on' to the heat of the exhaust from a jet engine, so that the missile always points towards the engine. This makes it harder for the aircraft to escape the missile by changing direction.

A heat-seeking missile is fired at an aircraft that is flying horizontally with constant speed v at a constant height of b. At the instant the missile is launched, the aircraft has coordinates (a, b) relative to the launch pad. At time t, the missile has position (x, y).

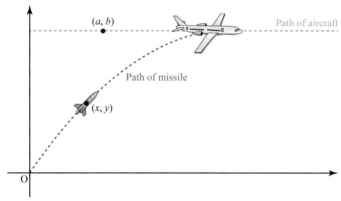

Figure 9.6

Formulate a differential equation for $\dfrac{dy}{dx}$, in terms of x, y and t, to model the path of the missile as it travels to intercept the aircraft.

⑮ Water is pumped into a conical tank at a rate of $0.3\,m^3$ per second. The tank has the dimensions shown in Figure 9.7, and the depth of the water is h metres at time t seconds.

Find an expression for $\dfrac{dh}{dt}$.

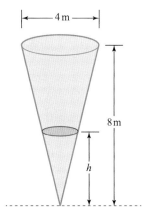

Figure 9.7

⑯ In a model (which applies up to heights of about $10\,\text{km}$) for the pressure p in the lower atmosphere at a height z metres above sea level, it is assumed that gravity is constant and that the air behaves as a perfect gas. This gives the differential equation

$\dfrac{\mathrm{d}p}{\mathrm{d}z} = -\dfrac{pg}{RT}$ where g is the acceleration due to gravity, R is the universal gas constant and T is the absolute temperature of the air in Kelvin (K). The value of R is $8.314\,\text{JK}^{-1}\,\text{mol}^{-1}$.

The temperature distribution in the lower atmosphere is given by $T = 300 - 0.006z$.

(i) Taking $g = 10\,\text{m}\,\text{s}^{-2}$, show that this differential equation may, to a good approximation, be written as:

$$\frac{1}{p}\frac{\mathrm{d}p}{\mathrm{d}z} = -\frac{200}{50\,000 - z}$$

(ii) Show that $p = p_0\left(1 - \dfrac{z}{50\,000}\right)^{200}$ satisfies this differential equation, given $p = p_0$, when $z = 0$.

2 Separation of variables

If a first order differential equation is written in the form:

$$\frac{\mathrm{d}y}{\mathrm{d}x} = \mathrm{f}(x)$$

(i.e. the RHS is a function of x only)

then you can solve it by directly integrating – assuming you can integrate $\mathrm{f}(x)$.

If, however, the differential equation is in the form:

$$\frac{\mathrm{d}y}{\mathrm{d}x} = \mathrm{f}(x)\,\mathrm{g}(y)$$

then you cannot directly integrate because of the y variable on the RHS. One method of proceeding is to use instead a technique covered in A Level Mathematics, called **separating the variables**. For example, look at the differential equation:

$$\frac{dy}{dx} = xy^2$$

Integrating directly is a problem because you do not know what y is in terms of x (that's precisely the final solution to the differential equation). However, since the RHS is just a product of a function of x and a function of y, you can separate them by writing:

$$\frac{1}{y^2}\frac{dy}{dx} = x$$

This is the process called **separating the variables**. Notice that the RHS is a function of x only, and the LHS is $\frac{dy}{dx}$ multiplied by a function of y only.

Integrating both sides with respect to x gives:

$$\int \frac{1}{y^2}\frac{dy}{dx}dx = \int x\,dx$$

It can be shown that the LHS simplifies, giving:

$$\int \frac{1}{y^2}dy = \int x\,dx$$

Since the LHS involves integrating a function involving y only, with respect to y, and the RHS involves integrating a function of x only, with respect to x, you can proceed by integrating each side separately.

This results in:

$$-\frac{1}{y} + c_1 = \frac{1}{2}x^2 + c_2$$

where c_1 and c_2 are the constants from the two integrations. It is simpler to combine these constants into one $c = c_2 - c_1$, giving:

$$-\frac{1}{y} = \frac{1}{2}x^2 + c$$

Tidying up the solution to make y the subject:

$$y = -\frac{1}{\frac{1}{2}x^2 + c}$$

The new constant $k = 2c$, but in practice it doesn't matter what you call it, as long as you keep track of where it fits.

which can be simplified further by writing:

$$y = -\frac{2}{x^2 + k}$$

This is the general solution of the differential equation $\frac{dy}{dx} = xy^2$.

You can use this method of separating the variables on any first order differential equation which can be written in the form:

$$\frac{dy}{dx} = f(x)\,g(y)$$

by rewriting it as:

$$\int \frac{1}{g(y)}\,dy = \int f(x)\,dx$$

The same approach can be used even if the equation is of the form

$$\frac{dy}{dx} = g(y)$$

as in the following example.

Example 9.5

The population, P (in millions), of a country grows so that $\frac{dP}{dt} = \frac{P}{50}$.

(i) Find the general solution of this differential equation.

(ii) Find the particular solution if $P = 100$ when $t = 0$.

Solution

(i) Dividing both sides by P gives:

$$\frac{1}{P}\frac{dP}{dt} = \frac{1}{50}$$

so:

$$\int \frac{1}{P}\,dP = \int \frac{1}{50}\,dt$$

$$\ln\left|P\right| = \frac{t}{50} + c$$

Since the population is always positive, $\left|P\right| = P$.

Rearranging gives:

$$P = e^{\left(\frac{t}{50} + c\right)}$$

which can be written as:

$$P = Ae^{\frac{t}{50}}$$

> This is a very common occurrence, where an additive constant becomes a multiplicative one.

where $A = e^c$.

The general solution is:

$$P = Ae^{\frac{t}{50}}$$

(ii) The particular solution needs $P = 100$ when $t = 0$, so substitute these in:

$$100 = Ae^0$$

$$A = 100$$

so the particular solution is $P = 100e^{\frac{t}{50}}$.

In the previous example the modulus sign that occurs when you integrated $\frac{1}{P}$ did not matter, but the following example shows that it is important not to neglect it.

Example 9.6

A drink is placed in a room where the ambient temperature is $20\,°C$. A model for the subsequent temperature T of the drink at time t, in hours, is given by Newton's law of cooling. This leads to the differential equation

$$\frac{dT}{dt} = -5(T - 20).$$

Find the solution of this differential equation, in the cases where

(i) $T = 80$ when $t = 0$

(ii) $T = 0$ when $t = 0$

Solution

Dividing both sides of the differential equation by $(T - 20)$, and rewriting it as two integrals gives

$$\int \frac{1}{T - 20}\, dT = \int -5\, dt$$

$$\ln|T - 20| = -5t + c$$

$$|T - 20| = e^{(-5t + c)}$$

$$|T - 20| = Ae^{-5t}$$

> This is the general solution of the differential equation, and since you don't know that $T - 20$ is necessarily positive you should not remove the modulus signs.

(i) $T = 80$ when $t = 0$. Substituting these gives:

$$|80 - 20| = A, \quad A = 60$$

> The RHS clearly always remains greater than zero, so the LHS must also never reach zero. Since $T - 20$ starts positive, it must always remain positive, so you can remove the modulus signs.

so:

$$|T - 20| = 60e^{-5t}$$

$T = 20 + 60e^{-5t}$ as the particular solution.

> The drink is cooling (the e^{-5t} term gets smaller as t gets bigger) towards the ambient room temperature of $20\,°C$. See the red curve in Figure 9.8.

(ii) $T = 0$ when $t = 0$. Substituting gives:

$$|0 - 20| = A \Rightarrow A = 20$$

so:

$$|T - 20| = 20e^{-5t}$$

$$-(T - 20) = 20e^{-5t} \quad -T + 20 = 20e^{-5t}$$

so:

$$T = 20 - 20e^{-5t}$$

is the particular solution.

> Again, the RHS is always greater than zero, so the LHS can never equal zero. Since $T - 20$ is negative to start with, it must remain negative. This means the modulus sign will always reverse the sign of the left-hand side and can be replaced with a negative sign.

> The drink is warming up (the e^{-5t} term gets smaller again, but it is being subtracted from the ambient temperature this time) to the ambient room temperature of $20\,°C$. See the blue curve in Figure 9.8.

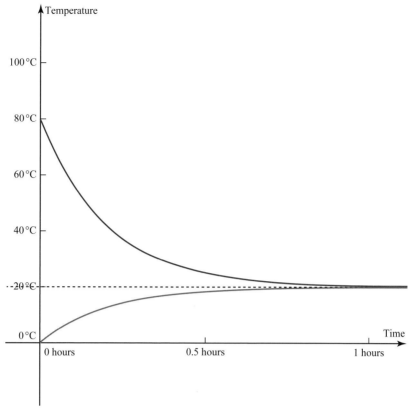

Figure 9.8 The red curve is the drink cooling down, starting at 80 °C, while the blue curve is the drink warming up having started at 0 °C. The ambient temperature can be seen as the horizontal asymptote at 20 °C which both curves approach.

 Historical note

The discovery of calculus is usually attributed to Isaac Newton with the invention of the **method of fluxions** in 1665, when he was 23 years old, although it is also claimed that Gottfried Leibniz discovered it first but did not publicise his work. The study of differential equations followed on naturally. Newton is known to have solved a differential equation in 1676 but he only published details of it in 1693, the same year in which a differential equation featured in Leibniz's work.

The **method of separation of variables** was developed by Johann Bernoulli between 1694 and 1697. He was a famous teacher and wrote the first calculus textbook in 1696; one of his students was Leonhard Euler. Johann Bernoulli was one of the older members of a quite remarkable family of Swiss mathematicians (three of them called Johann) spread over three generations, at least eight of whom may be regarded as famous.

① Which of the following differential equations can be solved by the method of separation of variables? Give the general solution of those which can be solved in this way.

(i) $\dfrac{dy}{dx} = y$ (ii) $\dfrac{dy}{dx} = xy$ (iii) $\dfrac{dy}{dx} = x^3 y$

(iv) $\dfrac{dy}{dx} = 3x^2 e^{-y}$ (v) $\dfrac{dy}{dx} = x + yx$ (vi) $\dfrac{dy}{dx} = x + y^2$

(vii) $\dfrac{dy}{dx} = e^{x+y}$ (viii) $\dfrac{dy}{dx} = \dfrac{y}{x(x-1)}$ (ix) $\dfrac{dy}{dx} = x^2 - y^2$

(x) $\dfrac{dy}{dx} = \dfrac{\sin x}{y^2}$ (xi) $\dfrac{dy}{dx} = \dfrac{y+2}{x-2}$ (xii) $\dfrac{dx}{dt} = x^2 - 8x$

② Find the particular solution of each of the differential equations below for the given conditions.

(i) $x\dfrac{dy}{dx} = y^2$ $y = 10$ when $x = 1$

(ii) $\dfrac{dz}{dt} = \dfrac{z^2}{t}$ $z = 2$ when $t = 1$

(iii) $\dfrac{dy}{dx} = \dfrac{x^2}{y}$ $y = 10$ when $x = 1$

(iv) $\dfrac{dp}{ds} = e^{-p} \sin 2s$ $p = 0$ when $s = 0$

(v) $\dfrac{dy}{dx} = x^2 e^{-y}$ $y = 10$ when $x = 0$

(vi) $\dfrac{dy}{dt} = \dfrac{e^y}{y}$ $y = 2$ when $t = 0$

③ The rate of radioactive decay of a chemical is given by:

$$\dfrac{dm}{dt} = -5m$$

(i) Find the general solution of this equation.

(ii) Given that $m = 10$ when $t = 0$, find the particular solution.

④ A bacterium reproduces such that the rate of increase (in bacteria per minute) of the population is given by:

$$\dfrac{dP}{dt} = 0.7P$$

(i) Find the general solution of this differential equation.

(ii) Find the particular solution if $P = 100$ when $t = 0$.

(iii) How long does it take for the population to double?

⑤ An object is projected horizontally on a surface with an initial speed of $20\,\text{m s}^{-1}$. Its speed is modelled by the differential equation:

$$\dfrac{dv}{dt} = -\dfrac{v^2}{10}$$

(i) Find the particular solution of this differential equation that corresponds to the initial conditions given.

(ii) How long does it take for the speed to drop to 10% of its original value?

⑥ Water evaporates from a conical tank such that the rate of change of the height of water is modelled by the differential equation:

$$\frac{dh}{dt} = -\frac{\pi h^2}{4}$$

(i) Find the general solution of this differential equation.

(ii) Initially the height of water is H. Find, in terms of H, how long it takes for the height of the water to decrease by 10%.

⑦ Water is leaving a tank through a pipe at the bottom, such that change in the height of water is modelled by the differential equation:

$$4\frac{dh}{dt} = -\sqrt{20h}, \text{ where } t \text{ is in minutes and } h \text{ in centimetres.}$$

(i) Find the particular solution of this differential equation given that the initial height is 4 cm.

(ii) How long does the tank take to empty?

⑧ The differential equation below models the motion of a body falling vertically subject to air resistance:

$$\frac{dv}{dt} = 10 - 0.2v, \text{ where } v \text{ is the downward vertical speed of the body in } ms^{-1}$$

and time t is measured in seconds.

(i) Find the general solution of the differential equation.

(ii) Find particular solutions when

(a) the body starts from rest

(b) the body has initial downward vertical speed of $80 \, ms^{-1}$

In each case state the terminal velocity.

⑨ The temperature of a hot body is initially 100 °C. It is placed in a tank of water at a temperature of 20 °C. The rate of change of temperature is modelled by the differential equation $\frac{dT}{dt} = -0.5(T - 20)$, where t is the time in minutes.

(i) Find the particular solution of this differential equation that fits the initial condition given.

(ii) After what time interval has the temperature of the body fallen to 50 °C?

⑩ A small particle moving in a fluid satisfies the differential equation $\frac{dv}{dt} = -0.2(v + v^2)$. Find the particular solution of this differential equation given that $v = 40$ when $t = 0$.

⑪ For the electrical circuit shown below, the current, i amperes, flowing once the switch is closed is given by:

$$L\frac{\mathrm{d}i}{\mathrm{d}t} + Ri = V$$

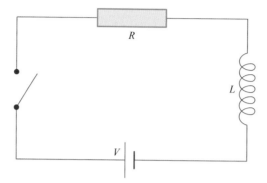

Figure 9.9

L is the inductance, measured in henries.

R is the resistance, measured in ohms.

V is the applied voltage, measured in volts.

(i) If $L = 0.04$, $V = 20$ and $R = 100$, find the general solution of the differential equation.

(ii) If $i = 0$ when $t = 0$, find the particular solution.

(iii) Find the general solution of the original differential equation in terms of R, V and L.

⑫ At 0300 in the morning the police were called to a house where the body of a murder victim had been found. The police doctor arrived at 0345 and took the temperature of the body, which was 34.5 °C. One hour later he took the temperature again and measured it to be 33.9 °C. The temperature of the room was fairly constant at 15.5 °C.

Take the normal body temperature of a human being as 37.0 °C, and use Newton's law of cooling

$$\frac{\mathrm{d}T}{\mathrm{d}t} = -k\left(T - T_{room}\right)$$

as a model to estimate the time of death of the victim.

⑬ A tank contains 2000 litres of salt solution with an initial concentration of $0.3\,\mathrm{kg\,l^{-1}}$. It is necessary to reduce the concentration to $0.2\,\mathrm{kg\,l^{-1}}$. In order to do this, pure water is pumped into the tank at a rate of 8 litres per minute, and the solution is pumped out of the tank at the same rate. The liquid in the tank is stirred so that perfect mixing may be assumed. At time t minutes the mass of salt in the tank is denoted by $M\,\mathrm{kg}$ and the concentration of salt in the solution by $\mathrm{kg\,l^{-1}}$.

(i) By considering the rate of change of mass of salt in the tank show that $\dfrac{\mathrm{d}M}{\mathrm{d}t} = -8C$ and also state the relationship between M and C.

(ii) Hence form a differential equation for C (with respect to time).

(iii) Find the time it will take to reduce the concentration to the required level (to the nearest minute).

3 Integrating factors

When a first order differential equation cannot be written in the form $\frac{dy}{dx} = f(x) g(y)$, it cannot be solved using the method of separation of variables.

If, however, it is a **linear** equation, you can multiply it by a special function called an **integrating factor** which converts it to a form which can be integrated.

Linear equations

Strictly, it is linear in y.

A linear differential equation is one in which the independent variable (y in these examples) only appears to the power of 1. So the differential equation $\frac{dy}{dx} = x^2 - xy$ is linear because the only terms that involve y are $\frac{dy}{dx}$ and $-xy$. There are no terms in y^2, y, $\sin y$, $\left(\frac{dy}{dx} \right)^2$, etc.

Any linear first order differential equation may be written in the form:

$$\frac{dy}{dx} + P(x)y = Q(x)$$

where P and Q are functions of x only.

For example, the equation:

$$\frac{dy}{dx} = x^2 - xy$$

can be rewritten as:

$$\frac{dy}{dx} + xy = x^2$$

This is in the form:

$$\frac{dy}{dx} + Py = Q$$

where the functions P and Q are given by $P = x$ and $Q = x^2$.

ACTIVITY 9.1

Which of the following differential equations

(i) can be written in the form $\frac{dy}{dx} = f(x) g(y)$ (i.e. which can be solved by separating variables)?

(ii) can be written in the form $\frac{dy}{dx} + P(x)y = Q(x)$ (i.e. which are **linear**)? Identify P and Q if so.

(a) $\frac{dy}{dx} = x^2 + x^2 y$

(b) $\frac{dy}{dx} = x^2 - xy$

(c) $\frac{dy}{dx} = x^2 + x + xy + y$

(d) $\frac{dy}{dx} = x + y^2$

Example 9.7

Find the general solution of the differential equation:

$$\cos x \frac{dy}{dx} - y \sin x = x^2$$

Solution

The equation looks forbidding until you notice that the left-hand side is a perfect derivative.

Since $\frac{d}{dx}(\cos x) = -\sin x$, it follows (using the product rule) that:

$$\frac{d}{dx}(y \cos x) = \cos x \frac{dy}{dx} - y \sin x.$$

So you can rewrite the differential equation as:

$$\frac{d}{dx}(y \cos x) = x^2$$

You may now integrate both sides to obtain:

$$y \cos x = \frac{x^3}{3} + c$$

Dividing both sides by $\cos x$, the general solution is:

$$y = \frac{x^3}{3 \cos x} + \frac{c}{\cos x}.$$

In the example above the left-hand side was already a perfect derivative. That is not the case in the next example but it is a simple matter to convert it into one.

Example 9.8

Find the general solution of the differential equation:

$$\frac{dy}{dx} + \frac{2}{x} y = \frac{4}{x^2} \text{ for } x \neq 0$$

Solution

First note that the equation is linear, because it can be written in the form:

$$\frac{dy}{dx} + Py = Q$$

where $P = \frac{2}{x}$ and $Q = \frac{4}{x^2}$.

> There are no terms in y^2, $\frac{1}{y}$, $\sqrt{y}$, etc.

If you now multiply through by x^2, the equation becomes

$$x^2 \frac{dy}{dx} + 2xy = 4.$$

The left-hand side of this equation is now the expression you obtain when you differentiate $x^2 y$ with respect to x, using the product rule:

$$\frac{d}{dx}(x^2 y) = x^2 \frac{dy}{dx} + 2xy$$

So the differential equation can be rewritten as:

$$\frac{d}{dx}(x^2 y) = 4$$

Now integrating both sides with respect to x gives:

$$x^2 y = \int 4 \, \mathrm{d}x = 4x + c$$

The general solution is:

$$y = \frac{4}{x} + \frac{c}{x^2}$$

In the previous two examples the differential equations could be rewritten in the form:

$$\frac{\mathrm{d}}{\mathrm{d}x}(Ry) = \text{function of } x$$

where R was some function of x. In Example 9.7 $R = \cos x$, and in the second, $R = x^2$. Once the differential equation was written in this form, it was a straightforward task to solve it, since all that remained was to integrate the function of x on the right-hand side.

However, in the Example 9.8 we had to multiply each term in the equation by a factor x^2 to bring the left-hand side into the required form. This factor of x^2 is an example of an **integrating factor**; multiplying by it made the left-hand side a perfect derivative.

To see how to calculate the integrating factor from a differential equation in the standard form, think about the general case:

$$\frac{\mathrm{d}y}{\mathrm{d}x} + Py = Q$$

You need a function R to multiply everything by:

$$R\frac{\mathrm{d}y}{\mathrm{d}x} + RPy = RQ$$

so that the LHS can be written as $\frac{\mathrm{d}}{\mathrm{d}x}(Ry)$.

This means you need:

$$\frac{\mathrm{d}}{\mathrm{d}x}(Ry) = R\frac{\mathrm{d}y}{\mathrm{d}x} + RPy$$

Differentiating the LHS (using the product rule and chain rule), remembering that R and y are functions of x) gives:

$$R\frac{\mathrm{d}y}{\mathrm{d}x} + \frac{\mathrm{d}R}{\mathrm{d}x}y = R\frac{\mathrm{d}y}{\mathrm{d}x} + RPy$$

Comparing the two sides, and realising that $y \neq 0$, you should be able to see that this is only true if:

$$\frac{\mathrm{d}R}{\mathrm{d}x} = RP$$

This is a first order differential equation, in R and x, but the variables are separable:

$$\frac{1}{R}\frac{\mathrm{d}R}{\mathrm{d}x} = P$$

$$\ln|R| = \int P \, \mathrm{d}x$$

so

$$R = e^{\int P\,dx}$$

This means that *any* first order linear differential equation written in the standard form can be multiplied by an **integrating factor** $R = e^{\int P\,dx}$ to convert it to the compact form: $\dfrac{d}{dx}(Ry) = RQ$.

You can then solve the equation by integrating the right-hand side, which is a function of x only.

Example 9.9

Solve the differential equation:

$$x\frac{dy}{dx} + 2y = \frac{4}{x} \text{ for } x \neq 0$$

Solution

Dividing through by x gives the standard form:

$$\frac{dy}{dx} + \frac{2y}{x} = \frac{4}{x^2}$$

and with the standard notation $P = \dfrac{2}{x}$ and $Q = \dfrac{4}{x^2}$.

The integrating factor is:

$$R = e^{\int \frac{2}{x}dx} = e^{2\ln|x|+c} = Ae^{\ln|x|^2} = Ax^2$$

> The constant of integration c becomes $A = e^c$

You will multiply the standard form equation by this. But the constant of integration will always become a multiplier, and since it cannot be zero (otherwise R would be zero) you can immediately divide by the constant again, eliminating it. In practice this means you can safely ignore this constant, so:

$$R = x^2$$

> One of the few times you can safely ignore a constant of integration is when calculating the integrating factor!

Multiplying the standard form of the differential equation by R gives:

$$x^2\frac{dy}{dx} + 2yx = 4$$

The left-hand side is now the derivative of a product, and can be written as $\dfrac{d}{dx}(Ry)$. In this case:

$$\frac{d}{dx}(x^2y) = 4$$

$$x^2y = 4x + c$$

Dividing by x^2 gives the general solution:

$$y = \frac{4}{x} + \frac{c}{x^2}$$

This is, of course, the same solution as you obtained in Example 9.8, but this time you didn't need to 'spot' a convenient form for the LHS, but used an explicit method to find an integrating factor.

Check you can follow the method in the following example, which also requests a particular solution to satisfy a condition.

Example 9.10	Find the particular solution of:

$$x^2 \frac{\mathrm{d}y}{\mathrm{d}x} + xy = \frac{2}{x}$$

that satisfies the condition $y = 1$ when $x = 2$.

Solution

Write the equation in the standard form:

$$\frac{\mathrm{d}y}{\mathrm{d}x} + \frac{y}{x} = \frac{2}{x^3}$$

Find the integrating factor R:

$$R = e^{\int \frac{1}{x} \mathrm{d}x} = e^{\ln|x|} = x$$

Multiply through by R:

$$x \frac{\mathrm{d}y}{\mathrm{d}x} + y = \frac{2}{x^2}$$

$$\frac{\mathrm{d}}{\mathrm{d}x}(xy) = \frac{2}{x^2}$$

Integrate with respect to x:

$$xy = \int \frac{2}{x^2} \mathrm{d}x$$

$$xy = -\frac{2}{x} + c$$

$$y = -\frac{2}{x^2} + \frac{c}{x}$$

which is the general solution.

The condition states that $y = 1$ when $x = 2$, so we need:

$$1 = -\frac{2}{2^2} + \frac{c}{2}$$

$$c = 3$$

Therefore, the particular solution is:

$$y = \frac{3}{x} - \frac{2}{x^2}$$

Exercise 9.3

① Find the integrating factor for each of the following differential equations:

(i) $\dfrac{dy}{dx} + x^2 y = x$

(ii) $\dfrac{dy}{dx} + y\sin(x) = x$

(iii) $4x\dfrac{dy}{dx} - y = x^2$

(iv) $x^2\dfrac{dy}{dx} + xy = 2$

(v) $\dfrac{dy}{dx} + 7y = 1$

(vi) $\cos(x)\dfrac{dy}{dx} + y\sin(x) = e^{-2x}$

② Consider the differential equation

$x^2\dfrac{dy}{dx} + xy = 1$, with condition $y = 0$ when $x = 1$.

(i) Rewrite the differential equation into the form $\dfrac{dy}{dx} + P(x)\, y = Q(x)$.

(ii) Find the integrating factor by calculating $e^{\int P\,dx}$.

(iii) Multiply your answer from part (i) by the integrating factor from part (ii).

(iv) Rewrite your answer from part (iii) in the form $\dfrac{d}{dx}(Ry) = RQ$.

(v) Integrate both sides with respect to x.

(vi) Rearrange your answer from part (v) to give the general solution.

(vii) Substitute the condition into the general solution to find the value of the constant.

(viii) Write down the particular solution.

③ Find the particular solution of each of the following differential equations:

(i) $x\dfrac{dy}{dx} + 2y = x^2$ $\qquad y = 0$ when $x = 1$

(ii) $\dfrac{dy}{dx} + xy = 4x$ $\qquad y = 2$ when $x = 0$

(iii) $6xy + \dfrac{dy}{dx} = 0$ $\qquad y = 3$ when $x = 1$

(iv) $\dfrac{dx}{dt} - 2tx = t$ $\qquad x = 1$ when $t = 0$

(v) $\dfrac{dy}{dx} - \dfrac{3}{x}y = x$ $\qquad y = 0$ when $x = 1$

(vi) $\dfrac{dv}{dt} + 3t^2 v = t^2$ $\qquad v = -1$ when $t = 0$

④ An object falling vertically experiences air resistance so that the velocity satisfies the differential equation:

$\dfrac{dv}{dt} = 10 - 0.4v$

(i) Use the integrating factor method to find the general solution of this differential equation.

(ii) Find the particular solution if, initially, $v = 0$.

(iii) Solve the original differential equation, with the condition, using the method of separation of variables.

(iv) Compare your solutions and comment on which method you would prefer to use, assuming that both were available (i.e. it is a first order, linear, differential equation).

⑤ A parachutist has a terminal speed of $30\,\text{m}\,\text{s}^{-1}$. The magnitude of the air resistance acting when the parachute is open is modelled by $F = kmv$ newtons, where k is a constant, v is the speed and m is the mass of the parachutist.

(i) By considering the forces acting at terminal velocity, find the value for k, taking the acceleration due to gravity to be $10\,\text{m}\,\text{s}^{-2}$.

(ii) Formulate a first order differential equation for the velocity, v, at a given time t, after the parachute opens.

(iii) Use the integrating factor method to find the general solution of the differential equation.

(iv) Find the particular solution if the parachutist is moving at $60\,\text{m}\,\text{s}^{-1}$ when the parachute opens.

⑥ A solution is sought to the differential equation:

$$\frac{dy}{dx} + \frac{y}{x} = \cos x \qquad (x > 0)$$

(i) Find the general solution for y in terms of x.

(ii) Given that $y = 0$, when $x = \pi$, find the particular solution.

(iii) Write down the function which approximates the solution as x gets very large.

(iv) State the behaviour of y as $x \to 0$ and sketch the shape of graph of y against x, focusing on the behaviour at large and small x.

⑦ The radioactive isotope uranium-238 decays into thorium-234, which in turn decays into protactinium-234. This can be summarised as

$${}^{238}_{92}\text{U} \xrightarrow{k_1} {}^{234}_{90}\text{Th} \xrightarrow{k_2} {}^{234}_{91}\text{Pa}$$ where k_1 and k_2 are reaction constants (i.e. the constants of proportionality by which the rates of decay occur).

The amounts of U-238, Th-234 and Pa-234 at time t are denoted by x, y and z, respectively.

You may assume that the rate of decay of an isotope is proportional to the amount present, and that the constant of proportionality is the relevant k-value.

An experiment begins with an amount a of U-238, but no Th-234 or Pa-234. The amount y of Th-234 present at time t satisfies the differential equation

$$\frac{dy}{dt} + k_2 y = k_1 a e^{-k_1 t}.$$

(i) Find the integrating factor for the differential equation, and hence its general solution.

(ii) Find the particular solution that satisfies the initial conditions.

(iii) Write down a differential equation for the variable x.

(iv) Solve the model you suggested in part (iii) and explain how its particular solution has been incorporated into the suggested differential equation for y.

(v) Find the particular solution of the differential equation for y, in the case where $k_1 = k_2 = k$.

(vi) Write down a differential equation for the variable z (still assuming $k_1 = k_2 = k$). By substituting in the solution you found in part (v) for y, solve this differential equation to find a particular solution for the variable z at any time t.

⑧ The differential equation

$$(1-x)\frac{dy}{dx} + \frac{2}{1+x}y = (1-x)^2 \text{ is to be solved for } |x| < 1.$$

(i) Solve the differential equation.

(ii) Find particular solutions when

(a) $y = 0$ when $x = 0$

(b) $y = 1$ when $x = 0$

in each case stating the behaviour of the solution as $x \to -1$.

(iii) Find a particular solution which tends to a finite limit as $x \to -1$. Sketch the graph of this solution.

⑨ The function $y = f(x)$ satisfies the differential equation

$$x\frac{dy}{dx} + 2y = \sqrt{1+x^2}.$$

(i) Using the integrating factor method, or otherwise, find the general solution of this differential equation.

(ii) Find the particular solution which satisfies the condition that $y = 1$ when $x = 1$. How does y behave when x becomes very small?

(iii) Write down the first three non-zero terms in the expansion of $\left(1 + x^2\right)^{\frac{3}{2}}$ in ascending powers of x.

(iv) Using the expansion in part (iii) and the general solution found in part (i), write down the power series expansion of the general solution y for small values of x up to and including the term in x^2. Hence find the particular solution of the differential equation which crosses the y-axis from the region $x > 0$ into the region $x < 0$.

LEARNING OUTCOMES

When you have completed this chapter you should be able to:

➤ formulate a differential equation from verbal descriptions involving rates of change

➤ use differential equations in modelling

➤ know the difference between a general solution and a particular solution

➤ find a particular solution to a differential equation by using initial or boundary conditions

➤ recognise differential equations which can be solved by separation of variables

➤ use separation of variables to solve a first order separable differential equation, to find both general and particular solutions

➤ recognise differential equations which can be solved using an integrating factor

➤ find an integrating factor and use it to solve a first order linear differential equation, to find both general and particular solutions.

KEY POINTS

1 A differential equation is an equation involving derivatives such as $\dfrac{dy}{dx}$, $\dfrac{dV}{dt}$ or $\dfrac{d^2x}{dt^2}$.

2 The order of a differential equation is the order of its highest derivative.

3 Differential equations are used to model situations which involve rates of change.

4 The solution of a differential equation gives the relationship between the variables themselves, not their derivatives.

5 The general solution of a first order differential equation satisfies the differential equation and has one constant of integration in the solution.

6 A particular solution of a differential equation is one in which additional information has been used to calculate the constant of integration.

7 The general solution may be represented by a family of curves and a particular solution is one member of that family.

8 The method of separation of variables can be used to solve first order differential equations of the form $\dfrac{dy}{dx} = f(x)g(y)$. Separating the variables gives $\displaystyle\int \dfrac{1}{g(y)}\,dy = \int f(x)\,dx$.

9 Any first order linear differential equation can be written in the form:

$\dfrac{dy}{dx} + Py = Q$ where P and Q are functions of x only.

10 To solve the equation $\dfrac{dy}{dx} + Py = Q$ you multiply throughout by the integrating factor $e^{\int P\,dx}$.

10 Complex numbers

Review: The modulus and argument of a complex number

Figure 10.1 shows the point representing $z = x + y\mathrm{i}$ on an Argand diagram.

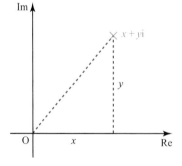

Prior knowledge

You need to be able to work with complex numbers and be familiar with using an Argand diagram to represent complex numbers. This is covered in Review: Complex numbers

Figure 10.1

The distance of this point from the origin is $\sqrt{x^2 + y^2}$. ← Using Pythagoras' theorem.

This distance is called the **modulus** of z, and is denoted by $|z|$.

So, for the complex number $z = x + y\mathrm{i}$, $|z| = \sqrt{x^2 + y^2}$.

Notice that since $zz^* = (x + \mathrm{i}y)(x - \mathrm{i}y) = x^2 + y^2$, then $|z|^2 = zz^*$.

Figure 10.2 shows the complex number z on an Argand diagram. The length r represents the modulus of the complex number and is denoted $|z|$. The angle θ is called the **argument** of the complex number.

The argument of $z = 0$ is undefined.

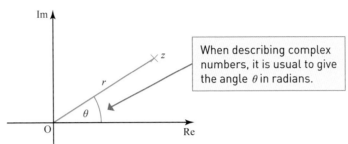

When describing complex numbers, it is usual to give the angle θ in radians.

Figure 10.2

The argument is measured anticlockwise from the positive real axis.

This angle is not uniquely defined since adding any multiple of 2π to θ gives the same direction. To avoid confusion, it is usual to choose that value of θ for which $-\pi < \theta \leq \pi$. This is called the **principal argument** of z and is denoted by arg z. Every complex number except zero has a unique principal argument.

Always draw a diagram when finding the argument of a complex number. This tells you which quadrant the complex number lies in.

In Figure 10.3, you can see the relationship between the components of a complex number and its modulus and argument.

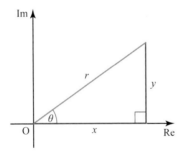

Discussion point

→ Give one similarity and one difference between modulus-argument form and polar coordinates.

Figure 10.3

Using trigonometry, you can see that $\sin\theta = \dfrac{y}{r}$ and so $y = r\sin\theta$.

Similarly, $\cos\theta = \dfrac{x}{r}$ so $x = r\cos\theta$.

Therefore, the complex number $z = x + y\mathrm{i}$ can be written

$$z = r\cos\theta + r\sin\theta\,\mathrm{i}$$

or $\quad z = r(\cos\theta + \mathrm{i}\sin\theta)$

This is called the **modulus–argument** or polar form of a complex number and is denoted (r, θ).

Example 10.1

For each of the following:

(a)   Show the complex number on Argand diagram.

(b)   Write the complex number in modulus-argument form.

(i) $z_1 = 3\mathrm{i}$ (ii) $z_2 = \sqrt{3} + \mathrm{i}$

(iii) $z_3 = -3 - 3\mathrm{i}$ (iv) $z_4 = -1 + \sqrt{3}\mathrm{i}$

Solution

(i) (a)

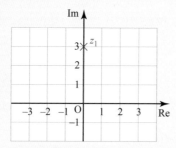

Figure 10.4

(b) z_1 has modulus 3 and argument $\frac{\pi}{2}$ so $z_1 = 3\left(\cos\frac{\pi}{2} + i\sin\frac{\pi}{2}\right)$

(ii) (a)

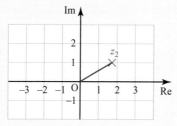

Figure 10.5

z_2 has modulus $\sqrt{\left(\sqrt{3}\right)^2 + 1^2} = 2$ and argument $\tan^{-1}\left(\frac{1}{\sqrt{3}}\right) = \frac{\pi}{6}$
so

(b) $z_2 = 2\left(\cos\frac{\pi}{6} + i\sin\frac{\pi}{6}\right)$.

(iii) (a)

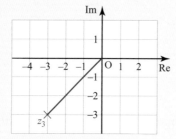

Figure 10.6

z_3 has modulus $\sqrt{3^2 + 3^2} = 3\sqrt{2}$

$\tan^{-1}\left(\frac{3}{3}\right) = \frac{\pi}{4}$ so the argument of z_3 is $-\frac{3\pi}{4}$

(b) $z_3 = 3\sqrt{2}\left(\cos\left(-\frac{3\pi}{4}\right) + i\sin\left(-\frac{3\pi}{4}\right)\right)$

(iv) (a)

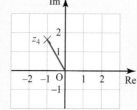

Figure 10.7

$$z_4 \text{ has modulus } \sqrt{(-1)^2 + \left(\sqrt{3}\right)^2} = 2$$

$$\tan^{-1}\left(\frac{\sqrt{3}}{1}\right) = \frac{\pi}{3} \text{ so the argument of } z_4 \text{ is } \frac{2\pi}{3}.$$

(b) $z_4 = 2\left(\cos\dfrac{2\pi}{3} + \mathrm{i}\sin\dfrac{2\pi}{3}\right)$

Multiplying and dividing complex numbers in modulus–argument form

To multiply complex numbers in modulus-argument form, you *multiply* their moduli and *add* their arguments.

$$\left|z_1 z_2\right| = \left|z_1\right|\left|z_2\right|$$

> You may need to add or subtract 2π to give the principal argument. For example, if $\arg(z_1) + \arg(z_2) = \dfrac{7\pi}{3}$ then $\arg(z_1 z_2) = \dfrac{\pi}{3}$.

$$\arg(z_1 z_2) = \arg z_1 + \arg z_2$$

You can prove these results using the compound angle formulae.

To divide complex numbers in modulus-argument form, you *divide* their moduli and *subtract* their arguments.

$$\left|\frac{z_1}{z_2}\right| = \frac{\left|z_1\right|}{\left|z_2\right|}$$

$$\arg\left(\frac{z_1}{z_2}\right) = \arg z_1 - \arg z_2$$

You can prove this easily from the multiplication results by letting $\dfrac{z_1}{z_2} = w$, so that $z_1 = w z_2$.

Example 10.2

The complex numbers w and z are given by $w = 3\left(\cos\dfrac{\pi}{3} + \mathrm{i}\sin\dfrac{\pi}{3}\right)$ and $z = 6\left(\cos\left(-\dfrac{\pi}{6}\right) + \mathrm{i}\sin\left(-\dfrac{\pi}{6}\right)\right)$.

Find (i) wz and (ii) $\dfrac{w}{z}$ in modulus-argument form. Illustrate each of these on a separate Argand diagram.

Solution

$$|w| = 3 \qquad \arg w = \frac{\pi}{3}$$

$$|z| = 6 \qquad \arg z = -\frac{\pi}{6}$$

(i) $\quad |wz| = |w||z| = 3 \times 6 = 18$

$$\arg wz = \arg w + \arg z = \frac{\pi}{3} + \left(-\frac{\pi}{6}\right) = \frac{\pi}{6}$$

So $wz = 18\left(\cos\dfrac{\pi}{6} + i\sin\dfrac{\pi}{6}\right).$

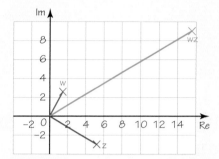

Figure 10.8

(ii) $\left|\dfrac{w}{z}\right| = \dfrac{|w|}{|z|} = \dfrac{1}{2}$

$\arg\dfrac{w}{z} = \arg w - \arg z = \dfrac{\pi}{3} - \left(-\dfrac{\pi}{6}\right) = \dfrac{\pi}{2}$

$\dfrac{w}{z} = \dfrac{1}{2}\left(\cos\dfrac{\pi}{2} + i\sin\dfrac{\pi}{2}\right)$

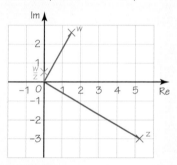

Figure 10.9

The effect of multiplication by a complex number in an Argand diagram

Much of the geometrical power of complex numbers comes from the result about multiplication of complex numbers in polar form: 'multiply the moduli, add the arguments'.

ACTIVITY 10.1

(i) Write the numbers i and −2 in modulus-argument form.

(ii) Using the result about multiplication of complex numbers in modulus-argument form investigate:

    (a)   multiplication of a complex number z by i

    (b)   multiplication of a complex number z by −2

You will have found in Activity 10.1 that multiplication of complex numbers in modulus-argument form gives rise to a simple geometrical interpretation of multiplication.

This combination of an enlargement followed by a rotation is called a **spiral dilation.**

To obtain the line representing z_1z_2 enlarge the line representing z_2 by the scale factor $|z_1|$ and rotate it through $\arg z_1$ anticlockwise about O (see Figure 10.10).

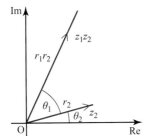

Figure 10.10

In modulus-argument form, you can say that multiplication by $r(\cos\theta + i\sin\theta)$ corresponds to enlargement with scale factor r with (anticlockwise) rotation through θ about the origin.

| Example 10.3 | Explain the geometrical effect of multiplying a complex number z by: |

(i) $-3i$

(ii) $-3\sqrt{3} + 3i$

Solution

(i) $-3i$ has modulus 3 and argument $-\dfrac{\pi}{2}$.

Multiplying by $-3i$ would enlarge the vector z by scale factor 3 and rotate it through $\dfrac{\pi}{2}$ radians clockwise (or $\dfrac{3\pi}{2}$ radians anticlockwise).

(ii) $-3\sqrt{3} + 3i$ has modulus 6 and argument $\dfrac{5\pi}{6}$.

Multiplying by $-3\sqrt{3} + 3i$ would enlarge the vector z by scale factor 6 and rotate it through $\dfrac{5\pi}{6}$ radians anticlockwise.

Loci in the Argand diagram

The complex number $z_2 - z_1$ can be represented on an Argand diagram by the vector from the point representing z_1 to the point representing z_2, as shown in Figure 10.11.

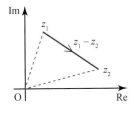

Figure 10.11

This means that:

■   **a locus of the form $|z - a| = r$ is a circle, centre a and radius r.**
    For example, the locus $|z - 3i| = 4$ represents a circle, centre 3i, radius 4 (see Figure 10.12).

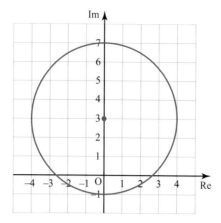

Figure 10.12

■ **a locus of the form arg(*z* − *a*) = *θ* is a half line of points from the point *a* and with angle *θ* from the positive horizontal axis.**

For example, the locus $\arg(z - 4) = \dfrac{\pi}{4}$ is the half line of points shown in Figure 10.13.

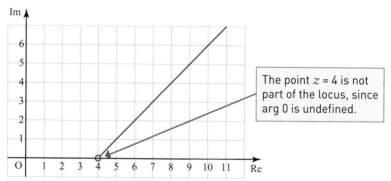

The point $z = 4$ is not part of the locus, since arg 0 is undefined.

Figure 10.13

■ **a locus of the form |*z* − *a*| = |*z* − *b*| represents the locus of all points which lie on the perpendicular bisector between the points represented by the complex numbers *a* and *b*.**

For example, the locus $|z - (2 + 3i)| = |z - (1 - i)|$ is the perpendicular bisector shown in Figure 10.14.

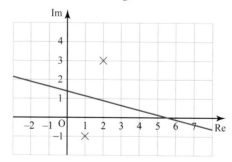

Figure 10.14

Example 10.4

Draw Argand diagrams showing the following sets of points z for which:

(i) $|z - 3i| \leqslant 3$

(ii) $0 < \arg(z - (2 - 3i)) < -\dfrac{2\pi}{3}$

(iii) $|z - 4| < |z + 4i|$ and $|z - (4 - 4i)| \leqslant 4$

Solution

(i)

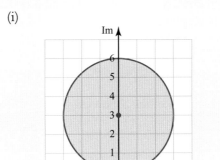

Figure 10.15

> $|z - 3i| = 3$ is a circle centre 3i, radius 3. The points for which $|z - 3i| \leqslant 3$ are the points inside the circle.

> The circle is shown as a solid line to indicate that it is included as part of the locus.

(ii)

> $\arg(z - (2 - 3i)) = 0$ and $\arg(z - (2 - 3i)) = -\dfrac{2\pi}{3}$ are represented by the two half lines. The points for which $0 < \arg(z - (2 - 3i)) < -\dfrac{2\pi}{3}$ lie in the shaded region between the two half lines.

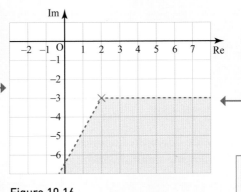

Figure 10.16

> The lines are shown dotted to indicate that they are not included as part of the locus

(iii)

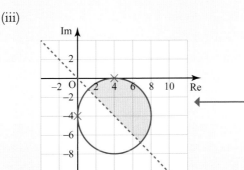

Figure 10.17

> The locus $|z - 4| = |z + 4i|$ is represented by the perpendicular bisector of line segment joining the points $z = 4$ and $z = -4i$. The locus $|z - (4 - 4i)| = 4$ is represented by the circle pictured.
>
> So the locus where both $|z - 4| < |z + 4i|$ and $|z - (4 - 4i)| \leqslant 4$ lies inside the circle and above the perpendicular bisector.

Review Exercise

① Write each of the following complex numbers in the form $x + y\text{i}$, giving surds in your answer where appropriate.

   (i) $6\left(\cos\left(-\dfrac{\pi}{4}\right) + \text{i}\sin\left(-\dfrac{\pi}{4}\right)\right)$
       (ii) $3\left(\cos\left(-\dfrac{\pi}{6}\right) + \text{i}\sin\left(-\dfrac{\pi}{6}\right)\right)$

   (iii) $2\left(\cos\dfrac{3\pi}{4} + \text{i}\sin\dfrac{3\pi}{4}\right)$
       (iv) $7\left(\cos\left(-\dfrac{5\pi}{6}\right) + \text{i}\sin\left(-\dfrac{5\pi}{6}\right)\right)$

② For each complex number, find the modulus and principal argument, and hence write the complex number in modulus-argument form. Give the argument in radians, as a multiple of π.

   (i) $2\sqrt{3} + 2\text{i}$
       (ii) $-2\sqrt{3} + 2\text{i}$

   (iii) $2\sqrt{3} - 2\text{i}$
       (iv) $-2\sqrt{3} - 2\text{i}$

③ Represent each of the following complex numbers on a separate Argand diagram, and write it in the form $x + y\text{i}$, giving surds in your answer where appropriate.

   (i) $|z| = 3$, $\arg z = \dfrac{\pi}{4}$
       (ii) $|z| = 5$, $\arg z = \dfrac{\pi}{2}$

   (iii) $|z| = 4$, $\arg z = -\dfrac{5\pi}{6}$
       (iv) $|z| = 6$, $\arg z = -\dfrac{\pi}{4}$

④ Given that $z = 5\left(\cos\left(-\dfrac{\pi}{3}\right) + \text{i}\sin\left(-\dfrac{\pi}{3}\right)\right)$ and $w = 3\left(\cos\dfrac{5\pi}{6} + \text{i}\sin\dfrac{5\pi}{6}\right)$, find the following complex numbers in modulus-argument form:

   (i) wz
       (ii) $\dfrac{w}{z}$
       (iii) $\dfrac{z}{w}$
       (iv) $\dfrac{1}{z}$

⑤ Explain the geometrical effect of multiplying a complex number z by:

   (i) $5 - 5\text{i}$
       (ii) $-\dfrac{1}{4} - \dfrac{1}{4}\sqrt{3}\text{i}$

⑥ Write down the complex number w such that the product wz represents the following transformations of z:

   (i)   an enlargement by scale factor 2 and a rotation of $\dfrac{\pi}{3}$ radians anticlockwise

   (ii)   an enlargement by scale factor $\dfrac{1}{3}$ and a rotation of $\dfrac{2\pi}{3}$ radians clockwise

⑦ For each of the parts (i) to (iii), draw an Argand diagram showing the set of points z for which the given condition is true.

   (i) $|z - 5\text{i}| \geqslant 3$
       (ii) $|z - 5| < 2$

   (iii) $|z + \sqrt{5} + \sqrt{5}\text{i}| \leqslant 5$

⑧ For each of the parts (i) to (iii), draw an Argand diagram showing the set of points z for which the given condition is true.

   (i) $\arg z = \dfrac{\pi}{2}$
       (ii) $0 < \arg(z + 2 + \text{i}) < \dfrac{\pi}{3}$

   (iii) $-\dfrac{2\pi}{3} \leqslant \arg(z + 2 + \text{i}) \leqslant \dfrac{2\pi}{3}$

⑨ For each of the parts (i) to (iii), draw an Argand diagram showing the set of points z for which the given condition is true.

   (i) $|z - 2| = |z - 4\text{i}|$
       (ii) $|z - 1 - \text{i}| < |z - 3 + 4\text{i}|$

   (iii) $|z + 3 + 2\text{i}| \geqslant |z + 3\text{i}|$

⑩ Write down, in terms of z, the loci for the regions that are represented in each of the Argand diagrams pictured below:

(i)

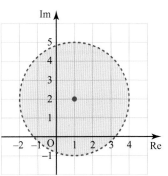

Figure 10.18

(ii)

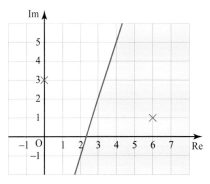

Figure 10.19

(iii)

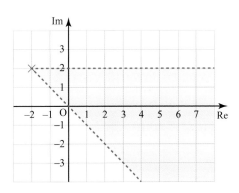

Figure 10.20

⑪ Sketch on the same Argand diagram:

(i)   The locus of points $|z - 4\mathrm{i}| = 1$

(ii) The locus of points $\arg(z - 4\mathrm{i}) = -\dfrac{\pi}{4}$

(iii) The locus of points $\arg(z - 4\mathrm{i}) = -\dfrac{\pi}{2}$

(iv) The locus of points $|z - 4\mathrm{i}| = |z|$. Shade the region defined by the inequalities $|z - 4\mathrm{i}| \geqslant 1$, $-\dfrac{\pi}{2} \leqslant \arg(z - 4\mathrm{i}) \leqslant -\dfrac{\pi}{4}$ and $|z - 4\mathrm{i}| < |z|$.

⑫ The complex number z is multiplied by the complex number $w*w$, where $w = a + b\mathrm{i}$ and $a, b \in \mathbb{Z}$. Find the possible values of a and b if the geometrical effect of the multiplication is to enlarge the vector z by a scale factor of 13.

⑬ (i) Given the point representing a complex number z on an Argand diagram, explain how to find the following points geometrically:

   (a) $3z$ (b) $2\mathrm{i}z$ (c) $(3 + 2\mathrm{i})z$

   (ii) Sketch an Argand diagram to represent the points O, $3z$, $2\mathrm{i}z$ and $(3 + 2\mathrm{i})z$ and explain the geometrical connection between the points.

1 De Moivre's theorem

On page 205 you saw that when you multiply two complex numbers in modulus–argument form you multiply their moduli and add their arguments.

ACTIVITY 10.2

For the complex number $z = 2(\cos 0.1 + \mathrm{i} \sin 0.1)$ write down z^2, z^3, z^4 and z^5.

Use your answers to write down an expression for z^n.

For the general complex number $z = r(\cos\theta + \mathrm{i}\sin\theta)$ write down an expression for z^n.

The product of:

$$z_1 = r_1\left(\cos\theta_1 + \mathrm{i}\sin\theta_1\right) \text{ and } z_2 = r_2\left(\cos\theta_2 + \mathrm{i}\sin\theta_2\right)$$

is: $\quad z_1 z_2 = r_1 r_2 \left(\cos\left(\theta_1 + \theta_2\right) + \mathrm{i}\sin\left(\theta_1 + \theta_2\right)\right)$

Using this result repeatedly with a single complex number z with modulus 1 allows you to concentrate on what happens to the argument.

If $z = \cos\theta + \mathrm{i}\sin\theta$

then $z^2 = \cos\left(\theta + \theta\right) + \mathrm{i}\sin\left(\theta + \theta\right) = \cos 2\theta + \mathrm{i}\sin 2\theta$

$z^3 = z^2 z = \cos\left(2\theta + \theta\right) + \mathrm{i}\sin\left(2\theta + \theta\right) = \cos 3\theta + \mathrm{i}\sin 3\theta$

and so on.

The diagram below shows $(\cos\theta + \mathrm{i}\sin\theta)^n$ for $n = 1, 2, 3, \ldots, 6$.

In this example $0 < \theta < \frac{\pi}{2}$.

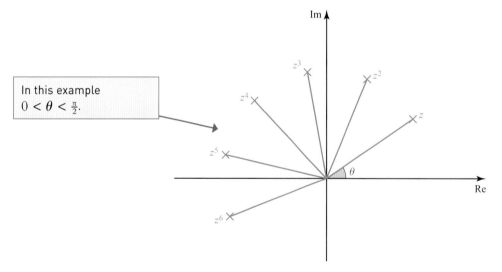

Figure 10.21

This suggests the following result which is called **de Moivre's theorem**.

If n is an integer then $(\cos\theta + i\sin\theta)^n = \cos n\theta + i\sin n\theta$

The proof of this result is in three parts, in which n is **(i)** positive **(ii)** zero **(iii)** negative.

(i) When n is a positive integer de Moivre's theorem can be proved by induction.

The theorem is obviously true if $n = 1$.

Assume the result is true for $n = k$, so

$$\left(\cos\theta + i\sin\theta\right)^k = \cos k\theta + i\sin k\theta$$

You want to prove that the result is true for $n = k + 1$ (if the assumption is true).

$$\left(\cos\theta + i\sin\theta\right)^{k+1} = (\cos k\theta + i\sin k\theta)(\cos\theta + i\sin\theta)$$
$$= \cos(k\theta + \theta) + i\sin(k\theta + \theta)$$
$$= \cos((k+1)\theta) + i\sin((k+1)\theta)$$

If the result is true for $n = k$, then it is true for $n = k + 1$. Since it is true for $n = 1$, it is true for all positive integer values of n.

(ii) By definition, $z^0 = 1$ for all complex numbers $z \neq 0$.

Therefore $\left(\cos\theta + i\sin\theta\right)^0 = 1 = \cos 0 + i\sin 0$.

(iii) For negative n the proof starts with the case $n = -1$.

As $(\cos\theta + i\sin\theta)(\cos(-\theta) + i\sin(-\theta)) = \cos(\theta - \theta) + i\sin(\theta - \theta) = 1$ it follows that

> If $a \times b = 1$ it follows that b is the reciprocal of a.

$$\left(\cos\theta + i\sin\theta\right)^{-1} = \cos(-\theta) + i\sin(-\theta). \quad †$$

If n is another negative integer, let $n = -m$ where m is a positive integer.

Then $\left(\cos\theta + i\sin\theta\right)^n = \left(\cos\theta + i\sin\theta\right)^{-m}$

> As m is a positive integer de Moivre's theorem holds using part (i).

$$= \left[\left(\cos\theta + i\sin\theta\right)^m\right]^{-1}$$
$$= \left[\cos m\theta + i\sin m\theta\right]^{-1}$$

> Using † with $m\theta$ in place of θ.

$$= \cos(-m\theta) + i\sin(-m\theta)$$
$$= \cos n\theta + i\sin n\theta$$

Therefore de Moivre's theorem holds for all integers n.

De Moivre's theorem can also be used for simplifying powers of complex numbers when the modulus is not 1, as in Figure 10.22. If $z = r\left(\cos\theta + i\sin\theta\right)$ then

$$z^n = \left[r(\cos\theta + i\sin\theta)\right]^n = r^n\left(\cos n\theta + i\sin n\theta\right)$$

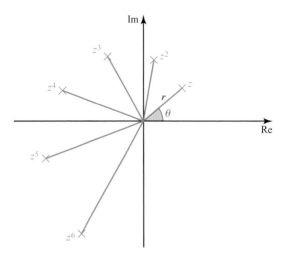

Figure 10.22

Example 10.5

Use de Moivre's theorem to simplify each of the following.

(i) $\left(\cos\dfrac{\pi}{12} + i\sin\dfrac{\pi}{12}\right)^{18}$

(ii) $\left(-1 + \sqrt{3}i\right)^{5}$

Solution

(i) $\left(\cos\dfrac{\pi}{12} + i\sin\dfrac{\pi}{12}\right)^{18} = \cos\left(18 \times \dfrac{\pi}{12}\right) + i\sin\left(18 \times \dfrac{\pi}{12}\right)$

$$= \cos\dfrac{3\pi}{2} + i\sin\dfrac{3\pi}{2}$$

$$= -i$$

(ii) First convert to modulus–argument form:

$$z = -1 + \sqrt{3}i \Rightarrow |z| = \sqrt{(-1)^2 + \left(\sqrt{3}\right)^2} = 2$$

$$\arctan\left(\dfrac{\sqrt{3}}{1}\right) = \dfrac{\pi}{3} \text{ so } \arg z = \dfrac{2\pi}{3}. \quad \longleftarrow \boxed{z \text{ is in the second quadrant.}}$$

So $\left(-1 + \sqrt{3}i\right)^{5} = 2^{5}\left(\cos\dfrac{2\pi}{3} + i\sin\dfrac{2\pi}{3}\right)^{5} = 2^{5}\left(\cos\dfrac{10\pi}{3} + i\sin\dfrac{10\pi}{3}\right)$

$$= 32\left(-\dfrac{1}{2} - \dfrac{\sqrt{3}}{2}i\right)$$

$$= -16 - 16\sqrt{3}i$$

Example 10.6

Simplify the expression $\dfrac{\left[4\left(\cos 5\theta + i\sin 5\theta\right)\right]^{5}}{\left[3\left(\cos 4\theta + i\sin 4\theta\right)\right]^{4}}.$

Use de Moivre's theorem with $\theta = -\phi$ and the facts:

$$\cos(-\theta) = \cos\theta$$
$$\sin(-\theta) = -\sin\theta$$

to show that:

$$\left(\cos\phi - i\sin\phi\right)^n = \cos n\phi - i\sin n\phi$$

Solution

$$\frac{\left[4\left(\cos5\theta + i\sin5\theta\right)\right]^5}{\left[3\left(\cos4\theta + i\sin4\theta\right)\right]^4} = \frac{1024\left(\cos25\theta + i\sin25\theta\right)}{81\left(\cos16\theta + i\sin16\theta\right)}$$

$$= \frac{1024}{81}\left(\cos9\theta + i\sin9\theta\right)$$

In Activity 10.3 you will prove the useful result:

$$\left(\cos\phi - i\sin\phi\right)^n = \cos n\phi - i\sin n\phi$$

Exercise 10.1

① Use de Moivre's theorem to simplify each of the following:
  (a) in the form $\cos\alpha + i\sin\alpha$
  (b) in the form $a + ib$

  (i) $\left(\cos\dfrac{\pi}{6} + i\sin\dfrac{\pi}{6}\right)^4$ (ii) $\left(\cos\dfrac{\pi}{3} + i\sin\dfrac{\pi}{3}\right)^{-8}$

  (iii) $\left(\cos\left(-\dfrac{\pi}{12}\right) + i\sin\left(-\dfrac{\pi}{12}\right)\right)^{10}$

② Given that $w = \cos\dfrac{\pi}{4} + i\sin\dfrac{\pi}{4}$, write each of the following complex numbers as a power of w:

  (i) $z_1 = \cos\dfrac{3\pi}{4} + i\sin\dfrac{3\pi}{4}$ (ii) $z_2 = \cos\dfrac{\pi}{2} + i\sin\dfrac{\pi}{2}$

  (iii) $z_3 = \cos\pi + i\sin\pi$

  Illustrate w, z_1, z_2 and z_3 on an Argand diagram.

③ Use de Moivre's theorem to simplify each of the following:

  (i) $\dfrac{\left(\cos3\theta + i\sin3\theta\right)^4}{\left(\cos5\theta + i\sin5\theta\right)^3}$

  (ii) $\dfrac{\left(\cos\dfrac{\pi}{4} + i\sin\dfrac{\pi}{4}\right)^3}{\left(\cos\dfrac{\pi}{6} + i\sin\dfrac{\pi}{6}\right)^2}$

  (ii) $\left(\cos\dfrac{\pi}{3} + i\sin\dfrac{\pi}{3}\right)^5\left(\cos\dfrac{\pi}{6} + i\sin\dfrac{\pi}{6}\right)^{-4}$

④ Given that $w = \cos\dfrac{\pi}{6} + i\sin\dfrac{\pi}{6}$, write each of the following complex numbers as a power of w.

  (i) $z_1 = \cos\dfrac{\pi}{6} - i\sin\dfrac{\pi}{6}$ (ii) $z_2 = \cos\dfrac{\pi}{2} - i\sin\dfrac{\pi}{2}$

  Illustrate w, z_1 and z_2 on an Argand diagram.

⑤ By converting to modulus–argument form and using de Moivre's theorem, find the following in the form $x + y$i giving x and y as exact expressions or correct to 3 significant figures.

(i) $\left(1 - \sqrt{3}i\right)^4$ (ii) $(-2 + 2i)^7$ (iii) $\left(\sqrt{27} + 3i\right)^6$

⑥ Without using a calculator write $\left(-\sqrt{3} - i\right)^7$ in the form $x + y$i where x and y are exact values.

⑦ Simplify the following expressions as far as possible.

(i) $\Big[3(\cos 2\theta + i\sin 2\theta)\Big]^4$ (ii) $\Big[i\left(\cos 3\theta + i\sin 3\theta\right)\Big]^5$

(iii) $\Big[2i\left(\cos 7\theta + i\sin 7\theta\right)\Big]^{-3}$

⑧ Show that

$$\frac{\Big[3(\cos 2\theta - i\sin 2\theta)\Big]^4 \Big[2(\cos\theta + i\sin\theta)\Big]^5}{\Big[4(\cos 3\theta + i\sin 3\theta)\Big]^2 \Big[\frac{1}{2}(\cos\theta - i\sin\theta)\Big]^8}$$

can be expressed in the form $k(\cos\theta - i\sin\theta)$ where k is a constant to be found.

⑨ The three complex numbers in Figure 10.23 below each have modulus 1. They form an equilateral triangle centred on the origin.

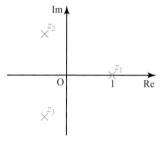

Figure 10.23

(i) Write down each of z_1, z_2 and z_3 in the form $\cos\theta + i\sin\theta$.

(ii) Use de Moivre's theorem to show that the cube of each of these complex numbers is the real number 1.

(iii) Draw an Argand diagram to show five complex numbers for which $z^5 = 1$, and write down these complex numbers in the form $\cos\theta + i\sin\theta$.

2 The nth roots of a complex number

The nth roots of unity

You already know that all quadratic equations have two roots (which may be a repeated root, or they may be complex roots). You have also solved cubic equations to find the three roots, some of which may be complex.

As early as 1629 Albert Girard stated that every polynomial equation of degree n has exactly n roots, including repeated roots. Some of these roots may be complex numbers. This was first proved by the 18-year-old Carl Freidrich Gauss 170 years later.

Therefore even the simple equation $z^n = 1$ has n roots. One of these roots is $z = 1$ and, if n is even, then $z = -1$ is another. All the other roots are complex numbers.

In this section you will look at methods for finding the other roots of the equation $z^n = 1$, and the relationship between them.

Example 10.7

(i) Write down the two roots of the equation $z^2 = 1$ and show them on an Argand diagram.

(ii) Use $z^3 - 1 = (z - 1)(z^2 + z + 1)$ to find the three roots of $z^3 = 1$. Show them on an Argand diagram.

(iii) Find the four roots of $z^4 = 1$ and show them on an Argand diagram.

Solution

(i) Using properties of real numbers $z = \pm 1$. These numbers are shown in Figure 10.24.

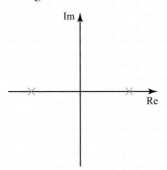

Figure 10.24

(ii) The equation $z^3 = 1$ can be rewritten as $z^3 - 1 = 0$.

So $(z - 1)\left(z^2 + z + 1\right) = 0$ ← This result is given in the question.

One of the roots is $z = 1$.

The equation $z^2 + z + 1$ has roots $z = \dfrac{-1 \pm \sqrt{3}i}{2}$.

The three roots

$z = 1, \quad z = \dfrac{-1 \pm \sqrt{3}i}{2}$

Using the quadratic formula.

are shown on an Argand diagram in Figure 10.25.

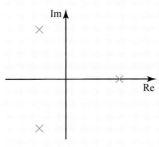

Figure 10.25

$z = 1, \quad z = \dfrac{-1 \pm \sqrt{3}i}{2}$

(iii) $z^4 = 1$ can be written in the form $z^4 - 1 = \left(z^2 - 1\right)\left(z^2 + 1\right)$ which has the four roots $z = \pm 1$, $z = \pm i$.

These roots are shown on the Argand diagram in Figure 10.26.

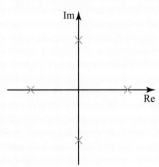

Figure 10.26

In the previous example you may have noticed that the roots of the equations $z^2 = 1$, $z^3 = 1$ and $z^4 = 1$ all lie on a unit circle, centred on the origin, with one root at the point 1.

In fact, every root of the equation $z^n = 1$ must have unit modulus, as otherwise the modulus of z^n would not be 1. So every root is of the form $z = \cos\theta + i\sin\theta$, and

> Using de Moivre's theorem.

$$z^n = 1 \Rightarrow \left(\cos\theta + i\sin\theta\right)^n$$
$$\Rightarrow \quad \cos n\theta + i\sin n\theta = 1$$
$$\Rightarrow \quad n\theta = 2k\pi \text{ where } k \text{ is any integer}$$

> Since in modulus–argument form 1 can be written $(1, 0)$ or $(1, 2\pi)$ or $(1, 4\pi)$ etc.

So, for example, in the case when $n = 3$ the result $n\theta = 2k\pi$ gives roots as follows:

when $k = 0$ $\quad 3\theta = 0 \Rightarrow \theta = 0$ $\qquad$ so $z = \cos 0 + i\sin 0 = 1$

when $k = 1$ $\quad 3\theta = 2\pi \Rightarrow \theta = \dfrac{2\pi}{3}$ $\quad$ so $z = \cos\dfrac{2\pi}{3} + i\sin\dfrac{2\pi}{3} = -\dfrac{1}{2} + \dfrac{\sqrt{3}i}{2}$

> Note that these are the same roots of $z^3 = 1$ as those obtained in Example 10.7.

when $k = 2$ $\quad 3\theta = 4\pi \Rightarrow \theta = \dfrac{4\pi}{3}$ $\quad$ so $z = \cos\dfrac{4\pi}{3} + i\sin\dfrac{4\pi}{3} = -\dfrac{1}{2} - \dfrac{\sqrt{3}i}{2}$

For larger values of k the same roots are obtained, so $k = 3$ gives the root $z = 1$, $k = 4$ gives the root $z = -\dfrac{1}{2} + \dfrac{\sqrt{3}i}{2}$, and so on.

So, when $n = 3$, the values of k that need to be considered are k = 0, 1 and 2.

Generally, as k takes values $0, 1, 2, 3, \ldots, (n - 1)$ the corresponding values of θ are:

$$0, \frac{2\pi}{n}, \frac{4\pi}{n}, \frac{6\pi}{n}, \ldots, \frac{2(n-1)\pi}{n}$$

giving n distinct values of z.

When $k = n$ then $\theta = 2\pi$, which gives the same z as $\theta = 0$. Similarly, any integer value of k larger than n will differ from one of $0, 1, 2, 3, \ldots, (n - 1)$ by a multiple of n, and so gives a value of θ differing by a multiple of 2π from one already listed; the same applies when k is any negative integer.

Therefore, the equation $z^n = 1$ has exactly n roots. These are:

$$z = \cos\frac{2k\pi}{n} + i\sin\frac{2k\pi}{n}, \quad k = 0, 1, 2, 3, \ldots, (n - 1)$$

These n complex numbers are called the **n^{th} roots of unity**. They include $z = 1$ when $k = 0$ and, if n is even, $z = -1$ when $k = \dfrac{n}{2}$.

It is usual to use ω (the Greek letter omega) for the root with the smallest positive argument:

$$\omega = \cos\frac{2\pi}{n} + i\sin\frac{2\pi}{n}$$

Then, by de Moivre's theorem:

$$\omega^k = \cos\frac{2k\pi}{n} + i\sin\frac{2k\pi}{n}$$

so the nth **roots of unity** can be written as:

$$1,\ \omega,\ \omega^2,\ \dots, \omega^{n-1}$$

These complex numbers can be represented on an Argand diagram by the vertices of a regular n-sided polygon inscribed in the unit circle, with one vertex at the point 1. Figure 10.27 shows the nth roots of unity when $n = 9$.

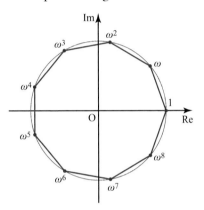

Figure 10.27

The sum of all of the nth roots of unity is a geometric series with common ratio ω:

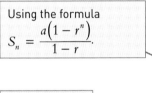

Using the formula $S_n = \dfrac{a\left(1 - r^n\right)}{1 - r}$.

$$1 + \omega + \omega^2 + \dots + \omega^{n-1}$$

$$= \frac{1 - \omega^n}{1 - \omega}$$

Since $\omega^n = 1$.

$$= 0$$

So the sum of the nth roots of unity is always zero.

ACTIVITY 10.4

Verify that the sum of the nth roots of unity is equal to zero in the cases where $n = 2$, $n = 3$ and $n = 4$.

Example 10.8

Solve the equation $z^6 = 1$. Show the roots on an Argand diagram.

Solution

The sixth roots of unity are given by:

$$z = \omega^k = \cos\frac{2k\pi}{6} + i\sin\frac{2k\pi}{6}$$

where $k = 0, 1, 2, 3, 4, 5$.

This gives the following roots z:

$k = 0 \qquad z = 1$

$k = 1 \qquad z = \omega = \cos\dfrac{2\pi}{6} + i\sin\dfrac{2\pi}{6} = \dfrac{1}{2} + \dfrac{\sqrt{3}}{2}i$

$k = 2 \qquad z = \omega^2 = \cos\dfrac{4\pi}{6} + i\sin\dfrac{4\pi}{6} = -\dfrac{1}{2} + \dfrac{\sqrt{3}}{2}i$

$k = 3 \qquad z = \omega^3 = \cos\dfrac{6\pi}{6} + i\sin\dfrac{6\pi}{6} = -1$

$k = 4 \qquad z = \omega^4 = \cos\dfrac{8\pi}{6} + i\sin\dfrac{8\pi}{6} = -\dfrac{1}{2} - \dfrac{\sqrt{3}}{2}i$

$k = 5 \qquad z = \omega^5 = \cos\dfrac{10\pi}{6} + i\sin\dfrac{10\pi}{6} = \dfrac{1}{2} - \dfrac{\sqrt{3}}{2}i$

The roots form a hexagon inscribed within a unit circle, as in Figure 10.28.

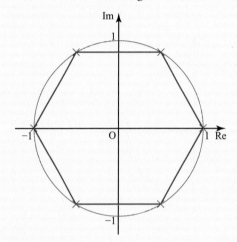

Figure 10.28

The nth roots of any complex number

To find the nth roots of any given non-zero complex number a you have to find z such that $z^n = a$. The method to find the nth roots is the same as that in the previous section on n^{th} roots of unity, adjusted to take account of the modulus s and argument ϕ of a.

Suppose, for example, you wanted to find the fifth roots of the complex number $a = -1 + i$.

> Writing $-1 + i$ in modulus–argument form.

You want to find the complex number z such that $z^5 = -1 + i$.

Let $z = r(\cos\theta + i\sin\theta)$, then:

$$\left[r\left(\cos\theta + i\sin\theta\right)\right]^5 = \sqrt{2}\left(\cos\dfrac{3\pi}{4} + i\sin\dfrac{3\pi}{4}\right)$$

> Using de Moivre's theorem.

$$\Leftrightarrow r^5\left(\cos 5\theta + i\sin 5\theta\right) = \sqrt{2}\left(\cos\dfrac{3\pi}{4} + i\sin\dfrac{3\pi}{4}\right)$$

Two complex numbers in modulus–argument form are equal only if they have the same moduli and their arguments are equal or differ by a multiple of 2π. Therefore:

$$r = 2^{\frac{1}{10}} \text{ and } 5\theta = \frac{3\pi}{4} + 2k\pi, \text{ where } k \text{ is an integer.}$$

Each of the roots will have the same modulus and so will lie on the circle $|z| = 2^{\frac{1}{10}}$.

The argument of z is $\theta = \dfrac{\dfrac{3\pi}{4} + 2k\pi}{5}$.

As k takes the values 0, 1, 2, 3 and 4 the arguments obtained are $\dfrac{3\pi}{20}$, $\dfrac{11\pi}{20}$, $\dfrac{19\pi}{20}$, $\dfrac{27\pi}{20}$ and $\dfrac{7\pi}{4}$.

Larger values of k generate the same set of arguments so, for example, $k = 5$ gives $\dfrac{43\pi}{20}$ which is equivalent to $\dfrac{3\pi}{20}$.

The five roots are shown in Figure 10.29 and form the five vertices of a regular pentagon.

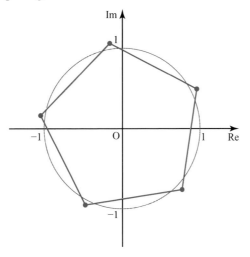

Figure 10.29

In a general case, suppose $z = r(\cos\theta + i\sin\theta)$ and $a = s(\cos\phi + i\sin\phi)$.
Then:

$$z^n = a \Leftrightarrow r^n(\cos\theta + i\sin\theta)^n = s(\cos\phi + i\sin\phi)$$

$$\Leftrightarrow r^n(\cos n\theta + i\sin n\theta) = s(\cos\phi + i\sin\phi)$$

$$\Leftrightarrow r^n = s \quad \text{and} \quad n\theta = \phi + 2k\pi, \text{ where } k \text{ is an integer}$$

Since r and s are positive real numbers the equation $r^n = s$ gives the *unique* value $r = s^{\frac{1}{n}}$ so all the roots lie on the circle $|z| = s^{\frac{1}{n}}$.

The argument of z is $\theta = \dfrac{\phi + 2k\pi}{n}$. As k can take the values 0, 1, 2, …, $n - 1$, this gives n distinct complex numbers z and, by the same argument as for the roots of unity, there are no other roots.

Therefore the non-zero complex number $s(\cos\phi + i\sin\phi)$ has precisely n different n^{th} roots, which are:

$$s^{\frac{1}{n}}\left(\cos\left(\frac{\phi + 2k\pi}{n}\right) + i\sin\left(\frac{\phi + 2k\pi}{n}\right)\right)$$

where $k = 0, 1, 2, …, n - 1$.

You may also express these n roots as α, $\alpha\omega$, $\alpha\omega^2$, ..., $\alpha\omega^{n-1}$ where:

$$\alpha = s^{\frac{1}{n}}\left(\cos\frac{\phi}{n} + i\sin\frac{\phi}{n}\right) \quad \text{and} \quad \omega = \cos\frac{2\pi}{n} + i\sin\frac{2\pi}{n}$$

The sum of these n^{th} roots of w is:

Since $\omega^n = 1$. ⟶ $\alpha + \alpha\omega + \alpha\omega^2 + ... + \alpha\omega^{n-1} = \dfrac{\alpha\left(1 - \omega^n\right)}{1 - \omega} = 0$

Example 10.9

Represent the five roots of the equation $z^5 = 32$ on an Argand diagram.

Hence, represent the five roots of the equation $(z - 3i)^5 = 32$ on an Argand diagram.

Solution

You know mod $32 = 32$ and $\arg(32) = 0$ so the fifth roots of 32 are given by:

$$32^{\frac{1}{5}}\left(\cos\left(\frac{0 + 2\pi k}{5}\right) + i\sin\left(\frac{0 + 2\pi k}{5}\right)\right)$$

Each root has modulus 2 and the arguments of the roots are $0, \dfrac{2\pi}{5}, \dfrac{4\pi}{5}, -\dfrac{4\pi}{5}$ and $-\dfrac{2\pi}{5}$. The roots form a regular pentagon inscribed in a circle, centre the origin and radius 2, as shown in by the blue points in Figure 10.30.

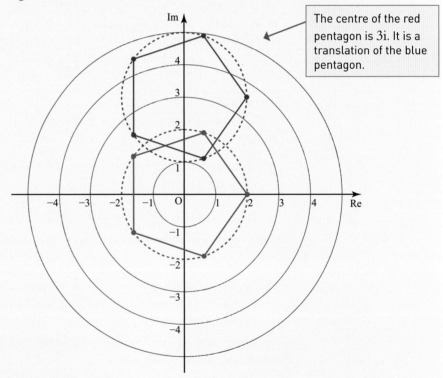

The centre of the red pentagon is 3i. It is a translation of the blue pentagon.

Figure 10.30

The roots of $(z - 3i) = 32$ are therefore represented by the same pentagon inscribed in a circle, centre 3i, radius 2 as shown by the red points in Figure 10.30.

① Find the roots of the equation $z^5 = 1$ giving your answers to 3 significant figures.

Show these roots on an Argand diagram.

Describe the polygon formed by the points representing the roots.

② Find the roots of the equation $z^8 = 1$ and show these on an Argand diagram.

Describe the polygon formed by the points representing the roots.

③ Find both square roots of $-7 + 5i$, giving your answers in the form $x + yi$ with x and y correct to 2 decimal places.

④ Find the fourth roots of -4, giving your answers in the form $x + yi$.

Show the roots on an Argand diagram.

⑤ In this question, give answers as exact values or to two decimal places where appropriate.

(i) Find the cube roots of $1 - i$.

(ii) Find the fourth roots of $2 + 3i$.

(iii) Find the fifth roots of $-3 + 4i$.

⑥ Explain geometrically why the set of tenth roots of unity is the same as the set of fifth roots of unity together with their negatives.

⑦ One fourth root of the complex number w is $2 + 3i$. Find w and its other fourth roots and represent all five points on an Argand diagram.

⑧ Solve the equation $z^3 - 4\sqrt{3} + 4i = 0$ giving your solutions in the form $r(\cos\theta + i\sin\theta)$, where $r > 0$ and $-\pi < \theta \leqslant \pi$.

⑨ A regular heptagon on an Argand diagram has centre $-1 + 3i$ and one vertex at $2 + 3i$.

Write down the equation whose solutions are represented by the vertices of this heptagon.

⑩ (i) Find the fourth roots of $-9i$ in the form $r(\cos\theta + i\sin\theta)$ where $r > 0$ and $0 < \theta < 2\pi$.

Illustrate the roots on an Argand diagram.

(ii) Let the points representing these roots, taken in order of θ increasing, be P, Q, R and S. The midpoints of the sides of the quadrilateral PQRS represent the fourth roots of a complex number w. Find the modulus and argument of the complex number w and mark the point representing w on your Argand diagram.

⑪ If ω is a complex cube root of unity, $\omega \neq 1$, prove that:

(i) $(1 + \omega)(1 + \omega^2) = 1$

(ii) $1 + \omega$ and $1 + \omega^2$ are complex cube roots of -1

(iii) $(a + b)(a + \omega b)(a + \omega^2 b) = a^3 + b^3$

(iv) $(a + b + c)(a + \omega b + \omega^2 c)(a + \omega^2 b + \omega c) = a^3 + b^3 + c^3 - 3abc$

3 Finding multiple angle identities using de Moivre's theorem

When they were first introduced, complex numbers were not generally accepted by mathematicians. However, during the eighteenth century, their usefulness in producing results involving only real numbers was recognised. The results could also be obtained without using complex numbers, but often only with considerably more effort. One of these results is finding expressions for the sine or cosine of multiple angles using de Moivre's theorem, as shown in the following example.

Example 10.10

Express $\cos 5\theta$ in terms of $\cos\theta$.

Solution

By de Moivre's theorem:

$$\cos 5\theta + i\sin 5\theta = \left(\cos\theta + i\sin\theta\right)^5$$

> Where c and s are used as abbreviations for $\cos\theta$ and $\sin\theta$ respectively.

$$= c^5 + 5ic^4 s - 10c^3 s^2 - 10ic^2 s^3 + 5cs^4 + is^5$$

Equating the real parts gives:

$$\cos 5\theta = c^5 - 10c^3 s^2 + 5cs^4$$

$$= c^5 - 10c^3\left(1 - c^2\right) + 5c\left(1 - c^2\right)^2 \quad\leftarrow$$

> Using $\sin^2\theta + \cos^2\theta \equiv 1$.

$$= c^5 - 10c^3 + 10c^5 + 5c - 10c^3 + 5c^5$$

Therefore $\cos 5\theta = 16\cos^5\theta - 20\cos^3\theta + 5\cos\theta$.

ACTIVITY 10.5

By equating the imaginary parts in Example 10.10, find $\sin 5\theta$ in terms of $\sin\theta$.

Example 10.10 expressed $\cos 5\theta$ in terms of powers of $\cos\theta$. Sometimes it is useful to do the reverse, for example if you wanted to integrate $\cos^5\theta$. The next example shows how this can be expressed in the form $a\cos 5\theta + b\cos 3\theta + c\cos\theta$, which is much easier to integrate.

To do this, you need expressions for $\cos n\theta$ and $\sin n\theta$ in terms of z^n and z^{-n}. You can deduce these expressions from de Moivre's theorem as follows:

If:

$$z = \cos\theta + i\sin\theta$$

then:

$$z^n = \cos n\theta + i\sin n\theta$$

and:

> As $\cos(-\theta) = \cos\theta$ and $\sin(-\theta) = \sin\theta$.

$$z^{-n} = \cos(-n\theta) + i\sin(-n\theta) = \cos n\theta - i\sin n\theta$$

Adding these two expressions gives $z^n + z^{-n} = 2\cos n\theta$ and so:

$$\cos n\theta = \frac{z^n + z^{-n}}{2}$$

Subtracting the two expressions gives $z^n - z^{-n} = 2\mathrm{i}\sin n\theta$ and so:

$$\sin n\theta = \frac{z^n - z^{-n}}{2\mathrm{i}}$$

Example 10.11

Express $\cos^5\theta$ in terms of multiple angles.

Solution

Let $z = \cos\theta + \mathrm{i}\sin\theta$.

Then $2\cos\theta = z + z^{-1}$.

> A rearrangement of $\cos n\theta = \dfrac{z^n + z^{-n}}{2}$ with $n = 1$.

$\Rightarrow \qquad (2\cos\theta)^5 = \left(z + z^{-1}\right)^5$

$\Rightarrow \qquad 32\cos^5\theta = z^5 + 5z^3 + 10z + 10z^{-1} + 5z^{-3} + z^{-5}$

> Expanding the right-hand side using the binomial expansion.

$\qquad\qquad = \left(z^5 + z^{-5}\right) + 5\left(z^3 + z^{-3}\right) + 10\left(z + z^{-1}\right)$

$\qquad\qquad = 2\cos 5\theta + 10\cos 3\theta + 20\cos\theta$

$\Rightarrow \qquad\qquad \cos^5\theta = \dfrac{\cos 5\theta + 5\cos 3\theta + 10\cos\theta}{16}$

> Using $\cos n\theta = \dfrac{z^n + z^{-n}}{2}$ three times, with $n = 1$, $n = 3$ and $n = 5$.

ACTIVITY 10.6

Use a similar method to that used in Example 10.11 to express $\sin^5\theta$ in terms of multiple angles.

Exercise 10.3

① (i) Using the result $(\cos\theta + \mathrm{i}\sin\theta)^3 = \cos 3\theta + \mathrm{i}\sin 3\theta$, compare real and imaginary parts to show that:
$$\cos 3\theta = 4\cos^3\theta - 3\cos\theta$$
and:
$$\sin 3\theta = 3\sin\theta - 4\sin^3\theta$$

   (ii) Hence express $\tan 3\theta$ in terms of $\tan\theta$.

② Let $z = \cos\theta + \mathrm{i}\sin\theta$.

   (i) Write down expressions for z^3 and z^{-3}.

   (ii) Use your expressions from (i) to show that $\cos 3\theta = \dfrac{z^3 + z^{-3}}{2}$ and $\sin 3\theta = \dfrac{z^3 - z^{-3}}{2\mathrm{i}}$.

③ Let $z = \cos\theta + \mathrm{i}\sin\theta$.

   (i) Write down an expression for z^{-1}.

   (ii) (a) Use your answer to (i) to show that $2\cos\theta = z + z^{-1}$.

   (b) Using the result in part (a), express $\cos^4\theta$ in terms of multiple angles.

   (iii) (a) Use your answer to (i) to show that $2\mathrm{i}\sin\theta = z - z^{-1}$.

   (b) Use the result in part (a), express $\sin^5\theta$ in terms of multiple angles.

④ Find $\cos 6\theta$ and $\dfrac{\sin 6\theta}{\sin \theta}$ in terms of $\cos\theta$.

⑤ Find an expression for $\sin^6\theta$ in terms of multiple angles.

Hence evaluate $\int \sin^6\theta \, d\theta$.

⑥ Express $\cos^4\theta \sin^3\theta$ in terms of multiple angles and hence find $\int \cos^4\theta \sin^3\theta \, d\theta$.

⑦ By first using de Moivre's theorem, evaluate:

$$\int_0^{\frac{\pi}{6}} \cos^5\theta \, d\theta$$

⑧ (i) Use de Moivre's theorem to show that:
$$\cos 5\theta = 16\cos^5\theta - 20\cos^3\theta + 5\cos\theta$$

(ii) Given $\cos 5\theta = 0$ but $\cos\theta \neq 0$, use your answer from (i) to find two possible values for $\cos^2\theta$. Give your answers in surd form.

(iii) Use (ii) to show that:

$$\cos 18° = \left(\frac{5 + \sqrt{5}}{8} \right)^{\frac{1}{2}}$$

and find, in a similar form, an expression for $\sin 18°$.

⑨ Use $\cos n\theta = \dfrac{z^n + z^{-n}}{2}$ to express:

$$\cos\theta + \cos 3\theta + \cos 5\theta + \ldots + \cos(2n-1)\theta$$

as a geometric series in terms of z.

Show that the sum of the geometric series can be expressed in the form $\dfrac{\sin(2n\theta)}{2\sin\theta}$.

⑩ (i) Given that $z = \cos\theta + i\sin\theta$, write down z^n and $\dfrac{1}{z^n}$ in the form $a + ib$.

Simplify $z^n + \dfrac{1}{z^n}$ and $z^n - \dfrac{1}{z^n}$.

(ii) Expand $\left(z^n + \dfrac{1}{z^n} \right)^2 \left(z^n - \dfrac{1}{z^n} \right)^4$ and hence find the constants p, q, r and s such that:
$$\sin^4\theta \cos^2\theta = p + q\cos 2\theta + r\cos 4\theta + s\cos 6\theta$$

(iii) Using a suitable substitution and your answer to part (ii), show that:
$$\int_1^2 x^4 \sqrt{4 - x^2} \, dx = \frac{4\pi}{3} + \sqrt{3}$$

4 The form $z = re^{i\theta}$

In Chapter 6 you saw that the series expansions of $\sin\theta$, $\cos\theta$ and e^x can be written as:

$$\sin\theta = \theta - \frac{\theta^3}{3!} + \frac{\theta^5}{5!} - \frac{\theta^7}{7!} + \ldots \frac{(-1)^r \theta^{2r+1}}{(2r+1)!}$$

$$\cos\theta = 1 - \frac{\theta^2}{2!} + \frac{\theta^4}{4!} - \frac{\theta^6}{6!} + \ldots \frac{(-1)^r \theta^{2r}}{(2r)!}$$

$$e^x = 1 + x + \frac{x^2}{2!} + \frac{x^3}{3!} + \frac{x^4}{4!} + \ldots + \frac{x^r}{r!} + \ldots$$

It can be shown that this series expansion is also true for complex powers.

→ Replacing x by $i\theta$ in the expansion e^x gives:

$$e^{i\theta} = 1 + i\theta + \frac{(i\theta)^2}{2!} + \frac{(i\theta)^3}{3!} + \frac{(i\theta)^4}{4!} + \frac{(i\theta)^5}{5!} + \frac{(i\theta)^6}{6!} + \ldots$$

$$= 1 + i\theta + \frac{i^2\theta^2}{2!} + \frac{i^3\theta^3}{3!} + \frac{i^4\theta^4}{4!} + \frac{i^5\theta^5}{5!} + \frac{i^6\theta^6}{6!} + \ldots$$

$$= 1 + i\theta - \frac{\theta^2}{2!} - \frac{i\theta^3}{3!} + \frac{\theta^4}{4!} + \frac{i\theta^5}{5!} - \frac{\theta^6}{6!} + \ldots$$

Collecting together real and imaginary terms.

$$= \left(1 - \frac{\theta^2}{2!} + \frac{\theta^4}{4!} - \frac{\theta^6}{6!} + \ldots\right) + i\left(\theta - \frac{\theta^3}{3!} + \frac{\theta^5}{5!} - \ldots\right)$$

Using the series expansions for $\cos\theta$ and $\sin\theta$.

Therefore:

$$e^{i\theta} = \cos\theta + i\sin\theta$$

and so:

$$z = r(\cos\theta + i\sin\theta)$$

can be rewritten as:

$$z = re^{i\theta}$$

This is called the **exponential form** of a complex number with modulus r and argument θ.

Discussion point

→ How would the result:
$$z = r(\cos\theta + i\sin\theta)$$
$$= re^{i\theta}$$
be adapted for a complex number of the form:
$$r(\cos\theta - i\sin\theta)$$
where $-\pi < \theta \leq \pi$?

This format is simply a more compact way of writing familiar expressions, as the modulus-argument form $z = r(\cos\theta + i\sin\theta)$ can now be abbreviated to $z = re^{i\theta}$.

This form allows you to derive de Moivre's theorem very easily for all rational n by using the laws of indices:

$$\cos\theta + i\sin\theta^n = (e^{i\theta})^n$$

$$= e^{in\theta}$$

$$= \cos n\theta + i\sin n\theta$$

Example 10.12

(i) Write $z = 6\left(\cos\frac{\pi}{6} + i\sin\frac{\pi}{6}\right)$ in the form $re^{i\theta}$.

(ii) Write $z = -1 + \sqrt{3}i$ in the form $re^{i\theta}$.

Solution

(i) $z = 6\left(\cos\frac{\pi}{6} + i\sin\frac{\pi}{6}\right)$ has modulus 6 and argument $\frac{\pi}{6}$.

Therefore $z = 6e^{\frac{i\pi}{6}}$.

(ii) $z = -1 + \sqrt{3}i$ has modulus 2 and argument $\frac{2\pi}{3}$.

Therefore $z = 2e^{\frac{2i\pi}{3}}$.

In the discussion point above, you should have noticed that since:

$$\cos(-\theta) = \cos\theta \text{ and } \sin(-\theta) = -\sin\theta$$

then:

$$r(\cos\theta - i\sin\theta) = r(\cos(-\theta) + i\sin(-\theta))$$

Therefore:

$$r(\cos\theta - i\sin\theta) = re^{-i\theta}$$

ACTIVITY 10.7

For a complex number $z = x + iy$, show that:

(i) $e^z = e^x(\cos y + i\sin y)$

(ii) $e^{z+2\pi n} = e^z$

(iii) $e^{i\pi} = -1$

The results in Activity 10.7 are useful when simplifying results involving exponential functions with complex exponents. Part (iii) is often written in the form $e^{i\pi} + 1 = 0$ which is a remarkable result that links the five numbers 0, 1, i, e and π.

The results from Activity 10.7 also give rise to two very interesting mathematical results that are useful when working with complex numbers:

$$\cos\theta = \frac{1}{2}\left(e^{i\theta} + e^{-i\theta}\right)$$

$$\sin\theta = \frac{1}{2i}\left(e^{i\theta} - e^{-i\theta}\right)$$

> Notice that these expressions for $\cos\theta$ and $\sin\theta$ are very similar to the definitions of the hyperbolic functions $\cosh\theta$ and $\sinh\theta$.

To prove these results, you can use:

$$e^{i\theta} = \cos\theta + i\sin\theta \qquad ①$$

and:

$$e^{-i\theta} = \cos\theta - i\sin\theta \qquad ②$$

Finding ① + ② gives:

$$e^{i\theta} + e^{-i\theta} = 2\cos\theta$$

so:

$$\cos\theta = \frac{1}{2}\left(e^{i\theta} + e^{-i\theta}\right)$$

Similarly, finding ① − ② gives:

$$e^{i\theta} - e^{-i\theta} = 2i\sin\theta$$

so:

$$\sin\theta = \frac{1}{2i}\left(e^{i\theta} - e^{-i\theta}\right)$$

Discussion point

→ These results are essentially the same as the ones given on pages 224 and 225. Explain why this is the case.

You need to learn the proofs of these results.

Prior knowledge

You need to be familiar with geometric sequences and series, including finding the sum to n terms and the sum to infinity of a geometric series.

Summing series using de Moivre's theorem

This section shows how complex numbers can be used to evaluate certain sums of real quantities. It may be possible to do these summations without using complex numbers, for example by induction if you already know the answer, but this is a lot more difficult.

Sometimes it is worth setting out to do more than is actually required, as shown in the next example.

Example 10.13

(i) Prove that:

$$1 + e^{i\theta} = 2\cos\frac{\theta}{2}\,e^{\frac{i\theta}{2}}$$

and:

$$1 - e^{i\theta} = -2i\sin\frac{\theta}{2}\,e^{\frac{i\theta}{2}}$$

(ii) Show that the sum of the series:

$$1 + {}_nC_1\cos\theta + {}_nC_2\cos 2\theta + {}_nC_3\cos 3\theta + \ldots + \cos n\theta$$

is:

$$2_n\cos_n\frac{\theta}{2}\cos\frac{n\theta}{2}$$

Solution

(i) The factor $e^{\frac{i\theta}{2}}$ on the right-hand side suggests writing each term on the left-hand side as a multiple of $e^{\frac{i\theta}{2}}$.

$$1 = e^{\frac{i\theta}{2}} \times e^{\frac{-i\theta}{2}}$$

$$e^{i\theta} = e^{\frac{i\theta}{2}} \times e^{\frac{i\theta}{2}}$$

So: $1 + e^{i\theta} = \left(e^{\frac{i\theta}{2}} \times e^{\frac{-i\theta}{2}}\right) + \left(e^{\frac{i\theta}{2}} \times e^{\frac{i\theta}{2}}\right)$

$$= e^{\frac{i\theta}{2}}\left(e^{\frac{-i\theta}{2}} + e^{\frac{i\theta}{2}}\right) \longleftarrow \boxed{\text{Taking out a factor of } e^{\frac{i\theta}{2}}.}$$

$$= e^{\frac{i\theta}{2}} \times 2\cos\frac{\theta}{2} \longleftarrow \boxed{\text{Using the earlier result } \cos\theta = \frac{1}{2}\left(e^{i\theta} + e^{-i\theta}\right).}$$

Similarly,

$$1 - e^{i\theta} = \left(e^{\frac{i\theta}{2}} \times e^{\frac{-i\theta}{2}}\right) - \left(e^{\frac{i\theta}{2}} \times e^{\frac{i\theta}{2}}\right)$$

$$= e^{\frac{i\theta}{2}}\left(e^{\frac{-i\theta}{2}} - e^{\frac{i\theta}{2}}\right)$$

$$= e^{\frac{i\theta}{2}} \times -2i\sin\frac{\theta}{2} \longleftarrow \boxed{\text{Using the earlier result } \sin\theta = \frac{1}{2i}\left(e^{i\theta} - e^{-i\theta}\right).}$$

(ii) At first sight this series seems to suggest the binomial expansion $(1 + \cos\theta)_n$. The binomial coefficients $1, {}_nC_1, {}_nC_2, \ldots, 1$ are correct, but there are multiple angles, $\cos r\theta$, instead of powers of cosines, $\cos^r\theta$. This suggests that de Moivre's theorem can be used.

The method involves introducing a corresponding sine series too.

Let: $\quad C = 1 + {}_nC_1\cos\theta + {}_nC_2\cos 2\theta + {}_nC_3\cos 3\theta + \ldots + \cos n\theta$

and: $\quad S = {}_nC_1\sin\theta + {}_nC_2\sin 2\theta + {}_nC_3\sin 3\theta + \ldots + \sin n\theta$

Then

$$C + iS = 1 + {}_nC_1(\cos\theta + i\sin\theta) + {}_nC_2(\cos 2\theta + i\sin 2\theta) + \ldots$$

$$+ (\cos n\theta + i\sin n\theta$$

$$= 1 + {}^nC_1 e^{i\theta} + {}^nC_2 e^{i2\theta} + \dots + e^{in\theta}$$

$$= 1 + {}^nC_1 e^{i\theta} + {}^nC_2 \left(e^{i\theta}\right)^2 + \dots + \left(e^{i\theta}\right)^n$$

> Using de Moivre's theorem and the fact that $e^{ir\theta} = \left(e^{i\theta}\right)^r$.

This is now recognisable as a binomial expansion, so that:

$$C + iS = \left(1 + e^{i\theta}\right)^n$$

To find C you need to find the real part of $\left(1 + e^{i\theta}\right)^n$ and here the results from part(i) are useful. Using the result:

$$1 + e^{i\theta} = 2\cos\frac{\theta}{2}\, e^{\frac{i\theta}{2}}$$

gives:

$$C + iS = \left(1 + e^{i\theta}\right)^n = \left(2\cos\frac{\theta}{2}\, e^{\frac{i\theta}{2}}\right)^n$$

$$= 2^n \cos^n \frac{\theta}{2}\, e^{\frac{in\theta}{2}}$$

$$= 2^n \cos^n \frac{\theta}{2}\left(\cos\frac{n\theta}{2} + i\sin\frac{n\theta}{2}\right)$$

Taking the real part:

$$C = 2^n \cos^n \frac{\theta}{2}\cos\frac{n\theta}{2}$$

ACTIVITY 10.8

For Example 10.13, state the corresponding result obtained by equating the imaginary parts.

Exercise 10.4

① Write the following complex numbers in the form $z = re^{i\theta}$ where $-\pi < \theta \leqslant \pi$.

(i) $\quad 4\left(\cos\frac{\pi}{3} + i\sin\frac{\pi}{3}\right)$

(ii) $\quad \sqrt{3}\left(\cos\left(-\frac{5\pi}{6}\right) + i\sin\left(-\frac{5\pi}{6}\right)\right)$

(iii) $\quad -5i$

(iv) $\quad -3 - 3i$

(v) $\quad \sqrt{3} - i$

② Write the following complex numbers in the form $x + yi$:

(i) $\quad 5e^{i\pi}$
(ii) $\quad \sqrt{2}e^{\frac{3\pi}{4}i}$
(iii) $\quad \sqrt{2}e^{-\frac{3\pi}{4}i}$
(iv) $\quad 5e^{\frac{23\pi}{4}i}$

③ Two complex numbers are given by $z = 2e^{\frac{3\pi}{4}i}$ and $w = 3e^{\frac{\pi}{3}i}$.

Find zw and $\dfrac{z}{w}$ giving your answers in the form $z = re^{i\theta}$, where $r > 0$ and $-\pi < \theta \leqslant \pi$.

④ (i) Write the complex number $w = 32i$ in exponential form.

(ii) Find the five fifth roots of w, giving your answers in exponential form.

⑤ Let:
$$C = 1 + \cos\theta + \cos 2\theta + \dots + \cos(n-1)\theta$$
and:
$$S = \sin\theta + \sin 2\theta + \dots + \sin(n-1)\theta$$

(i) Find $C + iS$ and show that this forms a geometric series with common ratio $e^{i\theta}$

(ii) Show that the sum of the series in part (i) is $\dfrac{1 - e^{in\theta}}{1 - e^{i\theta}}$

(iii) By multiplying the numerator and denominator of this sum by $1 - e^{-i\theta}$, show that:
$$C = \frac{1 - \cos\theta + \cos(n-1)\theta - \cos n\theta}{2 - 2\cos\theta}$$
and find S.

⑥ (i) Show that $1 + e^{i2\theta} = 2\cos\theta(\cos\theta + i\sin\theta)$.

(ii) The series C and S are defined as follows.
$$C = 1 + \binom{n}{1}\cos 2\theta + \binom{n}{2}\cos 4\theta + \dots \cos 2n\theta$$
$$S = \binom{n}{1}\sin 2\theta + \binom{n}{2}\sin 4\theta + \dots \sin 2n\theta$$

By considering $C + iS$, show that:
$$C = 2^n \cos^n\theta \cos n\theta$$
and find a corresponding expression for S.

⑦ (i) Use de Moivre's theorem to find the constants a, b, c in the identity
$$\cos 5\theta \equiv a\cos^5\theta + b\cos^3\theta + c\cos\theta.$$

(ii) Let:
$$C = \cos\theta + \cos\left(\theta + \frac{2\pi}{n}\right) + \cos\left(\theta + \frac{4\pi}{n}\right) + \dots \cos\left(\theta + \frac{(2n-2)\pi}{n}\right)$$
and:
$$S = \sin\theta + \sin\left(\theta + \frac{2\pi}{n}\right) + \sin\left(\theta + \frac{4\pi}{n}\right) + \dots \sin\left(\theta + \frac{(2n-2)\pi}{n}\right)$$
where n is an integer greater than 1.

Show that $C + iS$ forms a geometric series and hence show that $C = 0$, $S = 0$.

⑧ Use the result $e^{i\theta} = \cos\theta + i\sin\theta$ to prove that
$$e^{z^*} = \left(e^z\right)^*$$
for all complex numbers $z = a + b$.

⑨ Let $C = \displaystyle\int e^{3x}\cos 2x \, dx$ and $S = \displaystyle\int e^{3x}\sin 2x \, dx$.

(i) Find C and S by using integration by parts twice.

(ii) (a) Show that $C + iS = \dfrac{e^{(3+2i)x}}{3 + 2i} + A$ where A is a constant.

(b) Hence verify your answers for C and S from part (i).

⑩ The infinite series C and S are defined as follows:

$$C = \frac{\cos\theta}{2} - \frac{\cos 2\theta}{4} + \frac{\cos 3\theta}{8} - \frac{\cos 4\theta}{16} + \dots$$

$$S = \frac{\sin\theta}{2} - \frac{\sin 2\theta}{4} + \frac{\sin 3\theta}{8} - \frac{\sin 4\theta}{16} + \dots$$

Show that $C + iS = \dfrac{2e^{i\theta} + 1}{5 + 4\cos\theta}$ and hence find expressions for C and S in terms of $\cos\theta$ and $\sin\theta$.

LEARNING OUTCOMES

When you have completed this chapter you should be able to:

➤ find the modulus and argument of a complex number

➤ multiply and divide complex numbers in modulus-argument form

➤ understand the effect of multiplication by a complex number in an Argand diagram

➤ sketch loci of the form $|z - a| = r$, $\arg (z - a) = \theta$ and $|z - a| = |z - b|$ in an Argand diagram

➤ use de Moivre's theorem to simplify expressions involving powers of complex numbers

➤ find the nth roots of a complex number

➤ use de Moivre's theorem to find multiple angle identities

➤ use the exponential form of a complex number $z = re^{i\theta}$

➤ sum series using de Moivre's theorem.

KEY POINTS

1 The modulus r of $z = x + y\text{i}$ is $|z| = \sqrt{x^2 + y^2}$. This is the distance of the point z from the origin on the Argand diagram.

2 The argument of z is the angle θ, measured in radians, between the line connecting the origin and the point z and the positive real axis.

3 The principal argument of z, $\arg z$, is the angle θ, measured in radians, for which $-\pi < \theta \leqslant \pi$, between the line connecting the origin and the point z and the positive real axis.

4 For a complex number z, $zz^* = |z|^2$.

5 The modulus–argument form of z is $z = r(\cos\theta + \text{i}\sin\theta)$, where $r = |z|$ and $\theta = \arg z$. This is often written as (r, θ).

6 For two complex numbers z_1 and z_2:

$$|z_1 z_2| = |z_1||z_2| \qquad \arg(z_1 z_2) = \arg z_1 + \arg z_2$$

$$\left|\frac{z_1}{z_2}\right| = \frac{|z_1|}{|z_2|} \qquad \arg\left(\frac{z_1}{z_2}\right) = \arg z_1 - \arg z_2$$

7 Geometrically, to obtain the vector $z_1 z_2$ enlarge the vector z_2 by the scale factor $|z_1|$ and rotate it through $\arg z_1$ anticlockwise about O. In polar form, multiplication by $re^{i\theta}$ corresponds to enlargement with scale factor r with (anticlockwise) rotation through θ about the origin.

8 The distance between the points z_1 and z_2 in an Argand diagram is $|z_2 - z_1|$.

9 $|z - a| = r$ represents a circle, centre a and radius r. $|z - a| < r$ represents the interior of the circle, and $|z - a| > r$ represents the exterior of the circle.

10 $\arg(z - a) = \theta$ represents a half line starting at $z = a$ at an angle of θ from the positive real direction joining between the complex numbers $z = a$ and $z = b$

11 $|z - a| = |z - b|$ represents the perpendicular bisector of the points a and b.

12 De Moivre's theorem: If $z = r(\cos\theta + i\sin\theta)$ then:
$$z^n = \left[r(\cos\theta + i\sin\theta)\right]^n = r^n(\cos n\theta + i\sin n\theta)$$

13 The nth roots of unity can be written as
$$1, \omega, \omega^2, \ldots, \omega^{n-1}$$
where:
$$\omega = \cos\frac{2\pi}{n} + i\sin\frac{2\pi}{n}$$

14 The sum of all of the nth roots of unity is zero.

15 The non-zero complex number $r(\cos\theta + i\sin\theta)$ has precisely n different n^{th} roots, which are:
$$r^{\frac{1}{n}}\left(\cos\left(\frac{\theta + 2k\pi}{n}\right) + i\sin\left(\frac{\theta + 2k\pi}{n}\right)\right)$$
where $k = 0, 1, 2, \ldots, n - 1$.

These roots can also be written as $\alpha, \alpha\omega, \alpha\omega^2, \ldots, \alpha\omega^{n-1}$ where
$$\alpha = r^{\frac{1}{n}}\left(\cos\frac{\theta}{n} + i\sin\frac{\theta}{n}\right) \text{ and } \omega = \cos\frac{2\pi}{n} + i\sin\frac{2\pi}{n}.$$

16 The sum of the nth roots is zero.

17 If $z = \cos\theta + i\sin\theta$ then:
$$\cos n\theta = \frac{z^n + z^{-n}}{2}$$
and:
$$\sin n\theta = \frac{z^n - z^{-n}}{2i}$$

18 The exponential form of a complex number is:
$$z = r(\cos\theta + i\sin\theta) = re^{i\theta}$$
For a complex number $z = x + iy$ this can be written as:
$$e^z = e^x(\cos y + i\sin y)$$

19 For a complex number in exponential form
$$e^{z + 2\pi n} = e^z$$

20 For the complex number $z = r(\cos\theta + i\sin\theta)$
$$\cos\theta = \frac{1}{2}\left(e^{i\theta} + e^{-i\theta}\right)$$
$$\sin\theta = \frac{1}{2i}\left(e^{i\theta} - e^{-i\theta}\right)$$

FUTURE USES

■ Complex numbers will be needed for work on differential equations in A Level Further Mathematics, in particular in modelling oscillations (simple harmonic motion).

Vectors 2

Discussion point

→ The picture shows the ceiling of King's Cross Station in London.What mathematics might be involved in designing and building a structure like this?

1 The vector product

Prior knowledge

In Chapter 1 you met the scalar product $\mathbf{a.b}$ of two vectors $\mathbf{a}$ and $\mathbf{b}$:

$$\mathbf{a.b} = |\mathbf{a}||\mathbf{b}|\cos\theta$$

where θ is the angle between $\mathbf{a}$ and $\mathbf{b}$, the result being a scalar quantity.

Figure 11.1

You have already used the scalar product to find the angle between two vectors. It is particularly convenient if you want to test whether two vectors are perpendicular, as then the scalar product is zero.

The scalar product also enables you to write down the equation of a plane in the form $(\mathbf{r} - \mathbf{a}).\mathbf{n} = 0$ where $\mathbf{a}$ is the position vector of a specific point A in the plane, $\mathbf{r}$ is the position vector of a general point in the plane and $\mathbf{n}$ is a normal vector perpendicular to the plane (see Figure 11.2). This can be rearranged as $\mathbf{r}.\mathbf{n} = d$, where $d = \mathbf{n}.\mathbf{a}$; d is a scalar constant.

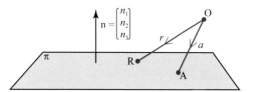

Figure 11.2

The **vector product** is a different method for 'multiplying' two vectors. As the name suggests, in this case the result is a vector rather than a scalar. The vector product of $\mathbf{a}$ and $\mathbf{b}$ is a vector perpendicular to both $\mathbf{a}$ and $\mathbf{b}$ and it is written $\mathbf{a} \times \mathbf{b}$.

It is given by

$$\mathbf{a} \times \mathbf{b} = |\mathbf{a}||\mathbf{b}| \sin \theta \, \hat{\mathbf{n}}$$

where θ is the angle between $\mathbf{a}$ and $\mathbf{b}$ and $\hat{\mathbf{n}}$ is a unit vector which is perpendicular to both $\mathbf{a}$ and $\mathbf{b}$.

This is often described as having opposite senses.

There are two unit vectors perpendicular to both $\mathbf{a}$ and $\mathbf{b}$, but they point in opposite directions.

The vector $\hat{\mathbf{n}}$ is chosen such that $\mathbf{a}, \mathbf{b}$ and $\hat{\mathbf{n}}$ (in that order) form a **right-handed set** of vectors, as shown in Figure 11.3. If you point the thumb of your right hand in the direction of $\mathbf{a}$, and your index finger in the direction of $\mathbf{b}$, then your second finger coming up from your palm points in the direction $\mathbf{a} \times \mathbf{b}$ as shown below.

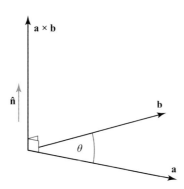

Figure 11.3

In component form, the vector product is expressed as follows:

$$\mathbf{a} \times \mathbf{b} = \begin{pmatrix} a_1 \\ a_2 \\ a_3 \end{pmatrix} \times \begin{pmatrix} b_1 \\ b_2 \\ b_3 \end{pmatrix} = \begin{pmatrix} a_2b_3 - a_3b_2 \\ a_3b_1 - a_1b_3 \\ a_1b_2 - a_2b_1 \end{pmatrix}$$

You will have the opportunity to prove this result in Exercise 11.1.

Notice that the first component of $\mathbf{a} \times \mathbf{b}$ is the value of the 2×2 determinant $\begin{vmatrix} a_2 & b_2 \\ a_3 & b_3 \end{vmatrix}$ obtained by covering up the top row of $\begin{pmatrix} a_1 \\ a_2 \\ a_3 \end{pmatrix} \times \begin{pmatrix} b_1 \\ b_2 \\ b_3 \end{pmatrix}$; the second component is the negative of the 2×2 determinant obtained by covering up the middle row; and the third component is the 2×2 determinant obtained by covering up the bottom row.

This means that the formula for the vector product can be expressed as a determinant:

$$\mathbf{a} \times \mathbf{b} = \begin{vmatrix} \mathbf{i} & a_1 & b_1 \\ \mathbf{j} & a_2 & b_2 \\ \mathbf{k} & a_3 & b_3 \end{vmatrix}$$

Expanding this determinant by the first column gives:

$$\mathbf{a} \times \mathbf{b} = \begin{vmatrix} a_2 & b_2 \\ a_3 & b_3 \end{vmatrix} \mathbf{i} - \begin{vmatrix} a_1 & b_1 \\ a_3 & b_3 \end{vmatrix} \mathbf{j} + \begin{vmatrix} a_1 & b_1 \\ a_2 & b_2 \end{vmatrix} \mathbf{k}$$

Note this sign.

Example 11.1

(i) Calculate $\mathbf{a} \times \mathbf{b}$ when $\mathbf{a} = 3\mathbf{i} + 2\mathbf{j} + 5\mathbf{k}$ and $\mathbf{b} = \mathbf{i} - 4\mathbf{j} + 2\mathbf{k}$.

(ii) Hence find $\hat{\mathbf{n}}$, a unit vector which is perpendicular to both $\mathbf{a}$ and $\mathbf{b}$.

Solution

(i) There are two possible methods:

Method 1

Using determinants:

$$\mathbf{a} \times \mathbf{b} = \begin{vmatrix} \mathbf{i} & 3 & 1 \\ \mathbf{j} & 2 & -4 \\ \mathbf{k} & 5 & 2 \end{vmatrix} \qquad \longleftarrow \qquad \mathbf{a} \times \mathbf{b} = \begin{vmatrix} \mathbf{i} & a_1 & b_1 \\ \mathbf{j} & a_2 & b_2 \\ \mathbf{k} & a_3 & b_3 \end{vmatrix}$$

$$= \mathbf{i} \begin{vmatrix} 2 & -4 \\ 5 & 2 \end{vmatrix} - \mathbf{j} \begin{vmatrix} 3 & 1 \\ 5 & 2 \end{vmatrix} + \mathbf{k} \begin{vmatrix} 3 & 1 \\ 2 & -4 \end{vmatrix}$$

$$= 24\mathbf{i} - \mathbf{j} - 14\mathbf{k}$$

Method 2

Using the result

$$\mathbf{a} \times \mathbf{b} = \begin{pmatrix} a_1 \\ a_2 \\ a_3 \end{pmatrix} \times \begin{pmatrix} b_1 \\ b_2 \\ b_3 \end{pmatrix} = \begin{pmatrix} a_2 b_3 - a_3 b_2 \\ a_3 b_1 - a_1 b_3 \\ a_1 b_2 - a_2 b_1 \end{pmatrix}$$

gives

$$\mathbf{a} \times \mathbf{b} = \begin{pmatrix} 3 \\ 2 \\ 5 \end{pmatrix} \times \begin{pmatrix} 1 \\ -4 \\ 2 \end{pmatrix} = \begin{pmatrix} 2 \times 2 - 5 \times (-4) \\ 5 \times 1 - 3 \times 2 \\ 3 \times (-4) - 2 \times 1 \end{pmatrix} = \begin{pmatrix} 24 \\ -1 \\ -14 \end{pmatrix}$$

(ii) So $\mathbf{a} \times \mathbf{b} = \begin{pmatrix} 24 \\ -1 \\ -14 \end{pmatrix}$, which is a vector perpendicular to the vectors $\mathbf{a}$ and $\mathbf{b}$.

$$|\mathbf{a} \times \mathbf{b}| = \sqrt{24^2 + (-1)^2 + (-14)^2} = \sqrt{773}$$

So a unit vector perpendicular to both $\mathbf{a}$ and $\mathbf{b}$ is $\hat{\mathbf{n}} = \dfrac{1}{\sqrt{773}} \begin{pmatrix} 24 \\ -1 \\ -14 \end{pmatrix}$.

Discussion point

➜ How can you use the scalar product to check that the answer to Example 11.1. is correct?

**TECHNOLOGY**

Investigate whether your calculator will find the vector product of two vectors.

If so, use your calculator to check the vector product calculated in Example 11.1.

Properties of the vector product

1 **The vector product is anti-commutative**

   The vector products $\mathbf{a} \times \mathbf{b}$ and $\mathbf{b} \times \mathbf{a}$ have the same magnitude but are in opposite directions, so $\mathbf{a} \times \mathbf{b} = -\mathbf{b} \times \mathbf{a}$. This is known as the **anti-commutative property.**

2 **The vector product of parallel vectors is zero**

   This is because the angle θ between two parallel vectors is $0°$ or $180°$, so $\sin\theta = 0$.

   In particular $\mathbf{i} \times \mathbf{i} = \mathbf{j} \times \mathbf{j} = \mathbf{k} \times \mathbf{k} = \mathbf{0}$.

3 **The vector product is compatible with scalar multiplication**

   For scalars m and n,

   $$(m\mathbf{a}) \times (n\mathbf{b}) = mn(\mathbf{a} \times \mathbf{b})$$

   This is because the vector $m\mathbf{a}$ has magnitude $|m||\mathbf{a}|$; $m\mathbf{a}$ and $\mathbf{a}$ have the same direction if m is positive, but opposite directions if m is negative.

4 **The vector product is distributive over vector addition**

   The result

   $$\mathbf{a} \times (\mathbf{b} + \mathbf{c}) = \mathbf{a} \times \mathbf{b} + \mathbf{a} \times \mathbf{c}$$

   enables you to change a product into the sum of two simpler products – in doing so the multiplication is 'distributed' over the two terms of the original sum.

> ## ACTIVITY 11.1
> In this activity you might find it helpful to take the edges of a rectangular table to represent the unit vectors **i**, **j** and **k** as shown in Figure 11.4.
>
>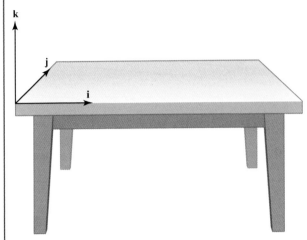
>
> **Figure 11.4**
>
> You could use pens to represent:
>
> **i**, the unit vector pointing to the right along the x-axis
>
> **j**, the unit vector pointing away from you along the y-axis
>
> **k**, the unit vector pointing upwards along the z-axis.
>
> The vector product of **a** and **b** is defined as
>
> $$\mathbf{a} \times \mathbf{b} = |\mathbf{a}||\mathbf{b}| \sin\theta \hat{\mathbf{n}}$$
>
> where θ is the angle between **a** and **b** and $\hat{\mathbf{n}}$ is a unit vector which is perpendicular to both **a** and **b** such that **a**, **b** and $\hat{\mathbf{n}}$ (in that order) form a right-handed set of vectors.
>
> Using this definition, check the truth of each of the following results.
>
> $$\mathbf{i} \times \mathbf{i} = 0 \qquad\qquad \mathbf{i} \times \mathbf{j} = \mathbf{k} \qquad\qquad \mathbf{i} \times \mathbf{k} = -\mathbf{j}$$
>
> Give a further six results for vector products of pairs of **i**, **j** and **k**.

Since the equation of a plane involves a vector which is perpendicular to the plane, the vector product is very useful in finding the equation of a plane.

Example 11.2

Find the cartesian equation of the plane which contains the points A(3, 4, 2), B(2, 0, 5) and C(6, 7, 8).

Solution

Start by finding two vectors in the plane, for example $\overrightarrow{AB}$ and $\overrightarrow{BC}$.

$$\overrightarrow{AB} = \begin{pmatrix} 2 \\ 0 \\ 5 \end{pmatrix} - \begin{pmatrix} 3 \\ 4 \\ 2 \end{pmatrix} = \begin{pmatrix} -1 \\ -4 \\ 3 \end{pmatrix} \text{ and } \overrightarrow{BC} = \begin{pmatrix} 6 \\ 7 \\ 8 \end{pmatrix} - \begin{pmatrix} 2 \\ 0 \\ 5 \end{pmatrix} = \begin{pmatrix} 4 \\ 7 \\ 3 \end{pmatrix}$$

You need to find a vector which is perpendicular to AB and BC.

You could find this result using your calculator.

$$\text{Then } \overrightarrow{AB} \times \overrightarrow{BC} = \begin{pmatrix} -1 \\ -4 \\ 3 \end{pmatrix} \times \begin{pmatrix} 4 \\ 7 \\ 3 \end{pmatrix} = \begin{pmatrix} -33 \\ 15 \\ 9 \end{pmatrix} \text{ which can be written as } -3\begin{pmatrix} 11 \\ -5 \\ -3 \end{pmatrix}.$$

So $\mathbf{n} = \begin{pmatrix} 11 \\ -5 \\ -3 \end{pmatrix}$ is a vector perpendicular to the plane containing A, B and C,

and the equation of the plane is of the form $11x - 5y - 3z = d$.

Substituting the coordinates of one of the points, say A, allows you to find the value of the constant d:

$$(11 \times 3) - (5 \times 4) - (3 \times 2) = 7$$

The plane has equation $11x - 5y - 3z = 7$.

Substituting for B and C provides a useful check of your answer.

Discussion point

→ Another way of finding the equation through three given points is to form three simultaneous equations and solve them.

Compare these two methods.

Exercise 11.1

In this exercise you should calculate the vector products by hand. You could check your answers using the vector product facility on a calculator.

① Calculate each of the following vector products:

(i) $\begin{pmatrix} 3 \\ 5 \\ 2 \end{pmatrix} \times \begin{pmatrix} 2 \\ 4 \\ -3 \end{pmatrix}$　(ii) $\begin{pmatrix} 7 \\ -4 \\ -5 \end{pmatrix} \times \begin{pmatrix} -4 \\ 5 \\ -3 \end{pmatrix}$

(iii) $(5\mathbf{i} - 2\mathbf{j} + 4\mathbf{k}) \times (\mathbf{i} + 5\mathbf{j} - 6\mathbf{k})$　(iv) $(3\mathbf{i} - 7\mathbf{k}) \times (2\mathbf{i} + 3\mathbf{j} + 5\mathbf{k})$

② Find a vector perpendicular to each of the following pairs of vectors:

(i) $\mathbf{a} = \begin{pmatrix} 2 \\ 0 \\ 5 \end{pmatrix}, \mathbf{b} = \begin{pmatrix} 3 \\ -1 \\ -2 \end{pmatrix}$　(ii) $\mathbf{a} = \begin{pmatrix} 12 \\ 3 \\ -2 \end{pmatrix}, \mathbf{b} = \begin{pmatrix} 7 \\ 1 \\ 4 \end{pmatrix}$

(iii) $\mathbf{a} = 2\mathbf{i} + 3\mathbf{j} + 4\mathbf{k}, \mathbf{b} = 3\mathbf{i} + 6\mathbf{j} + 7\mathbf{k}$

(iv) $\mathbf{a} = 3\mathbf{i} - 4\mathbf{j} + 6\mathbf{k}, \mathbf{b} = 8\mathbf{i} + 5\mathbf{j} - 3\mathbf{k}$

③ Three points A, B and C have coordinates $(1, 4, -2)$, $(2, 0, 1)$ and $(5, 3, -2)$ respectively.

(i) Find the vectors $\overrightarrow{AB}$ and $\overrightarrow{AC}$.

(ii) Use the vector product to find a vector that is perpendicular to $\overrightarrow{AB}$ and $\overrightarrow{AC}$.

(iii) Hence find the equation of the plane containing points A, B and C.

④ Find a unit vector perpendicular to both $\mathbf{a} = \begin{pmatrix} 1 \\ 2 \\ 7 \end{pmatrix}$ and $\mathbf{b} = \begin{pmatrix} 3 \\ -1 \\ 6 \end{pmatrix}$.

⑤ Find the magnitude of $\begin{pmatrix} 3 \\ 1 \\ -4 \end{pmatrix} \times \begin{pmatrix} 1 \\ -1 \\ 1 \end{pmatrix}$.

⑥ Find the cartesian equations of the planes containing the three points given:

(i) $A(1, 4, 2), B(5, 1, 3)$ and $C(1, 0, 0)$

(ii) $D(5, -3, 4), E(0, 1, 0)$ and $F(6, 2, 5)$

(iii) $G(6, 2, -2), H(1, 4, 3)$ and $L(-5, 7\ 1)$

(iv) $M(4, 2, -1), N(8, 2, 4)$ and $P(5, 8, -7)$

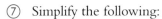

⑦ Simplify the following:

(i) $4\mathbf{i} \times 2\mathbf{k}$ (ii) $2\mathbf{i} \times (5\mathbf{i} - 2\mathbf{j} - 3\mathbf{k})$

(iii) $(6\mathbf{i} + \mathbf{j} - \mathbf{k}) \times 2\mathbf{k}$ (iv) $(3\mathbf{i} - \mathbf{j} + 2\mathbf{k}) \times (\mathbf{i} - \mathbf{j} - 4\mathbf{k})$

⑧ Prove algebraically that for two vectors $\mathbf{a} = a_1\mathbf{i} + a_2\mathbf{j} + a_3\mathbf{k}$ and $\mathbf{b} = b_1\mathbf{i} + b_2\mathbf{j} + b_3\mathbf{k}$

$$\mathbf{a} \times \mathbf{b} = \begin{pmatrix} a_2b_3 - a_3b_2 \\ a_3b_1 - a_1b_3 \\ a_1b_2 - a_2b_1 \end{pmatrix}$$

⑨ Two points A and B have position vectors $\mathbf{a} = 3\mathbf{i} + 5\mathbf{j} + 2\mathbf{k}$ and $\mathbf{b} = 2\mathbf{i} - \mathbf{j} + 4\mathbf{k}$ respectively.

(i) Find the lengths of each of the sides of the triangle OAB, and hence find the area of the triangle.

(ii) Find $|\mathbf{a} \times \mathbf{b}|$.

(iii) How does the definition $\mathbf{a} \times \mathbf{b} = |\mathbf{a}||\mathbf{b}| \sin\theta\hat{\mathbf{n}}$ explain the relationship between your answers to (i) and (ii)?

2 Finding distances

Sometimes you need to find the distance between points, lines and planes. In this section you will look at how to find:

■ the distance from a point to a line, in two or three dimensions;

■ the distance from a point to a plane;

■ the distance between parallel or skew lines.

Finding the distance from a point to a line

Figure 11.5 shows building works at an airport that require the use of a crane near the end of the runway. How far is it from the top of the crane to the flight path of the aeroplane?

plane
taking off

runway   crane

Figure 11.5

To answer this question you need to know the flight path and the position of the top of the crane.

Working in metres, suppose the position of the top of the crane is at P(70, 30, 22)

and the aeroplanes take off along the line $l \colon \mathbf{r} = \begin{pmatrix} -10 \\ 20 \\ 2 \end{pmatrix} + \lambda \begin{pmatrix} 5 \\ 4 \\ 3 \end{pmatrix}$ as illustrated in Figure 11.6.

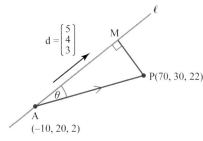

Figure 11.6

The shortest distance from P to the straight line l is measured along the line which is perpendicular to l. It is the distance PM in Figure 11.6. The vector product provides a convenient way of calculating such distances.

Since PM is perpendicular to l

$$PM = AP \sin P\hat{A}M$$

Compare this with the formula for the vector product of AP and AM:

$$\overrightarrow{AP} \times \overrightarrow{AM} = \left|\overrightarrow{AP}\right|\left|\overrightarrow{AM}\right| \sin P\hat{A}M \; \hat{\mathbf{n}}$$

$$\text{so } \left|\overrightarrow{AP}\right| \sin P\hat{A}M = \frac{\left|\overrightarrow{AP} \times \overrightarrow{AM}\right|}{\left|\overrightarrow{AM}\right|}$$

AM is the direction vector $\mathbf{d}$ for the line l, so

$$PM = \frac{\left|\overrightarrow{AP} \times \mathbf{d}\right|}{\left|\mathbf{d}\right|}$$

Returning to calculating the distance from the top of the crane to the flight path:

$$\mathbf{p} = \begin{pmatrix} 70 \\ 30 \\ 22 \end{pmatrix}, \quad \mathbf{a} = \begin{pmatrix} -10 \\ 20 \\ 2 \end{pmatrix} \text{ and } \mathbf{d} = \begin{pmatrix} 5 \\ 4 \\ 3 \end{pmatrix}$$

so that

$$\overrightarrow{AP} = \mathbf{p} - \mathbf{a} = \begin{pmatrix} 70 \\ 30 \\ 22 \end{pmatrix} - \begin{pmatrix} -10 \\ 20 \\ 2 \end{pmatrix} = \begin{pmatrix} 80 \\ 10 \\ 20 \end{pmatrix} = 10\begin{pmatrix} 8 \\ 1 \\ 2 \end{pmatrix}$$

and

$$\overrightarrow{AP} \times \mathbf{d} = 10\begin{pmatrix} 8 \\ 1 \\ 2 \end{pmatrix} \times \begin{pmatrix} 5 \\ 4 \\ 3 \end{pmatrix} = 10\begin{pmatrix} 1 \times 3 - 2 \times 4 \\ 2 \times 5 - 8 \times 3 \\ 8 \times 4 - 1 \times 5 \end{pmatrix} = 10\begin{pmatrix} -5 \\ -14 \\ 27 \end{pmatrix}$$

Therefore, $\left|\overrightarrow{AP} \times \mathbf{d}\right| = 10\sqrt{(-5)^2 + (-14)^2 + 27^2} = 50\sqrt{38}$ and $\left|\mathbf{d}\right| = \sqrt{5^2 + 4^2 + 3^2} = 5\sqrt{2}$.

So the shortest distance from P to l is:

$$PM = \frac{50\sqrt{38}}{5\sqrt{2}} = 10\sqrt{19} \approx 43.6 \text{ metres}$$

You might also need to find the point on l which is closest to P. The following example shows how you can do this for the scenario above.

Example 11.3

The line l has equation $\mathbf{r} = \begin{pmatrix} -10 \\ 20 \\ 2 \end{pmatrix} + \lambda \begin{pmatrix} 5 \\ 4 \\ 3 \end{pmatrix}$ and the point P has coordinates $(70, 30, 22)$.

The point M is the point on l that is closest to P.

(i) Express the position vector $\mathbf{m}$ of point M in terms of the parameter λ.

Hence find an expression for the vector $\overrightarrow{PM}$ in terms of the parameter λ.

(ii) By finding the scalar product of the vector $\overrightarrow{PM}$ with the direction vector $\mathbf{d}$, show that $\lambda = 10$ and hence find the coordinates of point M.

(iii) Verify that $PM = 10\sqrt{19}$ as found earlier.

Solution

(i)
$$\mathbf{m} = \begin{pmatrix} -10 + 5\lambda \\ 20 + 4\lambda \\ 2 + 3\lambda \end{pmatrix}$$

$$\overrightarrow{PM} = \begin{pmatrix} -10 + 5\lambda - 70 \\ 20 + 4\lambda - 30 \\ 2 + 3\lambda - 22 \end{pmatrix} = \begin{pmatrix} -80 + 5\lambda \\ -10 + 4\lambda \\ -20 + 3\lambda \end{pmatrix}$$

(ii) $\overrightarrow{PM} \cdot \mathbf{d} = 0$ ← Since M is the point on l closest to P, PM is perpendicular to l and so PM is perpendicular to $\mathbf{d}$.

$$\begin{pmatrix} -80 + 5\lambda \\ -10 + 4\lambda \\ -20 + 3\lambda \end{pmatrix} \cdot \begin{pmatrix} 5 \\ 4 \\ 3 \end{pmatrix} = 0$$

$$5(-80 + 5\lambda) + 4(-10 + 4\lambda) + 3(-20 + 3\lambda) = 0$$

$$-400 + 25\lambda - 40 + 16\lambda - 60 + 9\lambda = 0$$

$$50\lambda = 500$$

$$\lambda = 10$$

$$\mathbf{m} = \begin{pmatrix} -10 + 5\lambda \\ 20 + 4\lambda \\ 2 + 3\lambda \end{pmatrix} = \begin{pmatrix} 40 \\ 60 \\ 32 \end{pmatrix}$$

(iii)
$$\overrightarrow{PM} = \begin{pmatrix} -30 \\ 30 \\ 10 \end{pmatrix}$$

$$\overrightarrow{PM} = 10\sqrt{(-3)^2 + 3^2 + 1^2} = 10\sqrt{19}$$

As the vector product of vectors $\mathbf{a}$ and $\mathbf{b}$ is a vector perpendicular to both $\mathbf{a}$ and $\mathbf{b}$ the result $\dfrac{\left| \overrightarrow{AP} \times \mathbf{d} \right|}{\left| \mathbf{d} \right|}$ assumes that you are working in three dimensions. When you are working in two dimensions, in which case the vectors have only two components, you can use the following result, which you can prove in Activity 11.2 which follows.

The distance between a point $P(x_1, y_1)$ and the line $ax + by + c = 0$ is:

$$\frac{\left|ax_1 + by_1 + c\right|}{\sqrt{a^2 + b^2}}$$

ACTIVITY 11.2

In this activity, think of points $R(x, y)$ in two dimensional space as corresponding to the point $R'(x, y, 0)$ in three dimensional space.

Use the following steps to show that the distance between a point $P(x_1, y_1)$ and the line $ax + by + c = 0$ is:

$$\frac{\left|ax_1 + by_1 + c\right|}{\sqrt{a^2 + b^2}}$$

(i) Write down the coordinates (x, y) of the point A where the line $ax + by + c = 0$ meets the y-axis. Write down the corresponding coordinates of A′ in three dimensional space.

(ii) Write $ax + by + c = 0$ in the form $y = mx + c$. Find d, e, f so that $d\mathbf{i} + e\mathbf{j} + f\mathbf{k}$ is parallel to the line in three dimensional space which corresponds to the line $ax + by + c = 0$ in two dimensional space.

(iii) Use the formula $\dfrac{\left|(\mathbf{p} - \mathbf{a}) \times \mathbf{d}\right|}{\left|\mathbf{d}\right|}$ to show that the distance between a point $P(x_1, y_1)$ and the line $ax + by + c = 0$ is:

$$\frac{\left|ax_1 + by_1 + c\right|}{\sqrt{a^2 + b^2}}$$

Example 11.4

Find the shortest distance from the point $P(3, 5)$ to the line $5x - 3y + 4 = 0$.

Solution

In this case $x_1 = 3$, $y_1 = 5$ and $a = 5$, $b = -3$, $c = 4$ so the shortest distance from the point P to the line is

$$\frac{\left|ax_1 + by_1 + c\right|}{\sqrt{a^2 + b^2}} = \frac{(5 \times 3) + (-3 \times 5) + 4}{\sqrt{5^2 + (-3)^2}} = \frac{4}{\sqrt{34}}$$

The distance from a point to a plane

The distance from the point $P(x_1, y_1, z_1)$ to the plane π with equation $ax + by + cz + d = 0$ is PM, where M is the foot of the perpendicular from P to the plane (see Figure 11.7).

Notice that since PM is normal to the plane, it is parallel to the vector $\mathbf{n} = a\mathbf{i} + b\mathbf{j} + c\mathbf{k}$.

Take any point, other than M, on the plane and call it R, with position vector $\mathbf{r}$.

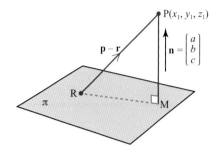

Figure 11.7

If the angle between the vectors $\mathbf{p} - \mathbf{r}$ and $\mathbf{n}$ is acute (as shown in Figure 11.7):

$$PM = RP\cos R\hat{P}M = \overrightarrow{RP}.\hat{\mathbf{n}} \leftarrow \boxed{\begin{array}{l} \text{Using the scalar product} \\ \mathbf{a}.\mathbf{b} = |\mathbf{a}|\,|\mathbf{b}|\cos\theta. \end{array}}$$

$$= (\mathbf{p} - \mathbf{r}).\hat{\mathbf{n}}$$

If the angle between $\mathbf{p} - \mathbf{r}$ and $\mathbf{n}$ is obtuse, $\cos R\hat{P}M$ is negative and

$$PM = -(\mathbf{p} - \mathbf{r}).\hat{\mathbf{n}}$$

Now you want to choose coordinates for the point R that will keep your

$\boxed{\left(0,0,-\dfrac{d}{c}\right) \text{ lies on the plane } ax + by + cz = d.}$ → working simple. A suitable point is $\left(0, 0, -\dfrac{d}{c}\right)$. For this point, $\mathbf{r} = \begin{pmatrix} 0 \\ 0 \\ -\dfrac{d}{c} \end{pmatrix}$.

$$\text{Then } (\mathbf{p} - \mathbf{r}).\mathbf{n} = \begin{pmatrix} x_1 \\ y_1 \\ z_1 + \dfrac{d}{c} \end{pmatrix} . \begin{pmatrix} a \\ b \\ c \end{pmatrix} = ax_1 + by_1 + cz_1 + d$$

and $PM = \left|(\mathbf{p} - \mathbf{r}).\hat{\mathbf{n}}\right|$

$$= \left|\frac{(\mathbf{p} - \mathbf{r}).\mathbf{n}}{|\mathbf{n}|}\right| = \frac{|ax_1 + by_1 + cz_1 + d|}{\sqrt{a^2 + b^2 + c^2}}$$

Notice how this formula for the distance from a point to a plane in three dimensions resembles the distance from a point to a line in two dimensions.

Example 11.5

Find the shortest distance from the point $(2, 4, -2)$ to the plane $6x - y - 3z + 1 = 0$.

Solution

The shortest distance from the point (x_1, y_1, z_1) to the plane $ax + by + cz + d = 0$ is:

$$\frac{|ax_1 + by_1 + cz_1 + d|}{\sqrt{a^2 + b^2 + c^2}}$$

In this case, $x_1 = 2$, $y_1 = 4$, $z_1 = -2$ and $a = 6$, $b = -1$, $c = -3$, $d = 1$

so the shortest distance from the point to the plane is

$$\frac{\left|ax_1 + by_1 + cz_1 + d\right|}{\sqrt{a^2 + b^2 + c^2}}$$

$$= \frac{(6 \times 2) + (-1 \times 4) + (-3 \times -2) + 1}{\sqrt{6^2 + (-1)^2 + (-3)^2}}$$

$$= \frac{15}{\sqrt{46}} \approx 2.21$$

Finding the distance between two parallel lines

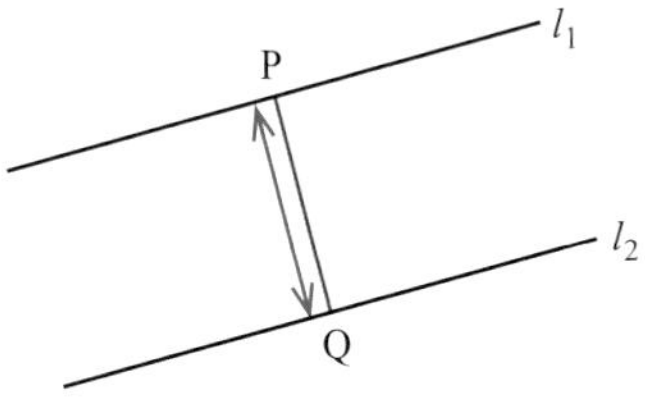

The distance between two parallel lines l_1 and l_2 is measured along a line PQ which is perpendicular to both l_1 and l_2, as shown in Figure 11.8.

You can find this distance by simply choosing a point P on l_1, say, and then finding the shortest distance from P to the line l_2.

Figure 11.8

Example 11.6

Two straight lines in three dimensions are given by the equations:

$$l_1 : \begin{pmatrix} 2 \\ -3 \\ 0 \end{pmatrix} + \lambda \begin{pmatrix} 1 \\ -3 \\ 2 \end{pmatrix} \text{ and } l_2 : \begin{pmatrix} 4 \\ 2 \\ 1 \end{pmatrix} + \mu \begin{pmatrix} -2 \\ 6 \\ -4 \end{pmatrix}$$

(i)   Show that the two lines are parallel.

(ii) Find the shortest distance between the two lines.

Solution

(i)   The direction vectors of the two lines are $\mathbf{d}_1 = \begin{pmatrix} 1 \\ -3 \\ 2 \end{pmatrix}$ and $\mathbf{d}_2 = \begin{pmatrix} -2 \\ 6 \\ -4 \end{pmatrix}$.

Since $\mathbf{d}_2 = -2\mathbf{d}_1$ the two lines are parallel.

You could use any value for λ.

(ii) Choose a point P on l_2 by setting $\lambda = 0$ which gives $\mathbf{p} = \begin{pmatrix} 4 \\ 2 \\ 1 \end{pmatrix}$.

To find the shortest distance of P from l_1, use $\mathbf{a} = \begin{pmatrix} 2 \\ -3 \\ 0 \end{pmatrix}$ and $\mathbf{d} = \begin{pmatrix} 1 \\ -3 \\ 2 \end{pmatrix}$.

$$\overrightarrow{AP} = \begin{pmatrix} 2 \\ 5 \\ 1 \end{pmatrix} \text{ and so } \overrightarrow{AP} \times \mathbf{d} = \begin{pmatrix} 2 \\ 5 \\ 1 \end{pmatrix} \times \begin{pmatrix} 1 \\ -3 \\ 2 \end{pmatrix} = \begin{pmatrix} 13 \\ -3 \\ -11 \end{pmatrix}.$$

$$\left|\overrightarrow{AP} \times \mathbf{d}\right| = \sqrt{13^2 + (-3)^2 + (-11)^2} = \sqrt{299}$$

$$\left|\mathbf{d}\right| = \sqrt{1^2 + (-3)^2 + 2^2} = \sqrt{14}$$

The shortest distance is $\dfrac{\left|\overrightarrow{AP} \times \mathbf{d}\right|}{\left|\mathbf{d}\right|} = \dfrac{\sqrt{299}}{\sqrt{14}} \approx 4.62.$

Finding the distance between skew lines

Two lines are **skew** if they do not intersect and are not parallel.

Figure 11.9 shows two skew lines l_1 and l_2. The shortest distance between the two lines is measured along a line that is perpendicular to both l_1 and l_2.

The common perpendicular of l_1 and l_2 is the perpendicular from l_1 that passes through the point Q, the point of intersection of l_2 and l_1'.

Drop perpendiculars from the points on l_1 to π to form l_1', which is the projection of l_1 on π.

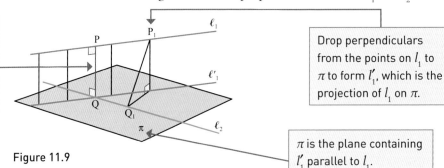

π is the plane containing l_1' parallel to l_1.

Figure 11.9

ACTIVITY 11.3

Explain why PQ is shorter than any other line, such as P_1Q_1 joining lines l_1 and l_2.

Figure 11.10 shows the lines l_1 and l_2 and two parallel planes. Then l_1 and l_2 have equations $\mathbf{r} = \mathbf{a}_1 + \lambda\mathbf{d}_1$ and $\mathbf{r} = \mathbf{a}_2 + \mu\mathbf{d}_2$ respectively. A_1 and A_2 are points on the lines l_1 and l_2 with position vectors $\mathbf{a}_1$ and $\mathbf{a}_2$ respectively.

π_1 contains l_1 and is parallel to l_2

π_2 contains l_2 and is parallel to l_1

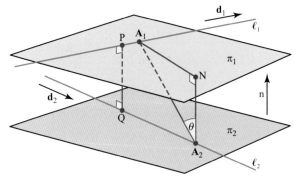

Figure 11.10

Then PQ, the common perpendicular of l_1 and l_2 has the same length as any other perpendicular between the planes such as A_2N. If angle $A_1A_2N = \theta$ then

$$PQ = A_2N = A_2A_1 \cos\theta = \left|\overrightarrow{A_2A_1}.\hat{\mathbf{n}}\right|$$

where $\hat{\mathbf{n}}$ is a unit vector parallel to A_2N, i.e. perpendicular to both planes.

Notice that the modulus function is used to ensure a positive answer: the vector $\hat{\mathbf{n}}$ may be directed from π_1 to π_2 making $\overrightarrow{A_2A_1}.\hat{\mathbf{n}}$ negative.

Since π_1 and π_2 are parallel to l_1 and l_2, which are parallel to $\mathbf{d}_1$ and $\mathbf{d}_2$ respectively, you can take $\mathbf{d}_1 \times \mathbf{d}_2$ as $\mathbf{n}$ with $\hat{\mathbf{n}} = \dfrac{\mathbf{d}_1 \times \mathbf{d}_2}{|\mathbf{d}_1 \times \mathbf{d}_2|}$.

Then:

$$PQ = A_2N = \left|\overrightarrow{A_2A_1}.\hat{\mathbf{n}}\right| = \dfrac{\overrightarrow{A_2A_1}.(\mathbf{d}_1 \times \mathbf{d}_2)}{|\mathbf{d}_1 \times \mathbf{d}_2|} = \dfrac{(\mathbf{a}_1 - \mathbf{a}_2).(\mathbf{d}_1 \times \mathbf{d}_2)}{|\mathbf{d}_1 \times \mathbf{d}_2|}$$

So, the distance between two skew lines is given by:

$$\left|\dfrac{\mathbf{d}_1 \times \mathbf{d}_2}{|\mathbf{d}_1 \times \mathbf{d}_2|}.(\mathbf{a}_1 - \mathbf{a}_2)\right|$$

where $\mathbf{a}_1$ is the position vector of a point on the first line and $\mathbf{d}_1$ is parallel to the first line, and similarly $\mathbf{a}_2$ is the position vector of a point on the second line and $\mathbf{d}_2$ is parallel to the second line.

Example 11.7

Find the shortest distance between the lines l_1: $\mathbf{r} = \begin{pmatrix} 8 \\ 9 \\ -2 \end{pmatrix} + \lambda \begin{pmatrix} 1 \\ 2 \\ -3 \end{pmatrix}$ and

l_2: $\mathbf{r} = \begin{pmatrix} 6 \\ 0 \\ -2 \end{pmatrix} + \mu \begin{pmatrix} 1 \\ -1 \\ -2 \end{pmatrix}$.

Solution

Line l_1 contains the point $A_1(8, 9, -2)$ and is parallel to the vector $\mathbf{d}_1 = \mathbf{i} + 2\mathbf{j} - 3\mathbf{k}$.

Line l_2 contains the point $A_2(6, 0, -2)$ and is parallel to the vector $\mathbf{d}_2 = \mathbf{i} - \mathbf{j} - 2\mathbf{k}$.

$$\mathbf{a}_1 - \mathbf{a}_2 = \begin{pmatrix} 8 \\ 9 \\ -2 \end{pmatrix} - \begin{pmatrix} 6 \\ 0 \\ -2 \end{pmatrix} = \begin{pmatrix} 2 \\ 9 \\ 0 \end{pmatrix}$$

and $\mathbf{d}_1 \times \mathbf{d}_2 = \begin{pmatrix} 1 \\ 2 \\ -3 \end{pmatrix} \times \begin{pmatrix} 1 \\ -1 \\ -2 \end{pmatrix} = \begin{pmatrix} -4-3 \\ -3+2 \\ -1-2 \end{pmatrix} = \begin{pmatrix} -7 \\ -1 \\ -3 \end{pmatrix}$

Then $(\mathbf{d}_1 \times \mathbf{d}_2).(\mathbf{a}_1 - \mathbf{a}_2) = \begin{pmatrix} -7 \\ -1 \\ -3 \end{pmatrix}.\begin{pmatrix} 2 \\ 9 \\ 0 \end{pmatrix} = -14 - 9 = -23$

Also $|\mathbf{d}_1 \times \mathbf{d}_2| = \sqrt{(-7)^2 + (-1)^2 + (-3)^2} = \sqrt{59}$

Therefore the shortest distance between the skew lines is:

$$\left|\dfrac{\mathbf{d}_1 \times \mathbf{d}_2}{|\mathbf{d}_1 \times \mathbf{d}_2|}.(\mathbf{a}_1 - \mathbf{a}_2)\right| = \dfrac{23}{\sqrt{59}} \approx 2.99 \text{ units}$$

① Calculate the distance from the point P to the line l:

(i) P$(1, -2, 3)$ $l: \dfrac{x-1}{2} = \dfrac{y-5}{2} = \dfrac{z+1}{-1}$

(ii) P$(2, 3, -5)$ $l: \mathbf{r} = \begin{pmatrix} 4 \\ 3 \\ 4 \end{pmatrix} + \lambda \begin{pmatrix} 6 \\ -7 \\ 6 \end{pmatrix}$

(iii) P$(8, 9, 1)$ $l: \dfrac{x-6}{12} = \dfrac{y-5}{-9} = \dfrac{z-11}{-8}$

② Find the distance from the point P to the line l:

(i) P$(8, 9)$ $l: 3x + 4y + 5 = 0$

(ii) P$(5, -4)$ $l: 6x - 3y + 3 = 0$

(iii) P$(4, -4)$ $l: 8x + 15y + 11 = 0$

③ Find the distance from the point P to the plane π:

(i) P$(5, 4, 0)$ $\pi: 6x + 6y + 7z + 1 = 0$

(ii) P$(7, 2, -2)$ $\pi: 12x - 9y - 8z + 3 = 0$

(iii) P$(-4, -5, 3)$ $\pi: 8x + 5y - 3z - 4 = 0$

④ A line l_1 has equation $\mathbf{r} = \begin{pmatrix} 2 \\ 0 \\ -1 \end{pmatrix} + \lambda \begin{pmatrix} 1 \\ -2 \\ -1 \end{pmatrix}$.

(i) Write down the equation of a line parallel to l_1 passing through the point $(3, 1, 0)$.

(ii) Find the distance between these two lines.

⑤ (i) Show that the lines $\mathbf{r} = \begin{pmatrix} 1 \\ 2 \\ 4 \end{pmatrix} + \lambda \begin{pmatrix} 3 \\ 0 \\ 2 \end{pmatrix}$ and $\mathbf{r} = \begin{pmatrix} 2 \\ 1 \\ 0 \end{pmatrix} + \mu \begin{pmatrix} 1 \\ 1 \\ -1 \end{pmatrix}$ are skew.

(ii) Find the shortest distance between these two lines.

⑥ Find the shortest distance between the lines l_1 and l_2.

In each case, state whether the lines are skew, parallel or intersect.

(i) $l_1: \dfrac{x-2}{1} = \dfrac{y-3}{2} = \dfrac{z-4}{2}$ and $l_2: \dfrac{x-2}{2} = \dfrac{y-9}{-2} = \dfrac{z-1}{1}$

(ii) $l_1: \dfrac{x-8}{4} = \dfrac{y+2}{3} = \dfrac{z-7}{5}$ and $l_2: \dfrac{x-2}{2} = \dfrac{y+6}{-6} = \dfrac{z-1}{-9}$

(iii) $l_1: \mathbf{r} = \begin{pmatrix} -5 \\ 6 \\ 1 \end{pmatrix} + \lambda \begin{pmatrix} 8 \\ 6 \\ 3 \end{pmatrix}$ and $l_2: \mathbf{r} = \begin{pmatrix} 5 \\ 8 \\ 3 \end{pmatrix} + \mu \begin{pmatrix} 5 \\ 1 \\ 1 \end{pmatrix}$

(iv) $l_1: \mathbf{r} = \begin{pmatrix} 2 \\ 3 \\ -1 \end{pmatrix} + \lambda \begin{pmatrix} 1 \\ 1 \\ 2 \end{pmatrix}$ and $l_2: \mathbf{r} = \begin{pmatrix} 4 \\ 0 \\ -1 \end{pmatrix} + \lambda \begin{pmatrix} -2 \\ -2 \\ -4 \end{pmatrix}$

⑦ (i) Find the shortest distance from the point P(13, 4, 2) to the line

$$l: \mathbf{r} = \begin{pmatrix} 2 \\ -8 \\ -21 \end{pmatrix} + \lambda \begin{pmatrix} 1 \\ -2 \\ 3 \end{pmatrix}.$$

(ii) Find the coordinates of the point M which is the foot of the perpendicular from P to the line l.

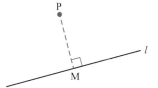

Figure 11.11

⑧ (i) Find the exact distance from the point A(2, 0, −5) to the plane
$\pi: 4x − 5y + 2z + 4 = 0$.

(ii) Write down the equation of the line l through the point A that is perpendicular to the plane π.

(iii) Find the exact coordinates of the point M where the perpendicular from the point A meets the plane π.

⑨ In a school production some pieces of the stage set is held in place by steel cables. The location of points on the cables can be measured, in metres, from an origin O at the side of the stage.

Cable 1 passes through the points A(2, −3, 4) and B(1, −3, 5) whilst cable 2 passes through the points C(0, 3, −2) and D(2, 3, 5).

(i) Find the vector equations of the lines AB and CD and determine the shortest distance between these two cables.

One piece of the stage set, with corner at E(1, 6, −1), needs to be more firmly secured with an additional cable. It is decided that the additional cable should be attached to cable 2.

(ii) If the additional cable available is three metres long, determine whether it will be long enough to attach point E to cable 2.

⑩ The point P has coordinates (4, k, 5) where k is a constant.

The line L has equation $\mathbf{r} = \begin{pmatrix} 1 \\ 0 \\ -4 \end{pmatrix} + \lambda \begin{pmatrix} 1 \\ 2 \\ -2 \end{pmatrix}$.

The line M has equation $\mathbf{r} = \begin{pmatrix} 4 \\ k \\ 5 \end{pmatrix} + \mu \begin{pmatrix} 7 \\ 3 \\ -4 \end{pmatrix}$.

(i) Show that the shortest distance from the point P to the line L is $\frac{1}{3}\sqrt{5(k^2 + 12k + 117)}$.

(ii) Find, in terms of k, the shortest distance between the lines L and M.

(iii) Find the value of k for which the lines L and M intersect.

(iv) When k = 12, show that the distances in parts (i) and (ii) are equal. In this case, find the equation of the line which is perpendicular to, and intersects, both L and M.

⑪ Four points have coordinates A($-2, -3, 2$), B($-3, 1, 5$), C($k, 5, -2$) and D($0, 9, k$).

(i) Find the vector product $\overrightarrow{AB} \times \overrightarrow{CD}$.

(ii) For the case when AB is parallel to CD:

(a) state the value of k

(b) find the shortest distance between the parallel lines AB and CD

(c) find, in the form $ax + by + cz + d = 0$, the equation of the plane containing AB and CD.

(iii) When AB is not parallel to CD, find the shortest distance between the lines AB and CD in terms of k.

(iv) Find the value of k for which the line AB intersects the line CD, and find the coordinates of the point of intersection in this case.

⑫ The point A($-1, 12, 5$) lies on the plane P with equation $8x - 3y + 10z = 6$.

The point B($6, -2, 9$) lies on the plane Q with equation $3x - 4y - 2z = 8$.

The planes P and Q intersect in the line L.

(i) (a) Show that the point $(0, -2, 0)$ lies on both planes.

(b) Use the vector product to find a vector perpendicular to both plane P and plane Q.

(c) Write down the equation of L, the line of intersection of planes P and Q.

(ii) Find the shortest distance between L and the line AB.

The lines M and N are both parallel to L, with M passing through A and N passing through B.

(iii) Find the distance between the parallel lines M and N.

The point C has coordinates $(k, 0, 2)$ and the line AC intersects the line N at the point D.

(iv) Find the value of k and the coordinates of D.

LEARNING OUTCOMES

When you have completed this chapter you should be able to:

➤ use the vector product in component form to find a vector perpendicular to two given vectors

➤ know that $\mathbf{a} \times \mathbf{b} = |\mathbf{a}||\mathbf{b}| \sin \theta \hat{\mathbf{n}}$, where $\mathbf{a}$, $\mathbf{b}$ and $\hat{\mathbf{n}}$, in that order, form a right-handed triple

➤ find the distance from a point to a line in two or three dimensions

➤ find the distance between two parallel lines

➤ find the shortest distance between two skew lines

➤ find the distance of a point from a plane.

KEY POINTS

1 The vector product $\mathbf{a} \times \mathbf{b}$ of $\mathbf{a}$ and $\mathbf{b}$ is a vector perpendicular to both $\mathbf{a}$ and $\mathbf{b}$

$$\mathbf{a} \times \mathbf{b} = |\mathbf{a}||\mathbf{b}|\sin\theta\,\hat{\mathbf{n}}$$

where θ is the angle between $\mathbf{a}$ and $\mathbf{b}$ and $\hat{\mathbf{n}}$ is a unit vector which is perpendicular to both $\mathbf{a}$ and $\mathbf{b}$ such that $\mathbf{a}$, $\mathbf{b}$ and $\hat{\mathbf{n}}$ (in that order) form a right-handed set of vectors.

2 $\mathbf{a} \times \mathbf{b} = \begin{pmatrix} a_1 \\ a_2 \\ a_3 \end{pmatrix} \times \begin{pmatrix} b_1 \\ b_2 \\ b_3 \end{pmatrix} = \begin{pmatrix} a_2b_3 - a_3b_2 \\ a_3b_1 - a_1b_3 \\ a_1b_2 - a_2b_1 \end{pmatrix} = \begin{vmatrix} \mathbf{i} & a_1 & b_1 \\ \mathbf{j} & a_2 & b_2 \\ \mathbf{k} & a_3 & b_3 \end{vmatrix}$

3 In three dimensions, the shortest distance from a point P with position vector $\mathbf{p}$ to a line with direction vector $\mathbf{d}$ and passing through the point A, with position vector $\mathbf{a}$, is given by:

$$\frac{\left|\overrightarrow{AP} \times \mathbf{d}\right|}{|\mathbf{d}|}$$

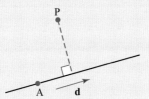

Figure 11.12

4 The shortest distance from a point $P(x_1, y_1, z_1)$ to the plane $ax + by + cz + d = 0$ is given by:

$$\frac{\left|ax_1 + by_1 + cz_1 + d\right|}{\sqrt{a^2 + b^2 + c^2}}$$

5 In three dimensions there are three possibilities for the arrangement of the lines. They are either parallel, intersecting or skew.

6 The shortest distance between two parallel lines can be found by choosing any point on one of the lines and finding the shortest distance from that point to the second line.

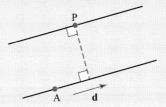

Figure 11.13

7 The distance between two skew lines is given by:

$$\left| \frac{\mathbf{d}_1 \times \mathbf{d}_2}{\left| \mathbf{d}_1 \times \mathbf{d}_2 \right|} \cdot (\mathbf{a}_1 - \mathbf{a}_2) \right|$$

where $\mathbf{a}_1$ is the position vector of a point on the first line and $\mathbf{d}_1$ is parallel to the first line, similarly for the second line.

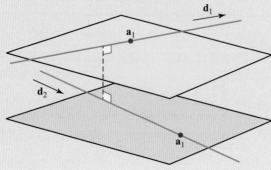

Figure 11.14

Second order differential equations

Discussion point

➜ The picture shows the Tacoma Narrows Bridge in the U.S. state of Washington, which collapsed on 7 November 1940. Find out what caused the collapse of the bridge.

1 Higher order differential equations

So far in this course you have mostly solved only first order differential equations. In this chapter you will extend these techniques to second (and higher) order differential equations. Second order equations are often needed in mechanics, particularly to model situations which involve acceleration.

A reasonable model of a parachutist (with parachute open) is provided by treating the parachutist as a particle of mass m, subject to two forces: weight, mg downwards, and resistance, kv, against the motion (i.e. upwards – where v is the velocity, and k is a constant).

Applying Newton's second law with the downward direction as positive gives:
$mg - kv = ma$

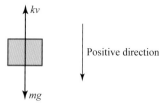

Using some standard notation

s = distance fallen

$v = \dfrac{ds}{dt}$ = velocity

$a = \dfrac{dv}{dt} = \dfrac{d^2 s}{dt^2}$ = acceleration

Figure 12.1

the equation can be written in several different ways:

1   $\dfrac{dv}{dt} = g - \dfrac{k}{m}v$ 2 $v\dfrac{dv}{ds} = g - \dfrac{k}{m}v$ 3 $\dfrac{d^2 v}{dt^2} = g - \dfrac{k}{m}v$

The first two equations are first order differential equations and can be solved using the method of separation of variables, covered in Chapter 9.

However, the third equation is of a type that is new to you. It is a second order differential equation, and this type of differential equation is the subject of this chapter.

Notation and vocabulary

As an example, the differential equation

$$\frac{d^3 y}{dx^3} - 7\frac{d^2 y}{dx^2} + 2\frac{dy}{dx} + 4y = 3\sin x$$

is described as:

- **third order** (because the highest derivative is a third derivative)
- **linear** (because where y and its derivatives appear they are to the power 1)
- having **constant coefficients** (because the coefficients of the terms involving y are constants – in this case 1, −7, 2 and 4)
- **non-homogeneous** (because the right-hand side, the part not containing y, is *not zero*). In cases where the right hand side *is zero* the differential equation is called **homogeneous**.

In this chapter you will meet **second order, linear differential equations, with constant coefficients**. In general they can be written as follows:

$$a\frac{d^2 y}{dx^2} + b\frac{dy}{dx} + cy = f(x)$$

with a, b and c constant. This may in fact be written without the a constant, by dividing through by a, but the form given is helpful as it is very similar to the form of quadratic equations which you will be familiar with. The rest of this chapter concentrates on the method of solving this type of differential equation and its applications.

The auxiliary equation method

Before solving second and higher order differential equations you will find it helpful to look at the form of first order, linear, homogeneous differential equation with constant coefficients. In general they are of the form:

First order. ⟶ $\dfrac{dy}{dx} + ky = 0$ ⟵ Homogeneous because the right-hand side is zero.

Constant coefficients because k is constant.

You can solve this by separating the variables:

$$\frac{dy}{dx} = -ky$$

$$\int \frac{1}{y}\,dy = \int -k\,dx$$

$$\ln|y| = -kx + c$$

$$y = \pm e^{-kx+c}$$

$$= Ae^{-kx}$$

So, in general, the solution of *any* first order equation of this type will contain an *exponential function* and *one unknown constant*.

Knowing the form of the solution allows you to find it without doing any integration, as shown in the next example.

Example 12.1

Solve the differential equation:

$$5\frac{dy}{dx} + y = 0$$

Solution

Since this is first order, linear, homogeneous and with constant coefficients, you know that the solution will be of the form

$$y = Ae^{\lambda x}$$

where λ and A are constants.

If this is to be a solution, it must satisfy the original differential equation.

Differentiating y with respect to x.

Substituting y and $\dfrac{dy}{dx}$ into the differential equation.

$$\frac{dy}{dx} = \lambda Ae^{\lambda x}$$

$$5\lambda Ae^{\lambda x} + Ae^{\lambda x} = 0$$

$$5\lambda + 1 = 0$$ ⟵ Dividing by $Ae^{\lambda x}$ (since it is not zero) gives an equation just involving λ.

$$\lambda = -\frac{1}{5}$$

So the general solution of the original differential equation is:

$$y = Ae^{-\frac{1}{5}x}$$

Note

In Example 12.1 there cannot be any other solutions because this one already contains the one necessary arbitrary constant to give it the generality it needs.

This method of assuming the form of the solution is extremely powerful and is called the **auxiliary equation method**. The equation $5\lambda + 1 = 0$ is the **auxiliary equation.**

ACTIVITY 12.1

All of the following differential equations can be solved by at least one of the methods:

- separation of variables
- integrating factor
- auxiliary equation.

(i) Which of the equations are linear? Which ones have constant coefficients?

(ii) For each equation, state which method (or methods) can be used and use it (or them) to solve the equation.

(a) $\dfrac{dy}{dx} - 17y = 0$

(b) $\dfrac{dy}{dx} - y^2 = 0$

(c) $\dfrac{dy}{dx} + xy = 0$

(d) $\dfrac{dy}{dx} - 3y = 0$

(e) $y\dfrac{dy}{dx} - y^2 = 0$

Second order homogenous differential equations

The ideas from the work above can be extended to cover second order equations.

Look at this differential equation. It is a second order, linear, homogeneous differential equation with constant coefficients.

$$\frac{d^2y}{dx^2} - 5\frac{dy}{dx} + 6y = 0$$

Suppose that you assume a solution of the form $y = Ae^{\lambda x}$ (just as before), where A and λ are constants. Then

$$\frac{dy}{dx} = A\lambda e^{\lambda x}$$

and

$$\frac{d^2y}{dx^2} = A\lambda^2 e^{\lambda x}$$

Substituting these into the differential equation gives:

$$A\lambda^2 e^{\lambda x} - 5A\lambda e^{\lambda x} + 6Ae^{\lambda x} = 0$$

Dividing through by $Ae^{\lambda x}$ you obtain the auxiliary equation. It is a quadratic equation in λ:

$$\lambda^2 - 5\lambda + 6 = 0$$

Notice that the form of the auxiliary equation is very close to the form of the original differential equation, and with experience you will often write it down straight away, without the intermediate working shown above.

Factorising the auxiliary equation gives:

$$(\lambda - 3)(\lambda - 2) = 0$$

You can see there are two different values for λ which satisfy the auxiliary equation: $\lambda = 3$ and $\lambda = 2$. So $y = Ae^{3x}$ and $y = Be^{2x}$ (with A and B constants) are two solutions of the differential equation.

Each of the two expressions Ae^{3x} and Be^{2x} is a **complementary function** of the differential equation. The sum of these expressions: $Ae^{3x} + Be^{2x}$, is usually called *the* complementary function, and is also a solution of the original differential equation. Since it has *two* arbitrary constants it is, in fact, the **general solution** of the original second order equation.

ACTIVITY 12.2

Verify that the complementary function $y = Ae^{3x} + Be^{2x}$ is a solution to the original differential equation, $\dfrac{d^2 y}{dx^2} - 5\dfrac{dy}{dx} + 6y = 0$.

This method can be used on any linear differential equation with constant coefficients, whatever its order. If $\lambda_1, \lambda_2, \lambda_3, \ldots$ are the roots of the auxiliary equation, then, assuming there are no repeated roots, the general solution of the homogeneous differential equation is

$$y = Ae^{\lambda_1 x} + Be^{\lambda_2 x} + Ce^{\lambda_3 x} + \ldots$$

There are three important points to notice about the auxiliary equation method:

- this method does not involve any integration;
- the number of terms in the complementary function is equal to the order of the differential equation;
- the number of unknown constants in the solution is equal to the order of the differential equation.

In general, a second order, linear, homogenous differential equation with constant coefficients

$$a\frac{d^2 y}{dx^2} + b\frac{dy}{dx} + cy = 0$$

has auxiliary equation

$$a\lambda^2 + b\lambda + c = 0$$

There are three types of solution to this equation and they each lead to different types of solution:

- When $b^2 > 4ac$, there are real, distinct roots.
  e.g. $x^2 + 3x - 10 = 0$ has roots $x = 2, \ x = -5$

- When $b^2 = 4ac$ there are real, repeated roots.
  e.g. $x^2 + 4x + 4 = 0$ has root $x = -2$ (twice)

- When $b^2 < 4ac$ there are complex, conjugate roots.
  e.g. $x^2 - 2x + 2 = 0$ has roots $x = 1 + i, \ x = 1 - i$

In this section you will look at the first two cases, and you will also see how you can find values for the arbitrary constants to give a **particular solution**.

Auxiliary equation with real distinct roots, $b^2 > 4ac$

$$a\frac{d^2y}{dx^2} + b\frac{dy}{dx} + cy = 0$$

You have already seen an example of this situation at the start of this section. In general, if $b^2 > 4ac$ there will be two distinct roots of the auxiliary equation, λ_1 and λ_2, which lead to the complementary function:

$$y = Ae^{\lambda_1 x} + Be^{\lambda_2 x}$$

This complementary function is the general solution in this case, but to find a particular solution you need to eliminate the two unknown constants. This is only possible with two extra pieces of information, which may come in two different ways:

- Initial conditions:
  If the two conditions are given for the same value of the independent variable (often x), you say that you have two **initial conditions**.
  For example: $y = 0$ when $x = 0$, and $\frac{dy}{dx} = 1$ when $x = 0$. Since both conditions are for $x = 0$, these are initial conditions, and the problem is called an **initial value problem**.

- Boundary conditions:
  If the two conditions are given for *different* values of the independent variable, you say that you have two **boundary conditions**.
  For example: $y = 0$ when $x = 0$, and $y = 1$ when $x = 1$. Since these use different x-values they are boundary conditions, and the problem is called a **boundary value problem**. Often the solutions to boundary value problems are restricted to the region between the boundary points in the conditions.

The following examples demonstrate one of each type of problem.

Example 12.2	

(i) Find the particular solution of $\dfrac{d^2y}{dx^2} + 4\dfrac{dy}{dx} + 3y = 0$

subject to the conditions $y = 0$ and $\dfrac{dy}{dx} = 1$ when $x = 0$.

(ii) Sketch the graph of the solution.

Solution

(i) Auxiliary equation:

$$\lambda^2 + 4\lambda + 3 = 0$$

Factorise:

$$(\lambda + 3)(\lambda + 1) = 0$$

Solving this gives two distinct roots, $\lambda_1 = -1$ and $\lambda_2 = -3$.

Complementary function:

$$y = Ae^{-x} + Be^{-3x}$$

and this is also the general solution.

To find the values of A and B you need to use the conditions $y = 0$ and $\dfrac{dy}{dx} = 1$ when $x = 0$.

> You need to differentiate your general solution in order to use the $\dfrac{dy}{dx}$ condition.

$$\frac{dy}{dx} = -Ae^{-x} - 3Be^{-3x}$$

When $x = 0$, $y = 0$ $\Rightarrow A + B = 0$

When $x = 0$, $\dfrac{dy}{dx} = 1$ $\Rightarrow -A - 3B = 1$

Solving these two equations simultaneously gives:

$$A = \tfrac{1}{2} \text{ and } B = -\tfrac{1}{2}$$

So the particular solution is:

$$y = \tfrac{1}{2}e^{-x} - \tfrac{1}{2}e^{-3x}$$

(ii)

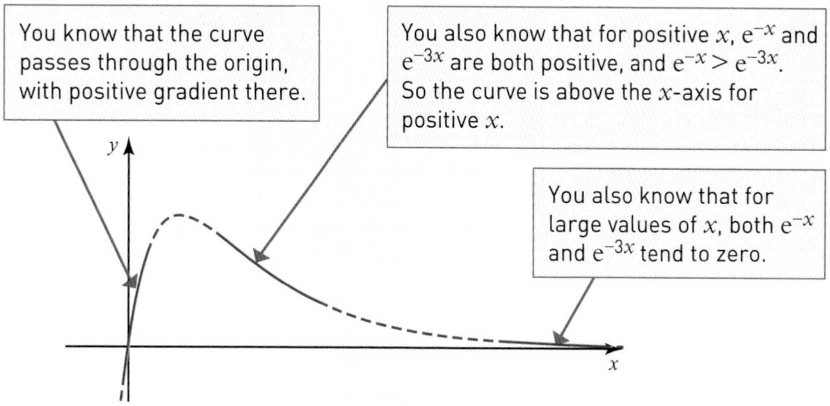

> You know that the curve passes through the origin, with positive gradient there.

> You also know that for positive x, e^{-x} and e^{-3x} are both positive, and $e^{-x} > e^{-3x}$. So the curve is above the x-axis for positive x.

> You also know that for large values of x, both e^{-x} and e^{-3x} tend to zero.

Note

You can check the shape of the graph using a graphical calculator or graphing software.

Figure 12.2

Example 12.3

The differential equation

$$\frac{d^2y}{dx^2} + 4\frac{dy}{dx} + 3y = 0$$

is used to model a situation in which the value of x lies between 0 and 2.

(i) Find the particular solution given the boundary conditions $y = 0$ at $x = 0$, and $y = 1$ when $x = 2$.

(ii) Sketch a graph of the solution for $0 \leqslant x \leqslant 2$.

Solution

(i) The general solution is the same as in the previous example:

$$y = Ae^{-x} + Be^{-3x}$$

Using the boundary conditions:

$y = 0, \ x = 0$ gives $\quad 0 = A + B$

$y = 1, \ x = 2$ gives $\quad 1 = Ae^{-2} + Be^{-6}$

Solving these simultaneously gives:

> Substituting for B from the first equation into the second.

$$1 = Ae^{-2} - Ae^{-6}$$

$$A = \frac{e^2}{1 - e^{-4}}$$

and then

$$B = -\frac{e^2}{1 - e^{-4}}$$

The particular solution then is:

$$y = \frac{e^2}{1 - e^{-4}}(e^{-x} - e^{-3x})$$

> **Note**
>
> Note that you have two points defined as the boundary conditions, but the same reasoning as before about the variable terms.

(ii)

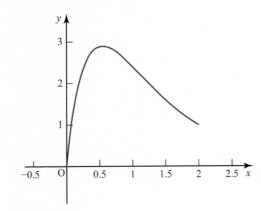

Figure 12.3

> **Discussion point**
> → The last two examples started off exactly the same. However, when it came to finding the particular solutions, they required different approaches because of the different types of conditions involved. Which is the easier to work with?

Auxiliary equation with repeated roots, $b^2 = 4ac$

$$a\frac{\mathrm{d}^2 y}{\mathrm{d}x^2} + b\frac{\mathrm{d}y}{\mathrm{d}x} + cy = 0$$

If $b^2 = 4ac$ then there is only one root from the auxiliary equation, and it is $\lambda = -\frac{b}{2a}$. (Check this with the quadratic formula).

This leads to a complementary function of $y = Ae^{\lambda x}$. But, since the original differential equation is second order, the general solution needs to have *two* arbitrary constants. So, you need to find another solution to the equation in order to provide this extra constant.

At this point, if someone suggested another possible solution of the form $y = Bxe^{\lambda x}$ (note the extra x in front), you could test to see if it satisfies the original differential equation:

Using the product rule

$$\frac{\mathrm{d}y}{\mathrm{d}x} = Be^{\lambda x} + B\lambda xe^{\lambda x}$$

and differentiating again

$$\frac{\mathrm{d}^2 y}{\mathrm{d}x^2} = B\lambda e^{\lambda x} + B\lambda e^{\lambda x} + B\lambda^2 xe^{\lambda x}$$

$$= 2B\lambda e^{\lambda x} + B\lambda^2 xe^{\lambda x}$$

If $y = Bxe^{\lambda x}$ is a solution, you should get 0 when you substitute these expressions into the left-hand side of the original equation. You get:

$$a\left(2B\lambda e^{\lambda x} + B\lambda^2 xe^{\lambda x}\right) + b\left(Be^{\lambda x} + B\lambda xe^{\lambda x}\right) + c\left(Bxe^{\lambda x}\right)$$

$$= Be^{\lambda x}\left(2a\lambda + b + x\left(a\lambda^2 + b\lambda + c\right)\right)$$

But, in this case, $\lambda = -\frac{b}{2a}$, which rearranges to give $2a\lambda = -b$, meaning that the first two terms in the above bracket sum to zero. (Notice this only works in this case because of the value of λ).

The last bracket (the coefficient of x) is $a\lambda^2 + b\lambda + c$, which is the left-hand side of the auxiliary equation, and therefore equal to zero too.

The whole expression is therefore zero, which means the expression $y = Bxe^{\lambda x}$ *does* satisfy the original differential equation, and so is a complementary function – and it is different from your first solution $y = Ae^{\lambda x}$.

So, you can combine them to form the complementary function (which is also the general solution), with two arbitrary constants:

$$y = Ae^{\lambda x} + Bxe^{\lambda x}$$

or

$$y = (A + Bx)e^{\lambda x}$$

See question 11 in Exercise 12.1 for a way to derive this result.

Example 12.4

(i) Find the particular solution of the equation

$$\frac{d^2 z}{dt^2} + 6\frac{dz}{dt} + 9z = 0$$

subject to the conditions $z = 0$ and $\frac{dz}{dt} = 5$ when $t = 0$.

(ii) Sketch the graph of the particular solution.

Solution

(i) Auxiliary equation:

$$\lambda^2 + 6\lambda + 9 = 0$$

$$(\lambda + 3)^2 = 0$$

$$\lambda = -3 \text{ (repeated)}$$

So, the general solution is:

$$z = (A + Bt)e^{-3t}$$

Differentiating this in order to use the condition gives:

$$\frac{dz}{dt} = Be^{-3t} - 3(A + Bt)e^{-3t}$$

And using the conditions:

$$0 = (A + B \times 0) \times 1$$

$$A = 0$$

and

$$5 = B - 3A$$

$$B = 5$$

So, the particular solution is:

$$z = 5te^{-3t}$$

(ii)

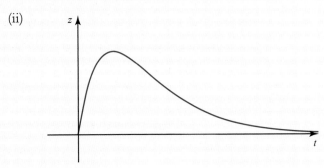

Figure 12.4

Note

For all graphs of the form $y = xe^{-kx}$, $y \to 0$ as $x \to \infty$. You are multiplying an increasing function, x, by a decreasing function e^{-kx}, and in this case the decreasing one wins.

To summarise: if the auxiliary equation has a repeated real root λ, then the complementary function is:

$$y = (A + Bx)e^{\lambda x}$$

① Use the auxiliary equation method to find the general solution of the following differential equations:

(i) $\dfrac{dy}{dx} - 3y = 0$ (ii) $\dfrac{dy}{dx} + 7y = 0$ (iii) $\dfrac{dx}{dt} + x = 0$

(iv) $\dfrac{dp}{dt} - 0.02p = 0$ (v) $5\dfrac{dz}{dt} - z = 0$

② Use the auxiliary equation method to find the particular solutions of the following differential equations:

(i) $\dfrac{dy}{dx} + 2y = 0$ $y = 3$ when $x = 0$

(ii) $2\dfrac{dy}{dx} - 5y = 0$ $y = 1$ when $x = 0$

(iii) $3\dfrac{dx}{dt} + x = 0$ $x = 2$ when $t = 1$

(iv) $\dfrac{dP}{dt} = kP$ $P = P_0$ when $t = 0$

(v) $\dfrac{dm}{dt} = -km$ $m = m_0$ when $t = 0$

③ Find the general solution of the following differential equations:

(i) $\dfrac{d^2y}{dx^2} - 16y = 0$ (ii) $\dfrac{d^2y}{dx^2} + \dfrac{dy}{dx} - 5y = 0$

(iii) $2\dfrac{d^2x}{dt^2} + 4\dfrac{dx}{dt} + x = 0$ (iv) $7\dfrac{d^2y}{dx^2} + 2\dfrac{dy}{dx} = 0$

(v) $9\dfrac{d^2y}{dx^2} - 12\dfrac{dy}{dx} + 4y = 0$

④ You are given the differential equation $\dfrac{d^2y}{dx^2} - 3\dfrac{dy}{dx} + 2y = 0$.

  (i) Write down the auxiliary equation in terms of λ.

  (ii) Solve the auxiliary equation to find its two real roots.

  (iii) Write down the complementary function in the form $y(x) = \ldots$

  (iv) Find the derivative of the complementary function.

  (v) You are also told that $y = 1$ and $\dfrac{dy}{dx} = 0$ when $x = 0$. Use these facts, the complementary function and its derivative, to eliminate the arbitrary constants and find the particular solution that satisfies these constraints.

⑤ Given $\dfrac{d^2y}{dx^2} - 8\dfrac{dy}{dx} + 16y = 0$

  (i) Write down the auxiliary equation, and solve it.

  (ii) Write down the complementary function (this is also the general solution).

  (iii) Find the particular solution which satisfies the initial conditions $y = 1$ and $\dfrac{dy}{dx} = -1$ when $x = 0$.

⑥ Find the particular solutions of the following differential equations that satisfy the conditions given. In each case sketch the graph of the solution (you may wish to use graphing software to help visualise this). Assuming each one models a real system, where x represents a variable changing over time, use the graph to describe the motion of each system in words.

(i) $\dfrac{d^2x}{dt^2} + 5\dfrac{dx}{dt} = 0$ $x = 0, \dfrac{dy}{dx} = 4$, when $t = 0$

(ii) $4\dfrac{d^2x}{dt^2} + 4\dfrac{dx}{dt} + x = 0$ $x = 1, t = 0$ and $x = 1, t = 1$

⑦ In an electrical circuit, the charge q coloumbs on a capacitor is modelled by the differential equation:

$$0.2\dfrac{d^2q}{dt^2} + \dfrac{dq}{dt} + 1.25q = 0$$

Initially $q = 2$ and $\dfrac{dq}{dt}$ (the current in amperes) is 4.

(i) Find an equation for the charge as a function of time.

(ii) Sketch the graphs of charge and current against time. Describe how the charge and current change.

(iii) What is the charge on the capacitor and the current in the circuit after a long period of time?

⑧ The temperature of a chemical undergoing a reaction is modelled by the differential equation

$$2\dfrac{d^2T}{dt^2} + \dfrac{dT}{dt} = 0$$

where T is the temperature in °C and t is the time in minutes.

For a particular experiment, the temperature is initially 50 °C, and it is 45 °C one minute later.

(i) Find an expression for the temperature T at any time.

(ii) What will the temperature be after two minutes?

(iii) Sketch a graph of T against t.

(iv) What is the steady state temperature?

⑨ The metal fins on a motorcycle engine help to cool it. A model for the change of temperature along a fin is given by $\dfrac{d^2T}{dx^2} - 4T = 0$, where T °C is the temperature and x m is the distance from the hot end.

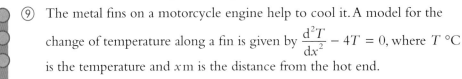

(i) Find the general solution of this differential equation.

When the engine has been running for some time the temperatures at the two ends of the fin are $100\,°C$ and $80\,°C$. The fin is $5\,cm$ long.

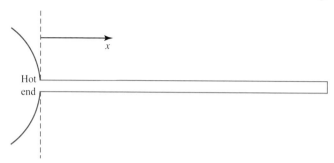

Figure 12.5

(ii) Find the particular solution.

(iii) Hence determine the temperature $3\,cm$ from the hot end.

⑩ A car of mass $800\,kg$ is travelling along a road at a constant velocity of $100\,km\,h^{-1}$. A catastrophic engine and brake failure renders the car unable to brake in any way other than natural slow-down from the friction from the air and road. The friction is modelled as a backwards force equal to $40\,000$ times the velocity (in $km\,h^{-1}$).

(i) Show that the equation of motion of the car as it slows (x in kilometres, t in hours), simplifies to $\dfrac{d^2x}{dt^2} + 50\dfrac{dx}{dt} = 0$.

(ii) Solve this differential equation, using appropriate initial conditions, to find a particular solution for x in terms of t as the distance (in km) travelled after the brake failure.

(iii) Find the time taken for the car to travel $1\,km$ further, and the speed of the car at this moment.

(iv) State at least one problem with the solution given by this model.

⑪ In this question you will prove the result for the general solution of a second order differential equation for which the auxiliary equation has repeated roots.

(i) Find the general solution of the differential equation $\dfrac{d^2y}{dx^2} = 0$.

(ii) Assume that the differential equation

$$\frac{d^2y}{dx^2} + 4\frac{dy}{dx} + 4y = 0$$

has a solution of the form $y = f(x)e^{-2x}$. Differentiate this twice and substitute it into the original differential equation, and hence find the form for $f(x)$.

(iii) Repeat for the differential equation $\dfrac{d^2y}{dx^2} + 2k\dfrac{dy}{dx} + k^2y = 0$.

2 Auxiliary equations with complex roots

$$a\frac{d^2y}{dx^2} + b\frac{dy}{dx} + cy = 0$$

If $b^2 < 4ac$ then the two roots will be complex (and conjugates of each other). For example, the differential equation

$$\frac{d^2y}{dx^2} + 4y = 0$$

has auxiliary equation

$$\lambda^2 + 4 = 0.$$

The roots of this quadratic are $\lambda_1 = 2i$ and $\lambda_2 = -2i$ (where $i = \sqrt{-1}$).

The general solution of the differential equation is therefore.

$$y = Ae^{2ix} + Be^{-2ix}$$

The terms in the solution contain complex exponentials, and it is important to recognise that these can always be written in terms of sine and cosine functions (see Chapter 10) by using the relationships:

$$e^{i\theta} = \cos\theta + i\sin\theta$$

$$e^{-i\theta} = \cos\theta - i\sin\theta$$

So you can rewrite the general solution as:

$$y = A(\cos 2x + i\sin 2x) + B(\cos 2x - i\sin 2x)$$
$$= (A + B)\cos 2x + i(A - B)\sin 2x$$

Notice now that the coefficients of $\cos 2x$ and $\sin 2x$ are constants, which you can call P and Q, such that $P = A + B$ and $Q = i(A - B)$. This means the general solution can be written simply as:

$$y = P\cos 2x + Q\sin 2x$$

It is important to note that this solution has an oscillating nature, and in fact all differential equations with complex roots from their auxiliary equation will give oscillating solutions.

> **Note**
>
> Notice that complex numbers are needed to find this form for the general solution, and yet this abstract result actually works as a model for oscillating systems in the real world.

Simple harmonic motion

Many real-life situations can be modelled by a differential equation of the form

$$\frac{d^2x}{dt^2} = -\omega^2 x.$$

For example:

- A mass fixed to the end of a spring is pulled down vertically and then released. The displacement x of the mass from its equilibrium position, at time t, can be modelled as

$$\frac{d^2x}{dt^2} = -\frac{k}{m}x$$

where k is the stiffness of the spring and m is the mass.

- A mass suspended from a fixed point is moved horizontally slightly from its equilibrium position and then released, forming a simple pendulum. The angle θ which the string makes with the vertical at time t, can be modelled as

$$\frac{\mathrm{d}^2\theta}{\mathrm{d}t^2} = -\frac{g}{l}\theta$$

where l is the length of the string and g is the acceleration due to gravity.

Both these equations have the same form:

$$\frac{\mathrm{d}^2 x}{\mathrm{d}t^2} = -\omega^2 x.$$

In the first case $\omega^2 = \frac{k}{m}$, in the second $\omega^2 = \frac{g}{l}$. In the second case the variable is θ rather than x.

In both cases the motion is called **simple harmonic motion** (SHM); it provides a model for many oscillations.

The general solution of the differential equation, as shown above, is

$$x = P\cos\omega t + Q\sin\omega t.$$

The constants of integration, P and Q, are unknown at this stage, but if you know suitable initial or boundary conditions you can calculate their values.

> The right-hand side of this equation should really involve $\sin\theta$ but the small angle approximation of $\sin\theta \approx \theta$ is very accurate and is used in this model.

Prior knowledge

You need to know how to write an expression of the form $a\cos\theta + b\sin\theta$ in the form $r\sin(\theta + \alpha)$. This is covered in Chapter 8 of the A Level Mathematics Year 2 textbook.

ACTIVITY 12.3

Given that the general solution $x = P\cos\omega t + Q\sin\omega t$ can be written in the form $x = a\sin(\omega t + \varepsilon)$, use the compound angle formulae to prove that $a = \sqrt{P^2 + Q^2}$ and $\tan\varepsilon = \frac{Q}{P}$.

Expressing the general solution $x = P\cos\omega t + Q\sin\omega t$ in the form $x = a\sin(\omega t + \varepsilon)$ tells you a lot about the solution.

- Since the sine function varies between $+1$ and -1, the solution varies between $-a$ and a.

- Since the sine function is periodic with period 2π, the solution is periodic with period $\frac{2\pi}{\omega}$.

ACTIVITY 12.4

Use graphing software to investigate the graph $x = a\sin(\omega t + \varepsilon)$ for different values of a, ω and ε.

In the activity above you should have noticed that the effect of ε is to translate the sine curve to the left by an amount $\frac{\varepsilon}{\omega}$, as shown in the diagram below. The quantity ε is called the **phase shift**.

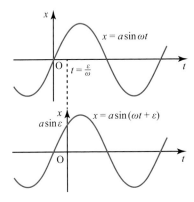

Figure 12.6

Example 12.5

(i) Find the general solution of the differential equation $\dfrac{d^2x}{dt^2} + 4x = 0$.

(ii) Find the particular solution in the case for which $x = 2$ and $\dfrac{dx}{dt} = -2$ when $t = 0$.

(iii) Find the period and amplitude of the oscillations, and sketch the graph of the particular solution.

Solution

(i) Auxiliary equation: $\lambda^2 + 4 = 0$

                                    $\lambda = \pm 2i$

     General solution: $x = A\cos 2t + B\sin 2t$

(ii) When $t = 0, x = 2$ $\Rightarrow 2 = A$

$$\frac{dx}{dt} = -2A\sin 2t + 2B\cos 2t$$

When $t = 0, \dfrac{dx}{dt} = -2$ $\Rightarrow -2 = 2B \quad \Rightarrow B = -1$

     Particular solution: $x = 2\cos 2t - \sin 2t$

> To sketch the graph and find the amplitude, it is helpful to write the solution in the form $a\sin(2t + \varepsilon)$.

> Using the compound angle formula.

(iii) The period of the oscillations is $\dfrac{2\pi}{\omega} = \dfrac{2\pi}{2} = \pi$

$2\cos 2t - \sin 2t = a\sin(2t + \varepsilon)$

$ = a\sin 2t\cos\varepsilon + a\cos 2t\sin\varepsilon$

Comparing coefficients of $\sin 2t$: $-1 = a\cos\varepsilon$

Comparing coefficients of $\cos 2t$: $2 = a\sin\varepsilon$

So $a = \sqrt{2^2 + 1^2} = \sqrt{5}$.

The amplitude of the oscillations is $\sqrt{5}$.

$\tan\varepsilon = -2 \quad \Rightarrow \varepsilon = -1.107$

The particular solution can be written in the form
$$x = \sqrt{5} \sin(2t - 1.107).$$

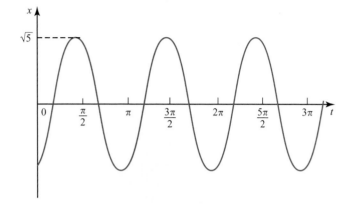

Figure 12.7

In the previous examples the roots of the auxiliary equation were purely imaginary. Often the roots also have a real part.

For example, suppose that the roots of the auxiliary equation are $2 \pm 3i$.

This gives the complementary function

$$y = Ae^{(2+3i)x} + Be^{(2+3i)x}$$

$$= e^{2x}\left(Ae^{3ix} + Be^{-3ix}\right)$$

But, as before, you can write $Ae^{3ix} + Be^{-3ix}$ as $P\cos 3x + Q\sin 3x$, so

$$y = e^{2x}(P\cos 3x + Q\sin 3x).$$

This is the standard form of the general solution.

Note

The four forms of complementary function, for the four cases of solutions of the auxiliary equation are:

■ Real distinct roots, λ_1 and λ_2: $y = Ae^{\lambda_1 x} + Be^{\lambda_2 x}$

■ Real, repeated root, λ: $y = (A + Bx)e^{\lambda x}$

■ Complex roots, $\alpha \pm \beta i$: $y = e^{\alpha x}(P\cos \beta x + Q\sin \beta x)$

■ Pure imaginary roots, $\pm \omega i$ $y = P\cos \omega x + Q\sin \omega x$

Example 12.6

(i) Find the particular solution of the differential equation

$$\frac{d^2 y}{dx^2} + 2\frac{dy}{dx} + 5y = 0$$

which satisfies the initial conditions $y = 0$ and $\dfrac{dy}{dx} = 1$ when $x = 0$.

(ii) Sketch the graph of this particular solution.

Solution

(i) The auxiliary equation is:

$$\lambda^2 + 2\lambda + 5 = 0$$

Completing the square gives:

> Alternatively you could use the quadratic formula.

$$(\lambda + 1)^2 - 1 + 5 = 0$$

$$\lambda + 1 = \pm\sqrt{-4}$$

$$\lambda = -1 \pm 2i$$

The general solution is $y = e^{-x}(P\cos 2x + Q\sin 2x)$.

To find the particular solution:

When $x = 0, y = 0 \Rightarrow P = 0$

So the general solution is now:

$$y = Qe^{-x}\sin 2x$$

$$\Rightarrow \frac{dy}{dx} = -Qe^{-x}\sin 2x + 2Qe^{-x}\cos 2x$$

When $x = 0, \dfrac{dy}{dx} = 1 \Rightarrow 1 = 2Q$

$$\Rightarrow Q = \tfrac{1}{2}$$

So the particular solution for these initial conditions is:

$$y = \tfrac{1}{2}e^{-x}\sin 2x$$

(ii) The graph of this particular solution is shown below. Notice that the oscillating solution moves between the curves $= \pm\frac{1}{2}e^{-x}$, which are indicated by the blue dotted lines.

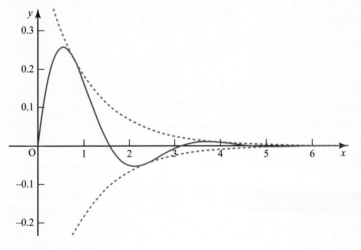

Figure 12.8

It is important to understand the relationship between the solution and its graph. In Example 12.6:

■ the e^{-x} factor tells you that the solution will decay as x increases (because of the negative sign)

■ the $\sin 2x$ factor tells you that there is an oscillation.

This form of solution can be described as **exponentially decaying oscillations** or **damped oscillations**, and it arises frequently when modelling real oscillating systems.

Damped oscillations

Simple harmonic motion has constant amplitude and goes on for ever. For many real oscillating systems, SHM is not a very good model: usually the amplitude of the oscillations gradually decreases, and the motion dies away.

Example 12.7	A damped oscillating system is modelled by the differential equation

$$\frac{d^2x}{dt^2} + k\frac{dx}{dt} + 25x = 0$$

where k is a constant that can be varied.

When $t = 0$, $x = 0$ and $\frac{dx}{dt} = 1$.

Solve the differential equation for each of the following values of k, and sketch the graph of the solution in each case.

(i) $k = 26$

(ii) $k = 6$

(iii) $k = 10$

Solution

(i)   Auxiliary equation: $\lambda^2 + 26\lambda + 25 = 0$

$(\lambda + 1)(\lambda + 25) = 0$

$\lambda = -1 \text{ or } -25$

General solution: $x = Ae^{-t} + Be^{-25t}$

When $t = 0$, $x = 0 \implies A + B = 0$

$$\frac{dx}{dt} = -Ae^{-t} - 25Be^{-25t}$$

When $t = 0$, $\frac{dx}{dt} = 1 \implies -A - 25B = 1$

Solving these equations simultaneously gives: $A = \frac{1}{24}$, $B = -\frac{1}{24}$

Particular solution is $x = \frac{1}{24}e^{-t} - \frac{1}{24}e^{-25t}$.

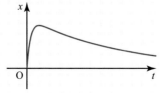

Figure 12.9

(ii)   Auxiliary equation: $\lambda^2 + 6\lambda + 25 = 0$

$$\lambda = \frac{-6 \pm \sqrt{36 - 100}}{2} = -3 \pm 4i$$

General solution: $x = e^{-3t}(A\sin 4t + B\cos 4t)$

When $t = 0, x = 0 \Rightarrow B = 0$

$x = Ae^{-3t}\sin 4t$

$\dfrac{dx}{dt} = -3Ae^{-3t}\sin 4t + 4Ae^{-3t}\cos 4t$

When $t = 0, \dfrac{dx}{dt} = 1 \Rightarrow 4A = 1 \Rightarrow A = \frac{1}{4}$

Particular solution is $x = \frac{1}{4}e^{-3t}\sin 4t$.

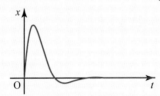

Figure 12.10

(iii)   Auxiliary equation: $\lambda^2 + 10\lambda + 25 = 0$

$(\lambda + 5) = 0$

$\lambda = -5$

General solution: $x = (A + Bt)e^{-5t}$

When $t = 0, x = 0 \Rightarrow A = 0 \quad x = Bte^{-5t}$

$\dfrac{dx}{dt} = Be^{-5t} - 5Bte^{-5t}$

When $t = 0, \dfrac{dx}{dt} = 1 \Rightarrow B = 1$

Particular solution is $x = te^{-5t}$.

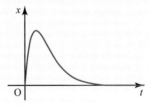

Figure 12.11

The three parts of Example 12.7 illustrate the three different types of damping. The type of damping that occurs depends on the nature of the roots of the auxiliary equation.

- Overdamping: the discriminant of the auxiliary equation is positive and the roots are negative; the system decays without oscillating.

Figure 12.12

- Underdamping: the discriminant of the auxiliary equation is negative and oscillations occur.

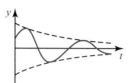

Figure 12.13

- Critical damping: the discriminant of the auxiliary equation is zero.

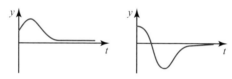

Figure 12.14

Critical damping is the borderline between overdamping and underdamping. It is not obvious in a physical situation when damping is critical, since the pattern of motion for critical damping can be very similar to that in the overdamped case.

Exercise 12.2

① For the differential equation

$$\frac{\mathrm{d}^2 x}{\mathrm{d}t^2} + 9x = 0$$

(i) Write down the general solution of the differential equation.

(ii) Given the initial conditions $x = 0$ and $\frac{\mathrm{d}x}{\mathrm{d}t} = 1$ when $t = 0$, find the particular solution.

(iii) Write down the period and amplitude of the oscillations, and sketch a graph of the solution.

② For the differential equation

$$4\frac{d^2x}{dt^2} + x = 0$$

(i) Write down the general solution of the differential equation.

(ii) Given the initial conditions $x = 4$ and $\frac{dx}{dt} = 0$ when $t = 0$, find the particular solution.

(iii) Write down the period and amplitude of the oscillations, and sketch a graph of the solution.

③ Find the general solution of each of the following differential equations:

(i) $\frac{d^2y}{dx^2} - 4\frac{dy}{dx} + 5y = 0$

(ii) $\frac{d^2y}{dx^2} - 2\frac{dy}{dx} + 5y = 0$

(iii) $\frac{d^2x}{dt^2} + 2\frac{dx}{dt} + 4x = 0$

(iv) $4\frac{d^2x}{dt^2} + 4\frac{dx}{dt} + 5x = 0$

④ Given $9\frac{d^2x}{dt^2} + 4x = 0$

(i) Find the general solution.

(ii) Find the particular solution which satisfies the conditions $x = 4$ and $\frac{dx}{dt} = 2$ when $t = 0$. Express the solution in the form $x = a\sin(t + \varepsilon)$.

(iii) Sketch the graph of the particular solution.

⑤ Find the particular solutions of the following differential equations that satisfy the conditions given. In each case sketch the graph of the solution (you may wish to use graphing software to help visualise this). Assuming each one models a real system, where x represents a variable changing over time, use the graph to describe the motion of each system in words.

(i) $\frac{d^2x}{dt^2} + 8x = 0$ $\qquad$ $x = 1, \frac{dx}{dt} = 1$ when $t = 0$

(ii) $\frac{d^2x}{dt^2} - \frac{dx}{dt} + x = 0$ $\qquad$ $x = 1, \frac{dx}{dt} = 0$ when $t = 0$

(iii) $\frac{d^2x}{dt^2} + 2\frac{dx}{dt} + 2x = 0$ $\qquad$ $x = 0, \frac{dx}{dt} = 2$ when $t = 0$

(iv) $4\frac{d^2x}{dt^2} - 8\frac{dx}{dt} + 5x = 0$ $\qquad$ $x = 2, \frac{dx}{dt} = 0$ when $t = 0$

(v) $\frac{d^2x}{dt^2} + 2\frac{dx}{dt} + 5x = 0$ $\qquad$ $x = 0, t = 0$ and $x = 3, t = \frac{\pi}{4}$

⑥ The motion of a spring-mass oscillator is modelled by the differential equation

$$\frac{d^2x}{dt^2} + 64x = 0$$

where x is the extension of the spring at time t.

(i) Find the general solution of the differential equation.

(ii)   Initially $x = 0.1$ and $\dfrac{dx}{dt} = 0$. Find the particular solution corresponding to these initial conditions.

(iii)   Write down the period and amplitude of the motion.

(iv)   Sketch a graph of the solution.

⑦   The angular displacement from its equilibrium position of a swing door is modelled by the differential equation

$$\frac{d^2\theta}{dt^2} + 4\frac{d\theta}{dt} + 5\theta = 0.$$

The door starts from rest at an angle of $\dfrac{\pi}{4}$ from its equilibrium position.

(i)   Find the general solution of the differential equation.

(ii)   Find the particular solution for the given initial conditions.

(iii)   Sketch a graph of the particular solution, and hence describe the motion of the door.

(iv)   What does your model predict as t becomes large?

⑧   A damped spring-mass oscillator consists of a spring and an object with mass m kg. The object is pulled down 10 cm from its equilibrium position and released. The motion of the system is modelled by the differential equation

$$m\frac{d^2x}{dt^2} + k\frac{dy}{dx} + 20x = 0$$

where x is the displacement from the equilibrium position at time t.

(i)   In the case where $m = 0.25$, find the value of k if the system is to be critically damped.

(ii)   If the value of k does not change, describe the motion of the system if:

(a) the mass of the object is increased to 0.3 kg

(b) the mass of the object is decreased to 0.2 kg

⑨   Prove that, if a, b and c are positive constants, then all possible solutions of

$$a\frac{d^2x}{dt^2} + b\frac{dx}{dt} + cx = 0$$

approach zero as $t \to \infty$.

⑩   Find the general solution of the differential equation

$$\frac{d^4y}{dx^4} - 16y = 0.$$

3 Non-homogeneous differential equations

So far you have seen that equations of the form

$$a\frac{d^2y}{dx^2} + b\frac{dy}{dx} + cy = 0$$

(linear, homogeneous, second order differential equations, with constant coefficients) have general solutions of the form

$$y = Au(x) + Bv(x)$$

where u(x) and v(x) may involve exponential and/or trigonometric functions (and possibly a factor x). In modelling real situations, equations like this often arise in which the right-hand side is non-zero. These equations are called **non-homogeneous** linear equations:

$$a\frac{d^2 y}{dx^2} + b\frac{dy}{dx} + cy = f(x)$$

There are many situations where such equations arise, such as the equation modelling the motion of the parachutist near the beginning of this chapter,

$\frac{d^2 x}{dt^2} + \frac{k}{m}\frac{dx}{dt} = g$, or when an external force causes a structure to vibrate (forced oscillations).

In order to learn how to deal with this type of second order differential equation, it is useful to look at similar first order equations, e.g.

| Linear, first order. |
| Constant coefficients. |

$$\frac{dy}{dx} + 2y = 3x - 1$$

| Non-homogeneous because the right-hand side is non-zero. |

You should remember how to solve this using an integrating factor method, but it is instructive to compare the solutions to several similar looking equations.

ACTIVITY 12.5

Verify that the following solutions satisfy their respective first order linear differential equations:

(i) $\dfrac{dy}{dx} + 2y = 0$ $\qquad\qquad$ $y = Ae^{-2x}$

(ii) $\dfrac{dy}{dx} + 2y = 3x - 1$ $\qquad$ $y = Ae^{-2x} + \frac{3}{2}x - \frac{5}{4}$

(iii) $\dfrac{dy}{dx} + 2y = e^{3x}$ $\qquad$ $y = Ae^{-2x} + \frac{1}{5}e^{3x}$

(iv) $\dfrac{dy}{dx} + 2y = \sin x$ $\qquad$ $y = Ae^{-2x} - \frac{1}{5}\cos x + \frac{3}{5}\sin x$

Compare the different solutions: what is the same and what is different?

You should notice that all the solutions have strong similarities, and in particular they consist of two distinct parts.

Notes
- The complementary function solves the homogeneous case, and contains an arbitrary constant.
- The particular integral solves the full non-homogeneous case, and does not contain an arbitrary constant.

- The first part Ae^{-2x} contains an arbitrary constant, and is the general solution for the homogeneous version (part (i) of the activity). It appears to depend only on the left-hand side of the differential equation. This part of the solution is the **complementary function**.

- The second part only exists in the non-homogeneous cases, contains no arbitrary constants, appears to depend only on the right-hand side of the differential equation – and is of a similar form to the original right-hand side (e.g. linear in part (ii), exponential in part (iii), trigonometric in part (iv)). This part of the solution is called the **particular integral**.

Since you already know how to find the complementary function, all that remains in order for you to use this method of constructing a solution is a way to find a particular integral. The following first order example demonstrates a way to approach this, using a **trial function**.

Example 12.8

Solve the differential equation

$$\frac{dy}{dx} + 2y = e^{3x}.$$

> This is one of the examples from the previous activity.

Solution

The auxiliary equation is $\lambda + 2 = 0$

$$\Rightarrow \lambda = -2$$

The complementary function is $y = Ae^{-2x}$.

To find the particular integral use a trial function of $y = ae^{3x}$, since the right-hand side is of the form ke^{3x} where k is a constant.

$$y = ae^{3x}$$

$$\frac{dy}{dx} = 3ae^{3x}$$

> Differentiate.

$$3ae^{3x} + 2ae^{3x} = e^{3x}$$

> Substitute into the differential equation.

Comparing coefficients of e^{3x} gives:

$$5a = 1$$

so $a = \frac{1}{5}$.

$$y = Ae^{-2x} + \frac{1}{5}e^{3x}$$

> Notice that you still have one arbitrary constant, as you would expect for a first order equation.

> The general solution is constructed by adding the complementary function and the particular integral.

The previous examples could also have been solved using first order methods, such as an integrating factor, but this method of complementary functions and particular integrals is more powerful since it can be used for higher order equations. The next example shows how this method is used for a second order differential equation, and also demonstrates a different form of particular integral.

Example 12.9

Find the general solution of the differential equation

$$\frac{d^2y}{dx^2} - 2\frac{dy}{dx} - 3y = 6x - 2.$$

Solution

The auxiliary equation is $\lambda^2 - 2\lambda - 3 = 0$

$$(\lambda - 3)(\lambda + 1) = 0$$

$$\lambda = 3 \text{ or } -1$$

> The right-hand side is a linear function, so your trial function should be a general linear function.

So the complementary function is $y = Ae^{3x} + Be^{-x}$.

To find a particular integral, use a trial function $y = ax + b$.

$$\frac{dy}{dx} = a \quad \text{and} \quad \frac{d^2y}{dx^2} = 0$$

> Differentiating twice.

> Substituting into the original differential equation.

$$0 - 2a - 3(ax + b) = 6x - 2$$

Comparing coefficients of x gives

$$-3a = 6$$
$$a = -2$$

and comparing constant terms

$$-2a - 3b = -2$$
$$4 - 3b = -2$$
$$-3b = -6$$
$$b = 2$$

The particular integral is $y = -2x + 2$.

So the general solution is:

$$y = Ae^{3x} + Be^{-x} - 2x + 2$$

Notice that the general solution has the properties you need:

■   it satisfies the full differential equation (you may wish to check this);

■   it has two arbitrary constants, which is consistent with it being the general solution to a second order differential equation.

Particular integrals

In order to find a suitable particular integral, the trial function needs to match the right-hand side of the differential equation.

Table 12.1 is a guide to what trial function to use for the differential equation

$$a\frac{d^2y}{dx^2} + b\frac{dy}{dx} + cy = f(x)$$

Right-hand side: $f(x)$	Trial function
linear function	$ax + b$
polynomial of order n	$a_n x^n + a_{n-1} x^{n-1} + \ldots + a_1 x + a_0$
trigonometric function involving $\cos px$ and/or $\sin px$	$a\cos px + b\sin px$
exponential function involving e^{px}	ae^{px}
sum of different functions	sum of matching functions

Table 12.1

You should note that this method finds a simple particular integral, but that there are infinitely many possible particular integrals, which can be constructed by adding on any term from the complementary function.

The following examples demonstrate this trial function approach, for the cases you have not yet seen.

Example 12.10

Find a particular integral of the differential equation

$$\frac{d^2z}{dt^2} - 2\frac{dz}{dt} - 3z = 6\cos 3t.$$

Solution

Use the trial function: $z = a\cos 3t + b\sin 3t$

Notice that both the $\cos 3t$ and $\sin 3t$ functions are used, since they differentiate to become each other.

Differentiating the trial function twice:

$$\frac{dz}{dt} = -3a\sin 3t + 3b\cos 3t$$

$$\frac{d^2z}{dt^2} = -9a\cos 3t - 9b\sin 3t$$

Substituting these into the differential equation gives

$$(-9a\cos 3t - 9b\sin 3t) - 2(-3a\sin 3t + 3b\cos 3t) - 3(a\cos 3t + b\sin 3t)$$
$$= 6\cos 3t$$

$$(-12a - 6b)\cos 3t + (-12b + 6a)\sin 3t = 6\cos(3t)$$

Equating coefficients of $\cos 3t$ gives $\quad -12a - 6b = 6$

Equating coefficients of $\sin 3t$ gives $\quad -12b + 6a = 0$

Solving for a and b gives $\qquad a = -\frac{2}{3}$ and $b = -\frac{1}{3}$

The particular integral is $-\frac{2}{5}\cos 3t - \frac{1}{5}\sin 3t$

Special cases

In some differential equations the function on the right-hand side has the same form as one of the complementary functions. For example, the complementary function of the differential equation

$$\frac{d^2y}{dx^2} - 5\frac{dy}{dx} + 6y = 4e^{3x}$$

is $Ae^{2x} + Be^{3x}$, and e^{3x} occurs on the right-hand side. In this situation it is no good using the trial function ae^{3x}, since upon substituting $y = ae^{3x}$, $\frac{dy}{dx} = 3ae^{3x}$ and $\frac{d^2y}{dx^2} = 9ae^{3x}$ into the differential equation, you obtain

$$9ae^{3x} - 5(3ae^{3x}) + 6(ae^{3x}) = 4e^{3x}$$

$$\Rightarrow 0 = 4e^{3x}$$

and so clearly this trial function does not work.

Instead $y = axe^{3x}$ is used as a trial function.

This gives $\qquad \frac{dy}{dx} = ae^{3x} + 3axe^{3x}$

and $\qquad \frac{d^2y}{dx^2} = 6ae^{3x} + 9axe^{3x}$

Substituting these in the differential equation gives:

$$(6ae^{3x} + 9axe^{3x}) - 5(ae^{3x} + 3axe^{3x}) + 6axe^{3x} = 4e^{3x}$$

$$\Rightarrow ae^{3x} = 4e^{3x}$$

$$\Rightarrow a = 4$$

A particular integral is $4xe^{3x}$.

This illustrates a general rule. If the function on the right-hand side of the differential equation has exactly the same form as one of the complementary functions, you multiply the usual trial function by the independent variable to give a new trial function. In order to recognise these special cases when they arise, it is worth getting into the habit of finding the complementary function before the particular integral.

Exercise 12.3

① Find a particular integral for each of the following differential equations.

(i) $\dfrac{d^2y}{dx^2} - 4\dfrac{dy}{dx} + y = -2x + 3$

(ii) $\dfrac{d^2x}{dt^2} + 4x = t + 2$

(iii) $\dfrac{d^2y}{dx^2} + 2\dfrac{dy}{dx} + y = \cos 3x$

(iv) $\dfrac{d^2x}{dt^2} + \dfrac{dx}{dt} + 2x = 3e^{-2t}$

② Find the general solutions of the following differential equations.

(i) $\dfrac{d^2y}{dx^2} + 4y = e^{2x}$

(ii) $\dfrac{d^2x}{dt^2} - 2\dfrac{dx}{dt} - 3x = 5e^{-2t}$

(iii) $\dfrac{d^2y}{dx^2} + 2\dfrac{dy}{dx} + 5y = \cos x$

(iv) $\dfrac{d^2y}{dx^2} + 2\dfrac{dy}{dx} + 5y = 3x + 2$

③ Find the general solutions of the following differential equations.

(i) $\dfrac{d^2x}{dt^2} + x = \sin t$

(ii) $\dfrac{d^2y}{dx^2} + 3\dfrac{dy}{dx} - 4y = e^x$

(iii) $\dfrac{d^2x}{dt^2} - 4\dfrac{dx}{dt} + 3x = e^{3t}$

(iv) $\dfrac{d^2x}{dt^2} - 6\dfrac{dx}{dt} + 9x = 4e^{3t}$

④ Find the particular solution of each of the following differential equations with the given initial conditions.

(i) $\dfrac{d^2y}{dx^2} - 5\dfrac{dy}{dx} + 6y = 36x$ $y = 0$ and $\dfrac{dy}{dx} = -10$ when $x = 0$

(ii) $\dfrac{d^2x}{dt^2} + 9x = 20e^{-t}$ $x = 0$ and $\dfrac{dx}{dt} = 1$ when $t = 0$

(iii) $\dfrac{d^2y}{dx^2} + 2\dfrac{dy}{dx} + 5y = 4e^{-x}$ $y = 0$ and $\dfrac{dy}{dx} = 0$ when $x = 0$

(iv) $\dfrac{d^2x}{dt^2} + \dfrac{dx}{dt} - 2x = 20\sin 2t$ $x = 2$ and $\dfrac{dx}{dt} = 0$ when $t = 0$

(v) $\dfrac{d^2x}{dt^2} - 3\dfrac{dx}{dt} + 2x = 1 - e^t$ $x = 0$ and $\dfrac{dx}{dt} = 1$ when $t = 0$

(vi) $\dfrac{d^2y}{dx^2} + 4y = 12\sin 2x$ $y = 0$ and $\dfrac{dy}{dx} = 1$ when $x = 0$

(vii) $\dfrac{d^2y}{dx^2} + 4\dfrac{dy}{dx} + 5y = 8\sin x$ $y = 1$ and $\dfrac{dy}{dx} = 0$ when $x = 0$

(viii) $\dfrac{d^2y}{dx^2} + 4\dfrac{dy}{dx} + 4y = 8e^{2x} + 4x$ $y = 0$ and $\dfrac{dy}{dx} = 1$ when $x = 0$

⑤ A biological population of size P at time t is growing in an environment which can support a maximum population which is subject to seasonal variation. The growth of the population is described by the first order linear differential equation

$$\frac{dP}{dt} + P = 100 + 50\sin t$$

Find:

(i) the complementary function of this differential equation

(ii) the particular integral

(iii) the complete solution given that initially $P = 20$

(iv) the mean size of the population after a long time has elapsed

(v) the amplitude of the oscillations of the population

⑥ The pointer on a set of kitchen scales oscillates before settling down at its final reading.

If x is the reading at time t then the oscillation of the pointer is modelled by the differential equation

$$\frac{d^2x}{dt^2} + 3\frac{dx}{dt} + 10x = 0.5$$

(i) Find the general solution of the equation for x.

(ii) Given that $x = 0.1$ and $\frac{dx}{dt} = 0$ when $t = 0$, find the particular solution. At what reading will the pointer settle?

(iii) What length of time will elapse before the amplitude of the oscillations of the pointer is less than 20% of the final value of x?

⑦ In an electrical circuit, the charge q coulombs stored in a capacitor at time t is given by the differential equation

$$\frac{d^2q}{dt^2} + 100\frac{dq}{dt} + 10\,000q = 1000\sin 10t$$

where t is the time in seconds.

(i) Find the complementary function of the differential equation.

(ii) Decide if the circuit is overdamped, critically damped or underdamped.

(iii) Find the particular integral and hence the general solution.

(iv) Initially both q and $\frac{dq}{dt}$, the current in the circuit, are zero. Find the particular solution.

(v) Use a graphical calculator or graphing software to sketch the solution and describe how the charge in the circuit changes with time.

⑧ A simple model for the motion of a delicate set of laboratory scales, when an object of mass m is placed gently on the scales, is given by the differential equation

$$m\frac{d^2x}{dt^2} + 7\frac{dx}{dt} + 5x = 2m$$

where x is the displacement from the initial position at time t.

(i) Find the particular solution that models the motion of the object in each of the cases:

(a) $m = 2$ (b) $m = 2.45$ (c) $m = 4$

(ii) Sketch a graph of the solution for each value of m. Describe the motion in each case.

⑨ (i) Solve the equation $\dfrac{dy}{dx} - ky = e^{pt}$, where $k \ne p$,

    (a) using the integrating factor method

    (b) by finding the complementary function and particular integral

(ii) Repeat part (i) for the equation $\dfrac{dy}{dx} - ky = e^{kt}$.

⑩ *You are advised to have a graphical calculator or graphing software to hand when doing this question.*

In normal running, a machine vibrates slightly and this is monitored by a marker on it. The marker moves along a straight line. Its displacement in mm from an origin is denoted by x.

The motion of the marker is subject to the differential equation

$$\dfrac{d^2x}{dt^2} + 9x = 0.$$

At the start of the motion, when $t = 0$, the marker is stationary and $x = 2$.

(i) Find the particular solution.

    Draw a graph of x against t.

One day the machine's operator attaches another mechanism to it with the result that a forced oscillation is applied to the machine and the differential equation governing x becomes

$$\dfrac{d^2x}{dt^2} + 9x = 5\cos 2t.$$

The same initial conditions apply.

(ii) Verify that the solution is $x = \cos 3t + \cos 2t$.

    Draw a graph of x against t for $0 \le t \le 2\pi$.

    Describe what has happened to the motion of the marker.

Another day the operator changes the frequency of the forced oscillation so that the differential equation becomes

$$\dfrac{d^2x}{dt^2} + 9x = 5\cos 3t.$$

Again the same initial conditions apply.

(iii) Solve the equation.

    Draw a sketch graph of x against t.

    Predict what happens to the machine when this forced oscillation is applied.

4 Systems of differential equations

Many situations that are modelled by differential equations involve a number of variables which may depend on one another, a **system**. Sometimes this means that the situation may be modelled by two or more differential equations.

Example 12.11

Figure 12.15 shows how lead from pollution can build up in the body. Biological research shows that it is reasonable to assume that the rate of transfer of lead from one part of the body to another is proportional to the amount of lead present. The symbols p, q, r and s are the constants of proportionality for the routes shown in the diagram.

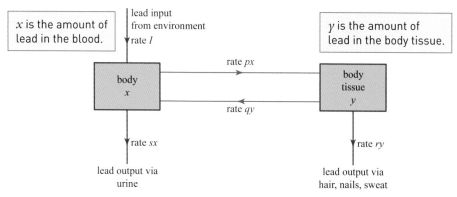

| x is the amount of lead in the blood. | | y is the amount of lead in the body tissue. |

Figure 12.15

Formulate a system of differential equations to model this situation.

Solution

For the blood, at any time t:

Rate of increase of lead in the bloodstream = rate of input − rate of output

$$\frac{\mathrm{d}x}{\mathrm{d}t} = (I + qy) - (px + sx)$$
$$= I - (p + s)x + qy$$

Similarly for the body tissue:

Rate of increase of lead in the body tissues = rate of input − rate of output

$$\frac{\mathrm{d}y}{\mathrm{d}t} = px - qy - ry$$
$$= px - (q + r)y$$

The example above resulted in two differential equations, both involving the dependent variables x and y and the independent value t. When two or more differential equations involve the same combination of variables they are called a **system of differential equations**. In this case, since both of the equations are linear, it is a **linear system**.

Solving linear simultaneous differential equations

To solve a pair of linear simultaneous differential equations, you need to eliminate one of the variables so that you end up with a differential equation involving just one dependent variable. This is shown in the following example.

Example 12.12

Solve the simultaneous equations

$$\frac{dx}{dt} = 2x + 4y \qquad ①$$

$$\frac{dy}{dt} = x - y \qquad ②$$

with initial conditions $x = 2$ and $y = -2$ when $t = 0$.

Solution

$$x = \frac{dy}{dt} + y \qquad ③$$

> Make x the subject in equation ②.

$$\frac{dx}{dt} = \frac{d^2y}{dy^2} + \frac{dy}{dt}$$

> Differentiate with respect to t.

$$\frac{d^2y}{dy^2} + \frac{dy}{dt} = 2\left(\frac{dy}{dt} + y\right) + 4y$$

> Substitute for x and $\frac{dx}{dt}$ in equation ①.

$$\frac{d^2y}{dy^2} - \frac{dy}{dt} - 6y = 0$$

> Simplify to give a second order differential equation which involves only y and not x.

Auxiliary equation: $\qquad \lambda^2 - \lambda - 6 = 0$

$$(\lambda - 3)(\lambda + 2) = 0$$

$$\lambda = 3 \text{ or } -2$$

General solution for y: $\quad y = Ae^{3t} + Be^{-2t}$

$$x = \frac{dy}{dt} + y$$

> Substituting into equation ③ to give the general solution for x.

$$= \left(3Ae^{3t} - 2Be^{-2t}\right) + \left(Ae^{3t} + Be^{-2t}\right)$$

$$= 4Ae^{3t} - Be^{-2t}$$

When $t = 0, x = 2 \quad \Rightarrow 2 = 4A - B$

When $t = 0, y = -2 \quad \Rightarrow -2 = A + B$

Solving for A and B gives $A = 0$ and $B = -2$.

The particular solution is $x = 2e^{-2t}$

$$y = -2e^{-2t}$$

In the example above, both x and y tend to zero as t tends to infinity. Figure 12.16 shows the behaviour of the solution graphically.

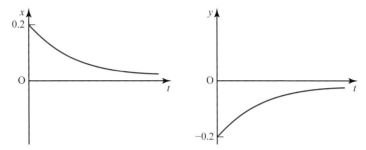

Figure 12.16

Sometimes it is important to understand the relationship between x and y, in which case the solution can be thought of as a pair of parametric equations with parameter t. In the example above, t can be eliminated simply by adding the two solutions:

$$x = 2e^{-2t}$$
$$\underline{y = -2e^{-2t}}$$
$$x + y = 0$$

This is illustrated in Figure 12.17. Notice that only part of the line $x + y = 0$ is required, since the starting point is $x = -2, y = 2$, and as $t \to \infty, (x, y) \to (0,0)$.

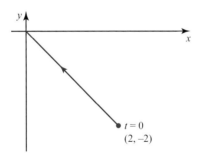

Figure 12.17

A graph like this is called a solution curve. It shows the relationship between the dependent variables (in this case x and y) as the independent variable (in this case t) increases through its permitted range.

Example 12.13

A system of differential equations is given by:

$$\frac{dx}{dt} = -x + y - 1 \quad ①$$

$$\frac{dy}{dt} = -x - y + 3 \quad ②$$

When $t = 0$, $x = 0$ and $y = 3$.

(i) Find expressions for x and y in terms of t.

(ii) Draw the graph of y against x for values of $t \geqslant 0$. Describe what happens as $t \to \infty$.

Solution

(i) Equation ① gives $y = x + \dfrac{dx}{dt} + 1 \quad ③$

Differentiating with respect to t: $\qquad \dfrac{dy}{dt} = \dfrac{dx}{dt} + \dfrac{d^2x}{dt^2}$

Substituting for y and $\dfrac{dy}{dt}$ into equation ②:

$$\frac{dx}{dt} + \frac{d^2x}{dt^2} = -x - \left(x + \frac{dx}{dt} + 1\right) + 3$$

Rearranging: $\quad \dfrac{d^2x}{dt^2} + 2\dfrac{dx}{dt} + 2x = 2$

Note

Verify this for yourself.

The general solution of this equation is $x = e^{-t}(A\sin t + B\cos t) + 1$.

Substituting for x and $\dfrac{dx}{dt}$ in equation ③ gives the general solution for y:

$$y = e^{-t}(A\cos t - B\sin t) + 2$$

The initial conditions give the values of A and B:

When $t = 0$, $x = 0 \qquad \Rightarrow 0 = B + 1 \quad \Rightarrow B = -1$

When $t = 0$, $y = 3 \qquad \Rightarrow 3 = A + 2 \quad \Rightarrow A = 1$

The particular solution that satisfies the initial conditions is therefore:

$$x = e^{-t}(\sin t - \cos t) + 1$$

$$y = e^{-t}(\cos t + \sin t) + 2$$

(ii) Figure 12.18 shows the graph of y against x.

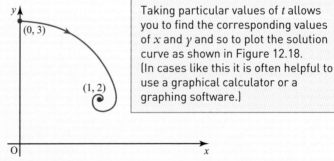

Taking particular values of t allows you to find the corresponding values of x and y and so to plot the solution curve as shown in Figure 12.18. (In cases like this it is often helpful to use a graphical calculator or a graphing software.)

Figure 12.18

As t increases, $e^{-t} \to 0$, so $x \to 1$ and $y \to 2$. In the long term the system approaches the point $(1, 2)$.

In each of questions 1–4:

(i) solve the equations to find expressions for x and y in terms of t

(ii) find the particular solutions for which $x = 1$ and $y = 2$ at $t = 0$

(iii) describe the long-term behaviour of the system

① $\dfrac{dx}{dt} = 3x - y$

$\dfrac{dy}{dt} = 2x$

② $\dfrac{dx}{dt} = 2x + 3y$

$\dfrac{dy}{dt} = 3x + 2y$

③ $\dfrac{dx}{dt} = x + 5y$

$\dfrac{dy}{dt} = -x - 3y$

④ $\dfrac{dx}{dt} = 2x - y - 1$

$\dfrac{dy}{dt} = 2y - 6$

⑤ A system of differential equations is given by:

$$\frac{dx}{dt} = x + y$$

$$\frac{dy}{dt} = x - y$$

When $t = 0$, $x = 0$ and $y = 1$.

(i) Find expressions for x and y in terms of t.

(ii) Describe the long-term behaviour of the system.

⑥ A system of differential equations is given by:

$$\frac{dx}{dt} = x + 2y - 3$$

$$\frac{dy}{dt} = -3x + y + 2$$

When $t = 0$, $x = 0$ and $y = 1$.

(i) Find expressions for x and y in terms of t.

(ii) Describe the long-term behaviour of the system.

⑦ Each of two competing species of insect reproduces at a rate proportional to its own number and is adversely affected by the other species at a rate proportional to the number of that other species. At time t, measured in centuries, the populations of these two species are x million and y million. The situation is modelled by the pair of simultaneous equations

$$\frac{dx}{dx} = 2x - 3y \text{ and } \frac{dy}{dx} = y - 2x$$

where, initially, at time $t = 0$, $x = 15$ and $y = 10$.

(i) Find the initial rates of changes of both x and y.

(ii) Differentiate the first differential equation with respect to t and use this together with the two original differential equations to eliminate y and obtain a second order differential equation for x.

(iii) Solve this second order equation to find x as a function of t.

(iv) Hence find y as a function of t.

(v) Show that one of the species becomes extinct and determine the time at which this occurs.

⑧ In a chemical process, compound P reacts with an abundant supply of a gas to form compound Q. The process is governed by two reaction rates, one for the forward reaction, and a lesser one for the reverse reaction, where Q decomposes into P and the gas. P is introduced into a reaction chamber at a constant rate of 21 kg per hour and Q is extracted at a rate proportional to the quantity of Q in the chamber. The equations which describe the process are

$$\frac{dp}{dt} = -5p + q + 21$$

$$\frac{dq}{dt} = -8q + 10p$$

where p and q are the quantities (in kg) of P and Q respectively and t is the time in hours.

(i) Calculate the values of p and q for which $\frac{dp}{dt}$ and $\frac{dq}{dt}$ are both zero.

(ii) Eliminate q to show that:

$$\frac{d^2p}{dt^2} + 13\frac{dp}{dt} + 30p = 168$$

(iii) Find the general solution of the differential equation in part (ii). Hence find the particular solutions for p and q for which $p = q = 0$ when $t = 0$.

(iv) Sketch the graphs of the solutions, showing the significance of your answers to part (i).

⑨ In a predator-prey environment the rate of growth of the predator population is found to be proportional to the size of the prey proportion. The rate of change of the prey proportion, however, is found to depend upon the sizes of both the predator and prey populations. The population dynamics are modelled by assuming that both populations vary continuously. The differential equations governing the relationships between the two populations are

$$100\frac{dx}{dt} = y \quad \text{and} \quad 100\frac{dy}{dt} = 2y - x$$

where x and y are the numbers of predator and prey respectively, and t is the time in years.

Initially the predator population is 10 thousand and the prey population 5 million.

(i) By eliminating y between the two equations show that the predator population, x, satisfies the second order differential equation:

$$10\,000\frac{d^2x}{dt^2} - 200\frac{dx}{dt} + x = 0$$

(ii) Solve this equation to find the predator population as a function of time.

(iii) Find the prey population, y, as a function of time.

(iv) Determine the size of each population after five years.

⑩ In a chemical decomposition a compound X produces a compound Y which in turn gives a compound Z. These decompositions are governed by the system of differential equations

$$\frac{dx}{dt} = -4x, \quad \frac{dy}{dt} = 4x - 2x, \quad \frac{dz}{dt} = 2y$$

where x, y and z are the masses in grams of X, Y and Z respectively, and time t is measured in hours.

Initially $x = 8$, $y = 0$ and $z = 0$.

(i) Find x, y and z in terms of t.

(ii) Determine the maximum value of y.

(iii) Determine the final value of z.

⑪ A population of cells consisting of a mixture of two-chromosome and four-chromosome cells is described approximately by the equations

$$\frac{dT}{dt} = (a - b)T, \quad \frac{dF}{dt} = bT + cF$$

where T is the number of two-chromosome cells and F is the number of four-chromosome cells. The variables T and F clearly cannot be negative, and a, b and c are constants with $a \neq b$ and $c \neq 0$.

(i) Show that the proportion of two-chromosone cells in the population can be written in the form $\dfrac{p}{q + re^{-(a+b+c)t}}$, where p, q are r are constants which depend on the values of a, b and c and the initial conditions.

(ii) Hence show that whatever the values of a, b and c, the proportion of two-chromosome cells in the population tends to a constant value in the long term, independent of the initial conditions.

(iii) Find conditions on the values of a, b and c which ensure that this limiting value of the proportion is non-zero, and find an expression for the limit when this is the case.

LEARNING OUTCOMES

➤ Be able to solve differential equations of the form $\dfrac{d^2y}{dx^2} + a\dfrac{dy}{dx} + by = 0$ using the auxiliary equation method.

➤ Understand and use the relationship between different cases of the solution and the nature of the roots of the auxiliary equation.

➤ Be able to solve differential equations of the form $\dfrac{d^2y}{dx^2} + a\dfrac{dy}{dx} + by = f(x)$

by solving the homogeneous case and adding a particular integral to the complementary function.

➤ Be able to find particular integrals for cases where $f(x)$ is a polynomial, trigonometry or exponential function, including cases where the form of the complementary function affects the form required for the particular integral.

➤ Be able to solve the equation for simple harmonic motion, $\dfrac{d^2x}{dt^2} = -\omega^2 x$, and be able to relate the solution to the motion.

➤ Be able to model damped oscillations using second order differential equations.

➤ Be able to interpret the solutions of equations modelling damped oscillations in words and graphically.

➤ Solve coupled first order simultaneous linear differential equations involving one independent variable and two dependent variables.

KEY POINTS

1 A second order linear differential equation with constant coefficients can be written in the form $\dfrac{d^2y}{dx^2} + a\dfrac{dy}{dx} + by = f(x)$, where a and b are constants.

2 The equation is homogeneous if $f(x) = 0$. Otherwise it is non-homogeneous.

3 When you are given a differential equation of the form above, you can immediately write down the auxiliary equation $\lambda^2 + a\lambda + b = 0$.

4 Each root of the auxiliary equation determines the form of one of the complementary functions.

5 If the auxiliary equation has two real, distinct roots λ_1 and λ_2, then the complementary function of the differential equation is:

$$y = Ae^{\lambda_1 x} + Be^{\lambda_2 x}$$

6 If the auxiliary equation has a repeated root α, then the complementary function of the differential equation is:

$$y = e^{\alpha x}(A + Bx)$$

7 If the auxiliary function has complex roots $\lambda = \alpha \pm \beta i$, then the complementary function of the differential equation is:

$$y = e^{\alpha x}(A\sin\beta x + B\cos\beta x)$$

8 Motion for which the differential equation is $\dfrac{d^2x}{dt^2} + \omega^2 x = 0$ is called

simple harmonic motion (SHM). The solution of this differential equation is of the form:

$$x = A\sin\omega t + B\cos\omega t \text{ or } x = a\sin(\omega t + \varepsilon)$$

The period of this motion is $\dfrac{2\pi}{\omega}$ and the amplitude is $\sqrt{A^2 + B^2}$.

9 Motion for which the differential equation is $\dfrac{d^2x}{dt^2} + \alpha\dfrac{dx}{dt} + \omega^2 x = 0$, where $\alpha > 0$, is called damped harmonic motion. The following table shows the features of damped harmonic motion.

$\alpha^2 - 4\omega^2$	Type of damping
$\alpha = 0$	no damping
$\alpha^2 - 4\omega^2 > 0$	overdamping
$\alpha^2 - 4\omega^2 = 0$	critical damping
$\alpha^2 - 4\omega^2 < 0$	underdamping

Table 12.2

10 The general solution of the non-homogeneous linear differential equation with constant coefficients

$$\frac{d^2y}{dx^2} + a\frac{dy}{dx} + by = f(x)$$

is the sum of the complementary function and a particular integral.

11 The number of unknown constants is the same as the order of the equation.

12 A particular integral is any function that satisfies the full equation; it does not contain any arbitrary constants.

13 To find the particular integral, use the trial function shown in the following table:

Function	Trial function
linear function	$ax + b$
polynomial of order n	$a_n x^n + a_{n-1} x^{-1} + \ldots + a_1 x + a_0$
trigonometric function involving $\sin px$ and/or $\cos px$	$a \sin px + b \cos px$
exponential function involving e^{px}	$a e^{px}$
sum of different functions	sum of matching functions

Table 12.3

14 If the trial function for a particular integral is the same as one of the complementary functions, you multiply the trial function by x.

15 A system of differential equations involves two or more dependent variables and one independent variable.

16 In a linear system

$$\frac{dx}{dt} = a_1 x + b_1 y + f_1(t)$$

$$\frac{dy}{dt} = a_2 x + b_2 y + f_2(t)$$

x and y are the dependent variables, and t is the independent variable.

17 Solving such a system of equations involves finding x in terms of t and y in terms of t. This is done by differentiating one equation and substituting into the other, and then solving the resulting second order linear differential equation.

① A computer program has been used to investigate the family of solutions to a differential equation. In the following graph, the line segments are tangents to the solution curves.

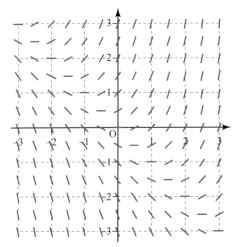

Figure 1

(i) Write down the equation of the straight line solution to the differential equation. [2 marks]

(ii) Now consider the solution curve for which $y = -0.5$ when $x = 0$. Estimate the value of x for which $y = 0$ on this curve. [2 marks]

② Software has been used to plot the locus of a complex number z in an Argand diagram with the following result.

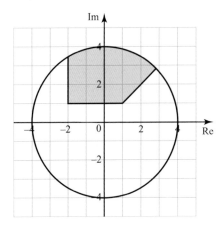

Figure 2

Write down the four inequalities, one for $\text{Re}(z)$, one for $\text{Im}(z)$, one for $\arg(z)$ and one for $|z|$, satisfied by the points in the region shaded. (Note: the boundaries are included in the region.) [4 marks]

③ Find the particular solution of the differential equation

$$\frac{\mathrm{d}y}{\mathrm{d}x} - 2xy = \mathrm{e}^{2x+x^2}$$

for which $y = 1$ when $x = 0$. [5 marks]

④ (i) Find, in the form $a + ib$, the complex number with modulus 1 and argument $\dfrac{2\pi}{3}$.

The point P represents the complex number $2 + i\sqrt{3}$ in an Argand diagram. P is one vertex of an equilateral triangle centred on the origin. [2 marks]

(ii) Find, in the form $a + ib$, the complex numbers represented by the other two vertices. [3 marks]

(iii) The midpoints of the sides of the equilateral triangle represent the cube roots of a complex number w. Find w in the form $a + ib$. [3 marks]

MP ⑤ (i) Prove that $e^{2i\theta} - 1 = 2ie^{i\theta}\sin\theta$. [3 marks]

(ii) Given that $\theta \neq n\pi$ for any integer n, find the sum of the series

$$e^{i\theta} + e^{3i\theta} + \ldots + e^{(2n-1)i\theta}.$$

Hence prove that $\cos\theta + \cos 3\theta + \ldots \cos(2n-1)\theta = \dfrac{\sin n\theta \cos n\theta}{\sin\theta}$ provided $\theta \neq n\pi$. [7 marks]

(iii) Find the corresponding expression for $\sin\theta + \sin 3\theta + \ldots + \sin(2n-1)\theta$. [2 marks]

⑥ The current flowing through a particular component in an electrical circuit is x amps at time t seconds, where x and t are modelled by the differential equation

$$\frac{d^2x}{dt^2} + 6\frac{dx}{dt} + 34x = 25\sin 4t.$$

(i) Find the general solution of this differential equation. [8 marks]

(ii) Write down an expression, in terms of t, for the current in the component when the circuit has been operating for a long time. Explain why this expression does not depend on the initial current in the component. [2 marks]

(iii) Discuss briefly whether or not the model will break down if the coefficient of $\dfrac{dx}{dt}$ in the differential equation is -6 instead of 6. [2 marks]

⑦ An area of moorland is modelled as a flat (but not horizontal) plane Π. Two paths on the moorland are AB and AC, where the coordinates of A, B, C, relative to a convenient origin, are as follows. A(0, 3, 1), B(10, 12, 2), C(4, −5, 3). The units are metres.

(i) Find the vector product of $\overrightarrow{AB}$ and $\overrightarrow{AC}$. Hence show that the equation of the plane Π may be written as $13x - 8y - 58z = -82$. [5 marks]

A drone is hovering above the moorland at the point P with coordinates (5, 2, 51). Charlie is standing on the path AB at the point Q that is closest to the drone.

(ii) Find the coordinates of Q and hence the distance from Charlie's position to the drone. [6 marks]

In order to get a better look, Charlie walks to the point on the moorland that is closest to the drone.

(iii) Find the distance from Charlie's new position to the drone. Hence determine how far Charlie has walked. [4 marks]

Answers

Chapter 1

Review Exercise (Page 8)

1. (i) $40.2°$ (ii) $90°$ (iii) $25.6°$
2. (i) $90°$ (ii) $134.2°$ (iii) $0°$
3. Vector equation $\left(\mathbf{r} - \begin{pmatrix} 5 \\ -2 \\ 0 \end{pmatrix} \right) \cdot \begin{pmatrix} 3 \\ -2 \\ 1 \end{pmatrix} = 0$

   Cartesian equation $3x - 2y + z - 19 = 0$
4. $2x + y - 7x + 31 = 0$
5. (i) $87.1°$ (ii) $85.3°$
6. (i) $k = 2, k = 3$

   (ii) $2x - 5y + 3z - 5 = 0$ or

   $3x - 5y + 3z - 7 = 0$
7. $-3x + y + 7z - 30 = 0$
8. (i) $\overrightarrow{AB} = \begin{pmatrix} 2 \\ 1 \\ 2 \end{pmatrix}$, $\overrightarrow{AC} = \begin{pmatrix} -5 \\ -1 \\ 1 \end{pmatrix}$

   (iii) $x - 4y + z - 1 = 0$
9. $6x + 27y + 22z + 2 = 0$
11. π_1 is parallel to π_3; π_2 and π_4 are not parallel to any of the other planes. π_1 and π_3 are perpendicular to π_2; π_1 and π_3 are perpendicular to π_4.
12. (i) $\cos(A - B) = \cos A \cos B + \sin A \sin B$

    (ii) **a.b**

Activity 1.1 (Page 10)

(ii) $\begin{pmatrix} -2 \\ -9 \end{pmatrix} \begin{pmatrix} 0 \\ -5 \end{pmatrix} \begin{pmatrix} 2 \\ -1 \end{pmatrix} \begin{pmatrix} 3 \\ 1 \end{pmatrix} \begin{pmatrix} 3.5 \\ 2 \end{pmatrix} \begin{pmatrix} 6 \\ 7 \end{pmatrix} \begin{pmatrix} 8 \\ 11 \end{pmatrix}$

(iii) The points join to form the straight line $y = 2x - 5$

(iv) (a) It lies between the point A$(2, -1)$ and the point B$(4, 3)$

(b) It lies beyond the point B$(4, 3)$

(c) It lies beyond A in the opposite direction to the point B.

Activity 1.2 (Page 12)

(i) $x = a_1 + \lambda d_1$ $y = a_2 + \lambda d_2$

(ii) $\lambda = \dfrac{x - a_1}{d_1}$ $\lambda = \dfrac{y - a_2}{d_2}$

$y = \dfrac{d_2}{d_1} x + \dfrac{a_2 d_1 - a_1 d_2}{d_1}$ so $m = \dfrac{d_2}{d_1}$ and

$c = \dfrac{a_2 d_1 - a_1 d_2}{d_1}$

Activity 1.3 (Page 18)

$y = \dfrac{1}{2}x + 3$ and $y = -3x + 8$. You can find the angle between them by finding the difference between arctan (gradient) for each line.

Exercise 1.1 (Page 19)

1. (i) $\mathbf{r} = \begin{pmatrix} 3 \\ 1 \end{pmatrix} + \lambda \begin{pmatrix} 5 \\ -2 \end{pmatrix}$

   (ii) $\mathbf{r} = \begin{pmatrix} 5 \\ -1 \end{pmatrix} + \lambda \begin{pmatrix} 0 \\ 4 \end{pmatrix}$

   (iii) $\mathbf{r} = \begin{pmatrix} -2 \\ 4 \end{pmatrix} + \lambda \begin{pmatrix} 5 \\ 5 \end{pmatrix}$

   (iv) $\mathbf{r} = \begin{pmatrix} 0 \\ 8 \end{pmatrix} + \lambda \begin{pmatrix} -2 \\ -11 \end{pmatrix}$

2. (i) $\mathbf{r} = \begin{pmatrix} 2 \\ 4 \\ -1 \end{pmatrix} + \lambda \begin{pmatrix} 3 \\ 6 \\ 4 \end{pmatrix}$

   (ii) $\mathbf{r} = \begin{pmatrix} 1 \\ 0 \\ -1 \end{pmatrix} + \lambda \begin{pmatrix} 1 \\ 0 \\ 0 \end{pmatrix}$

   (iii) $\mathbf{r} = \begin{pmatrix} 1 \\ 0 \\ 4 \end{pmatrix} + \lambda \begin{pmatrix} 5 \\ 3 \\ -6 \end{pmatrix}$

   (iv) $\mathbf{r} = \begin{pmatrix} 0 \\ 0 \\ 1 \end{pmatrix} + \lambda \begin{pmatrix} 2 \\ 1 \\ 3 \end{pmatrix}$

3. (i) $\dfrac{x - 2}{3} = \dfrac{y - 4}{6} = \dfrac{z + 1}{4}$

   (ii) $x - 1 = \dfrac{y}{3} = \dfrac{z + 1}{4}$

   (iii) $x - 3 = \dfrac{z - 4}{2}$ and $y = 0$

   (iv) $\dfrac{x}{2} = \dfrac{z - 1}{4}$ and $y = 4$

4. (i) $\mathbf{r} = \begin{pmatrix} 3 \\ -2 \\ 1 \end{pmatrix} + \lambda \begin{pmatrix} 5 \\ 3 \\ 4 \end{pmatrix}$

   (ii) $\mathbf{r} = \begin{pmatrix} 0 \\ 0 \\ -1 \end{pmatrix} + \lambda \begin{pmatrix} 1 \\ 2 \\ 3 \end{pmatrix}$

(iii) $\mathbf{r} = \lambda \begin{pmatrix} 1 \\ 1 \\ 1 \end{pmatrix}$

(iv) $\mathbf{r} = \begin{pmatrix} 2 \\ 0 \\ 0 \end{pmatrix} + \lambda \begin{pmatrix} 0 \\ 1 \\ 1 \end{pmatrix}$

5 $\mathbf{r} = \begin{pmatrix} 3 \\ -5 \\ 2 \end{pmatrix} + \lambda \begin{pmatrix} 0 \\ 1 \\ 0 \end{pmatrix}$;

$x = 3, z = 2$ and $y - 5(= \lambda)$

6 (i) $\begin{pmatrix} 4 \\ 1 \end{pmatrix}$ (ii) $\begin{pmatrix} 5 \\ 5 \end{pmatrix}$ (iii) $\begin{pmatrix} -5 \\ 6 \end{pmatrix}$

7 (i) Intersect at $(3, 2, -13)$
   (ii) Lines are skew
   (iii) Lines are parallel
   (iv) Intersect at $(4, -7, 11)$
   (v) Lines are skew

8 (i) $45°$ (ii) $56.3°$
   (iii) $53.6°$ (iv) $81.8°$
   (v) $8.7°$

9 Do not meet

10 $6, 9, \sqrt{77}$

11 (i) $\begin{pmatrix} -0.25 \\ 0 \\ 0 \end{pmatrix}$

   (ii) $C(0, 0.05, 1.1)$

   (iii) DE: $\mathbf{r} = \begin{pmatrix} 0 \\ 0 \\ 1 \end{pmatrix} + \lambda \begin{pmatrix} 1 \\ 0 \\ 0 \end{pmatrix}$

   EF: $\mathbf{r} = \begin{pmatrix} 0.25 \\ 0 \\ 1 \end{pmatrix} + \mu \begin{pmatrix} 0 \\ 1 \\ 2 \end{pmatrix}$

12 (i) $d = 33$

   (iii) $\mathbf{r} = \begin{pmatrix} 2 \\ 3 \\ -1 \end{pmatrix} + \lambda \begin{pmatrix} 7 \\ 8 \\ 5 \end{pmatrix}$

   (iv) $p = -5, q = -6; AC = \sqrt{138}$
   (v) $3\sqrt{161}$

13 (i) $(1, 0.5, 0)$ (ii) $41.8°$
   (iii) $027°$ (iv) $(2, 2.5, 2)$
   (v) $t = 2; \sqrt{5}$ km

Exercise 1.2 (Page 24)

2 (i) $(0, 1, 3), 67.8°$ (ii) $(1, 1, 1), 3.01°$
   (iii) $(8, 4, 2), 12.6°$ (iv) $(0, 0, 0), 70.5°$

3 (i) $\mathbf{r} = \begin{pmatrix} 4 \\ 1 \\ 3 \end{pmatrix} + \lambda \begin{pmatrix} 2 \\ 3 \\ 5 \end{pmatrix}$ or

   $\mathbf{r} = \begin{pmatrix} 6 \\ 4 \\ 8 \end{pmatrix} + \mu \begin{pmatrix} 2 \\ 3 \\ 5 \end{pmatrix}$ or equivalent

   (ii) Intersect at $(0, -5, -7)$
   (iii) $11.5°$

4 (i) $\mathbf{r} = \begin{pmatrix} 13 \\ 5 \\ 0 \end{pmatrix} + \lambda \begin{pmatrix} 3 \\ 1 \\ -2 \end{pmatrix}$

   (ii) $(4, 2, 6)$
   (iii) $\sqrt{126}$

5 (i) $\overrightarrow{AB} = \begin{pmatrix} 2 \\ -3 \\ 2 \end{pmatrix}$

   $\overrightarrow{AC} = \begin{pmatrix} -5 \\ 2 \\ -1 \end{pmatrix}$ $x + 8y + 11z = 15$

   (ii) $P\left(\dfrac{24}{13}, -1, \dfrac{25}{13}\right)$ $Q\left(\dfrac{267}{40}, \dfrac{-67}{40}, \dfrac{79}{40}\right)$
   (iii) $R(6, -1, 4)$ (iv) $22.4°$ (v) 0.89

6 (i) $\overrightarrow{AA'} = \begin{pmatrix} 1 \\ 2 \\ -3 \end{pmatrix}$ so AA′ is perpendicular to
   $x + 2y - 3z = 0$
   M has coordinates $(1.5, 3, 2.5)$ and
   $1.5 + (2 \times 3) - (3 \times 2.5) = 0$ so lies in
   the plane.

   (ii) $(0, 3, 2); \mathbf{r} = \begin{pmatrix} 2 \\ 4 \\ 1 \end{pmatrix} + \lambda \begin{pmatrix} -2 \\ -1 \\ 1 \end{pmatrix}$

   (iii) $80.4°$

7 (i) $AB = \sqrt{29}$ $AC = 5$; $56.1°$; 11.2
   (ii) (b) $4x - 3y + 10z = -12$

   (iii) $\mathbf{r} = \begin{pmatrix} 0 \\ 4 \\ 5 \end{pmatrix} + \lambda \begin{pmatrix} 4 \\ -3 \\ 10 \end{pmatrix}$;

   Intersect at $(-1.6, 5.2, 1)$
   (iv) 16.7

Review: Matrices and transformations

Exercise R.1 (Page 31)

1. (i) **A**: 2×2 **B**: 1×3 **C**: 2×1 **D**: 2×3
   E: 3×2 **F**: 3×3 **G**: 1×1 **H**: 1×5

   (ii) (a) $\begin{pmatrix} -10 \\ -28 \end{pmatrix}$ (b) $\begin{pmatrix} 3 \\ 27 \\ -5 \end{pmatrix}$

   (c) $(19 \quad 14 \quad 3)$
   (d) non-conformable
   (e) $(-4 \quad 8 \quad -12 \quad 16 \quad -20)$
   (f) non-conformable
   (g) $\begin{pmatrix} -5 & 11 & -3 \\ 6 & 1 & 2 \end{pmatrix}$

   (iii) (a) $\begin{pmatrix} -4 & 3 \\ 1 & 10 \end{pmatrix}$

   (b) $\begin{pmatrix} 8 & 3 \\ -1 & 4 \end{pmatrix}$

   (c) $\begin{pmatrix} 8 \\ -17 \end{pmatrix}$

   (d) non-conformable

   (e) $\begin{pmatrix} 1 & -12 & 7 \\ 3 & -4 & -1 \\ -4 & 5 & 2 \end{pmatrix}$

   (f) non-conformable
   (g) non-conformable

2. (i) $\mathbf{MN} = \begin{pmatrix} 6 & 0 \\ -3 & 3 \end{pmatrix}$, $\mathbf{NM} = \begin{pmatrix} 5 & 1 \\ 2 & 4 \end{pmatrix}$

   (ii) Matrix multiplication is not commutative

3. (i) $\mathbf{PQ} = \begin{pmatrix} 3 & 0 & -3 \\ -3 & 2 & -1 \end{pmatrix}$

   $(\mathbf{PQ})\mathbf{R} = \begin{pmatrix} 3 & 3 & 3 \\ -7 & -9 & 3 \end{pmatrix}$

   (ii) $\mathbf{QR} = \begin{pmatrix} -2 & -3 & 3 \\ 7 & 9 & -3 \end{pmatrix}$

   $\mathbf{P(QR)} = \begin{pmatrix} 3 & 3 & 3 \\ -7 & -9 & 3 \end{pmatrix}$

   (iii) Matrix multiplication is associative

4. $a = -\frac{1}{5}$ or 3, $b = \pm 4$

5. (i) $\begin{pmatrix} 25 & 6 \\ 0 & 1 \end{pmatrix}$

   (ii) $\begin{pmatrix} 125 & 31 \\ 0 & 1 \end{pmatrix}$

   (iii) $\begin{pmatrix} 625 & 156 \\ 0 & 1 \end{pmatrix}$

   (iv) Top right entry is the sum of powers of

   5 from 0 to $n - 1$, so $A^n = \begin{pmatrix} 5^n & \sum\limits_{r=0}^{n-1} 5^r \\ 0 & 1 \end{pmatrix}$

   (v) $\begin{pmatrix} 15625 & 3906 \\ 0 & 1 \end{pmatrix}$

6. (i) $\begin{pmatrix} 10 + 3x & -15 + x^2 \\ -5 + 3x & 18 - x \end{pmatrix}$

   (ii) $x = -2$ or 5

   (iii) $\begin{pmatrix} 4 & -11 \\ -11 & 20 \end{pmatrix}$ or $\begin{pmatrix} 25 & 10 \\ 10 & 13 \end{pmatrix}$

7. $a = -7, b = 2$

Exercise R.2 (Page 38)

1. (i) $\begin{pmatrix} 0 & 1 \\ 1 & 0 \end{pmatrix}$

   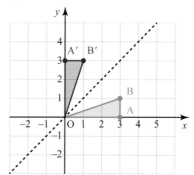

   (ii) $\begin{pmatrix} -1 & 0 \\ 0 & -1 \end{pmatrix}$

   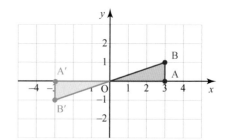

   (iii) $\begin{pmatrix} 4 & 0 \\ 0 & 4 \end{pmatrix}$

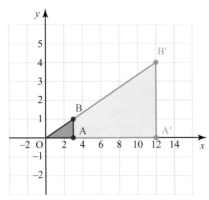

(iv) $\begin{pmatrix} 1 & 2 \\ 0 & 1 \end{pmatrix}$

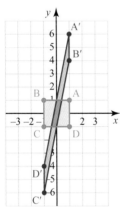

2 (i) Rotation of 180° about the origin
(ii) Stretch, scale factor 4, parallel to the x-axis
(iii) Enlargement scale factor 4, centre the origin
(iv) Reflection in the x-axis
(v) Rotation of 90° clockwise about the origin

3 (i)

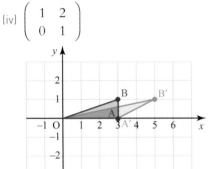

(ii) Shear with the y-axis fixed; A is mapped from $(1, 1)$ to $(1, 6)$

4 (i) $\begin{pmatrix} 1 & 0 & 0 \\ 0 & 0 & 1 \\ 0 & -1 & 0 \end{pmatrix}$ (ii) $\begin{pmatrix} 1 & 0 & 0 \\ 0 & 1 & 0 \\ 0 & 0 & -1 \end{pmatrix}$

(iii) $\begin{pmatrix} -1 & 0 & 0 \\ 0 & 1 & 0 \\ 0 & 0 & -1 \end{pmatrix}$

5 (i) Enlargement scale factor 2, centre $(0, 0)$
(ii) Reflection in the plane $z = 0$

(iii) Rotation of 90° clockwise about the x-axis
(iv) Three way stretch of factor 4 in the x-direction, factor 0.5 in the y-direction and factor 3 in the z-direction

6 (i) $\mathbf{P} = \begin{pmatrix} 0 & -1 \\ 1 & 0 \end{pmatrix}$, $\mathbf{Q} = \begin{pmatrix} 0 & -1 \\ -1 & 0 \end{pmatrix}$

(ii) $\mathbf{PQ} = \begin{pmatrix} 1 & 0 \\ 0 & -1 \end{pmatrix}$ Reflection in the x-axis

(iii) $\mathbf{QP} = \begin{pmatrix} -1 & 0 \\ 0 & 1 \end{pmatrix}$ Reflection in the y-axis

(iv) $\begin{pmatrix} 1 & 0 \\ 0 & -1 \end{pmatrix}\begin{pmatrix} x \\ y \end{pmatrix} = \begin{pmatrix} x \\ -y \end{pmatrix}$ and

$\begin{pmatrix} -1 & 0 \\ 0 & 1 \end{pmatrix}\begin{pmatrix} x \\ y \end{pmatrix} = \begin{pmatrix} -x \\ y \end{pmatrix}$

so if the points have the same image,

$\begin{pmatrix} x \\ -y \end{pmatrix} = \begin{pmatrix} -x \\ y \end{pmatrix}$ which is only true

when $x = y = 0$. The point that has the same image under both transformations is the origin $(0, 0)$.

7 (i) $\mathbf{A} = \begin{pmatrix} -\dfrac{1}{2} & -\dfrac{\sqrt{3}}{2} \\ \dfrac{\sqrt{3}}{2} & -\dfrac{1}{2} \end{pmatrix}$

(ii) $\mathbf{B} = \begin{pmatrix} \dfrac{\sqrt{3}}{2} & -\dfrac{1}{2} \\ \dfrac{1}{2} & \dfrac{\sqrt{3}}{2} \end{pmatrix}$, $\mathbf{C} = \begin{pmatrix} 0 & -1 \\ 1 & 0 \end{pmatrix}$

$\mathbf{BC} = \begin{pmatrix} \dfrac{\sqrt{3}}{2} & -\dfrac{1}{2} \\ \dfrac{1}{2} & \dfrac{\sqrt{3}}{2} \end{pmatrix}\begin{pmatrix} 0 & -1 \\ 1 & 0 \end{pmatrix}$

$= \begin{pmatrix} -\dfrac{1}{2} & -\dfrac{\sqrt{3}}{2} \\ \dfrac{\sqrt{3}}{2} & -\dfrac{1}{2} \end{pmatrix} = \mathbf{A}$

(iii) $\mathbf{B}^3 = \begin{pmatrix} \dfrac{\sqrt{3}}{2} & -\dfrac{1}{2} \\ \dfrac{1}{2} & \dfrac{\sqrt{3}}{2} \end{pmatrix}\begin{pmatrix} \dfrac{\sqrt{3}}{2} & -\dfrac{1}{2} \\ \dfrac{1}{2} & \dfrac{\sqrt{3}}{2} \end{pmatrix}$

$\begin{pmatrix} \dfrac{\sqrt{3}}{2} & -\dfrac{1}{2} \\ \dfrac{1}{2} & \dfrac{\sqrt{3}}{2} \end{pmatrix} = \begin{pmatrix} 0 & -1 \\ 1 & 0 \end{pmatrix}$.

This verifies that three successive anticlockwise rotations of 30° about the origin is equivalent to a single anticlockwise rotation of 90° about the origin.

8 (i) $\mathbf{J} = \begin{pmatrix} 1 & 0 & 0 \\ 0 & -1 & 0 \\ 0 & 0 & 1 \end{pmatrix}$ $\mathbf{K} = \begin{pmatrix} 0 & -1 & 0 \\ 1 & 0 & 0 \\ 0 & 0 & 1 \end{pmatrix}$

$\mathbf{L} = \begin{pmatrix} -1 & 0 & 0 \\ 0 & 1 & 0 \\ 0 & 0 & 1 \end{pmatrix}$ $\mathbf{M} = \begin{pmatrix} 1 & 0 & 0 \\ 0 & -1 & 0 \\ 0 & 0 & -1 \end{pmatrix}$

(ii) (a) $\mathbf{LJ} = \begin{pmatrix} -1 & 0 & 0 \\ 0 & -1 & 0 \\ 0 & 0 & 1 \end{pmatrix}$

(b) $\mathbf{KJ} = \begin{pmatrix} 0 & 1 & 0 \\ 1 & 0 & 0 \\ 0 & 0 & 1 \end{pmatrix}$

(c) $\mathbf{K}^2 = \begin{pmatrix} -1 & 0 & 0 \\ 0 & -1 & 0 \\ 0 & 0 & 1 \end{pmatrix}$

(d) $\mathbf{JLM} = \begin{pmatrix} 1 & 0 & 0 \\ 0 & 1 & 0 \\ 0 & 0 & -1 \end{pmatrix}$

9 (i) $\begin{pmatrix} 1 & 0 \\ 0 & 3 \end{pmatrix}$

(ii) A reflection in the x-axis and a stretch of scale factor 2 parallel to the x-axis

(iii) $\begin{pmatrix} 2 & 0 \\ 0 & -3 \end{pmatrix}$ reflection in the x-axis; stretch scale factor 2 parallel to the x-axis; stretch factor 3 parallel to the y-axis. The outcome of these three transformations would be the same regardless of the order in which they are applied. There are 6 different possible orders.

(iv) $\begin{pmatrix} \dfrac{1}{2} & 0 \\ 0 & -\dfrac{1}{3} \end{pmatrix}$

10 (i) $\mathbf{A} = \begin{pmatrix} \cos\theta & -\sin\theta \\ \sin\theta & \cos\theta \end{pmatrix}$

$\mathbf{B} = \begin{pmatrix} \cos\phi & -\sin\phi \\ \sin\phi & \cos\phi \end{pmatrix}$

(ii)

$\mathbf{BA} = \begin{pmatrix} \cos\theta\cos\phi - \sin\theta\sin\phi & -\sin\theta\cos\phi - \cos\theta\sin\phi \\ \sin\theta\cos\phi + \cos\theta\sin\phi & -\sin\theta\sin\phi + \cos\theta\cos\phi \end{pmatrix}$

(iii) $\mathbf{C} = \begin{pmatrix} \cos(\theta+\phi) & -\sin(\theta+\phi) \\ \sin(\theta+\phi) & \cos(\theta+\phi) \end{pmatrix}$

(iv) $\sin(\theta+\phi) = \sin\theta\cos\phi + \cos\theta\sin\phi$
$\cos(\theta+\phi) = \cos\theta\cos\phi - \sin\theta\sin\phi$

(v) A rotation through angle θ followed by rotation through angle ϕ has the same effect as a rotation through angle ϕ followed by angle θ.

11 A reflection in a line followed by a second reflection in the same line returns a point to its original position.

12 (i) $\mathbf{A}$ represents an anticlockwise rotation through 90° about the origin; $\mathbf{A}^4$ represents four rotations each of 90°, totalling 360° which leaves an object unchanged – this is equivalent to the identity matrix $\mathbf{I}$.

(ii) $\mathbf{B} = \begin{pmatrix} 0 & 1 \\ -1 & 0 \end{pmatrix}$ which represents a rotation of 90° clockwise about the origin.

(iii) $\mathbf{C} = \begin{pmatrix} \dfrac{1}{2} & -\dfrac{\sqrt{3}}{2} \\ \dfrac{\sqrt{3}}{2} & \dfrac{1}{2} \end{pmatrix}$

(iv) $m = 2, n = 3$. $\mathbf{A}^2 = \mathbf{C}^3$ because both represent a rotation through 180°.

(v) $\mathbf{AC} = \begin{pmatrix} -\dfrac{\sqrt{3}}{2} & -\dfrac{1}{2} \\ \dfrac{1}{2} & -\dfrac{\sqrt{3}}{2} \end{pmatrix}$ Rotations of 60° and 90° can be carried out in either order, both result in a rotation of 150°.

Exercise R.3 (Page 43)

1 (i) $(0, 0)$ is the only invariant point
(ii) $(0, 0)$ is the only invariant point
(iii) Invariant points have the form $(\lambda, -\lambda)$
(iv) Invariant points have the form $(2\lambda, 3\lambda)$

2 (i) x-axis, y-axis, lines of the form $y = mx$
(ii) x-axis, y-axis, lines of the form $y = mx$
(iii) No invariant lines
(iv) $y = x$, lines of the form $y = -x + c$
(v) $y = -x$, lines of the form $y = x + c$
(vi) x-axis

3 (ii) $y = \pm x$

4 (i) $y = \dfrac{3}{2}x$ (ii) $y = \dfrac{3}{2}x$

(iii)

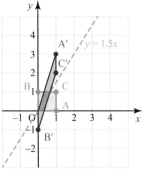

5   (i) $(3, 6); y = 2x$
    (ii) $(-2, 3);$ Rotation 90° anticlockwise about
          the origin
    (iii) $\begin{pmatrix} -0.8 & -0.6 \\ -0.6 & 0.8 \end{pmatrix}$
    (iv) $3x + y = 0$

Chapter 2

Review Exercise (Page 52)

1   $x = -7, \dfrac{3}{2}$

2   (i) $\mathbf{A} = \begin{pmatrix} -1 & 0 \\ 0 & 1 \end{pmatrix}$ $\mathbf{B} = \begin{pmatrix} 0 & -1 \\ -1 & 0 \end{pmatrix}$

    (iii)

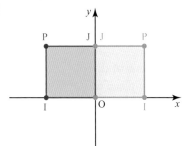

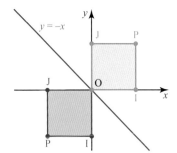

3   (i) $\begin{pmatrix} 1 & 0 \\ 4 & 1 \end{pmatrix}$
    (ii)   determinant = 1 so area is preserved
4   (i) $(11, -3)$
    (ii) $\dfrac{1}{20}\begin{pmatrix} 5 & 5 \\ -1 & 3 \end{pmatrix}$
    (iii) $(1, 0)$

5   $k = 1, 6$
6   (i) $\det(\mathbf{M}) = -9, \det(\mathbf{N}) = -67$
    (ii) $\mathbf{MN} = \begin{pmatrix} 19 & -17 \\ 50 & -13 \end{pmatrix}$
          $\det(\mathbf{MN}) = 603$ and $603 = -9 \times -67$
7   $\begin{pmatrix} 3 & -1 & 3 & 6 \\ -1 & 0 & 0 & 0 \end{pmatrix}$
8   (i)

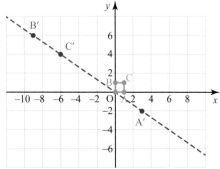

    (ii)   The image of all points lie on the line
          $y = -\dfrac{2}{3}x.$ The determinant of the matrix is
          zero which shows that the image will have
          zero area.
9   (i) $x = 2, y = -1$ (ii) $x = -3, y = 4$
10   $k = \pm 6; k = 6$ gives the same line so an infinite
     number of solutions

     $k = -6$ gives parallel lines so there are no
     solutions
11   (ii) $\mathbf{M}^n = (a + d)^{n-1}\mathbf{M}$

12   $\begin{pmatrix} x' \\ y' \end{pmatrix} = \begin{pmatrix} a & b \\ c & d \end{pmatrix}\begin{pmatrix} x \\ y \end{pmatrix} \Rightarrow$

     $\begin{pmatrix} x' \\ y' \end{pmatrix} = \begin{pmatrix} ax + by \\ cx + dy \end{pmatrix} \Rightarrow \begin{array}{l} x' = ax + by \\ y' = cx + dy \end{array}$

     Solving simultaneously and using the fact
     $ad - bc = 0$ gives the result.
13   (i) $(4, 1), (2, 2)$ and $(-12, -3)$

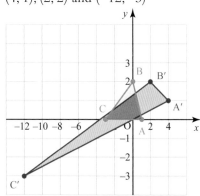

(ii) Area of T = 4; Area of T′ = 12

Ratio 12:4 or 3:1. 3 is the determinant of **M**.

(iii) $\mathbf{M}^{-1} = \dfrac{1}{3}\begin{pmatrix} 1 & -1 \\ -1 & 4 \end{pmatrix}$

14 (i) (b) $\mathbf{S}^{-1} = \dfrac{1}{2}\begin{pmatrix} 4 & -2 \\ 3 & -1 \end{pmatrix}$

(ii) $\mathbf{T}\begin{pmatrix} x \\ y \end{pmatrix} = \begin{pmatrix} x \\ y \end{pmatrix}$

$\Rightarrow \mathbf{T}^{-1}\mathbf{T}\begin{pmatrix} x \\ y \end{pmatrix} = \mathbf{T}^{-1}\begin{pmatrix} x \\ y \end{pmatrix}$

$\Rightarrow \begin{pmatrix} x \\ y \end{pmatrix} = \mathbf{T}^{-1}\begin{pmatrix} x \\ y \end{pmatrix}$ so (x, y)

is invariant under $\mathbf{T}^{-1}$

15 (i) $\mathbf{M} = \begin{pmatrix} 2 & -1 \\ 3 & k \end{pmatrix}$

(ii) $k = -\dfrac{3}{2}$; $x = 2, y = 3$

(iii) There are no unique solutions.

(iv) (a) Lines intersect at a unique point

(b) Lines are parallel

(c) Lines are coincident

16 (i) $\mathbf{M}^{-1} = \dfrac{1}{3}\begin{pmatrix} 1 & -2 \\ 0 & 3 \end{pmatrix}$, $\mathbf{N}^{-1} = \dfrac{1}{7}\begin{pmatrix} 4 & 3 \\ -1 & 1 \end{pmatrix}$

(ii) $\mathbf{MN} = \begin{pmatrix} 5 & -1 \\ 1 & 4 \end{pmatrix}$

$(\mathbf{MN})^{-1} = \dfrac{1}{21}\begin{pmatrix} 4 & 1 \\ -1 & 5 \end{pmatrix} = \mathbf{N}^{-1}\mathbf{M}^{-1}$

Activity 2.1 (Page 55)

$\det(\mathbf{A}) = -17$ $\mathbf{A}^{-1} = \begin{pmatrix} -\dfrac{1}{17} & -\dfrac{2}{17} & \dfrac{5}{17} \\ -\dfrac{8}{17} & \dfrac{1}{17} & \dfrac{6}{17} \\ \dfrac{4}{17} & \dfrac{8}{17} & -\dfrac{3}{17} \end{pmatrix}$

B is singular.

$\mathbf{C}^{-1} = \begin{pmatrix} -\dfrac{1}{15} & -\dfrac{2}{3} & \dfrac{1}{5} \\ \dfrac{2}{5} & 0 & -\dfrac{1}{5} \\ -\dfrac{1}{15} & \dfrac{1}{3} & \dfrac{1}{5} \end{pmatrix}$

$\mathbf{D}^{-1} = \begin{pmatrix} -1 & -3 & \dfrac{4}{3} \\ 1 & 2 & -\dfrac{2}{3} \\ 1 & 3 & -1 \end{pmatrix}$

Exercise 2.1 (Page 59)

1 (i) (a) 5 (b) 5

(ii) (a) −5 (b) −5

Interchanging the rows and columns has not changed the determinant.

(iii) (a) 0 (b) 0

If a matrix has a repeated row or column the determinant will be zero.

2 (i) $\dfrac{1}{3}\begin{pmatrix} 3 & 0 & 6 \\ -4 & 2 & 3 \\ 2 & -1 & 0 \end{pmatrix}$

(ii) Matrix is singular

(iii) $\begin{pmatrix} -0.06 & -0.1 & -0.1 \\ 0.92 & 0.2 & 0.7 \\ 0.66 & 0.1 & 0.6 \end{pmatrix}$

(iv) $\dfrac{1}{21}\begin{pmatrix} 34 & 11 & 32 \\ 9 & 6 & 6 \\ -38 & -16 & -37 \end{pmatrix}$

3 $\dfrac{1}{7}\begin{pmatrix} 2 & 18 & -11 \\ 2 & 39 & -25 \\ 3 & 41 & -27 \end{pmatrix}$; $x = 8, y = 4, z = -3$

4 $\mathbf{M}^{-1} = \dfrac{1}{28 - 10k}\begin{pmatrix} 4 & -10 & 12 \\ -(4k + 8) & 8 & -(4 + 2k) \\ -k & 7 & -3k \end{pmatrix}$;

$k = 2.8$

5 (i) The columns of the matrix have been moved one place to the right, with the final column moving to replace the first. This is called **cyclical interchange** of the columns.

(ii) $\det(\mathbf{A}) = \det(\mathbf{B}) = \det(\mathbf{C}) = -26$

Cyclical interchange of the columns leaves the determinant unchanged.

6 $x = \dfrac{-1 \pm \sqrt{41}}{2}$

7 $x = 1, x = 4$

8 $1 < k < 5$

9 (i) Let $\mathbf{X} = (\mathbf{PQ})^{-1}$ so $\mathbf{X}(\mathbf{PQ}) = \mathbf{I}$.

$\Rightarrow \mathbf{X}(\mathbf{PQ})\mathbf{Q}^{-1} = \mathbf{IQ}^{-1} = \mathbf{Q}^{-1}$

$\Rightarrow \mathbf{XP}(\mathbf{QQ}^{-1}) = \mathbf{XP} = \mathbf{Q}^{-1}$

$\Rightarrow \mathbf{XPP}^{-1} = \mathbf{Q}^{-1}\mathbf{P}^{-1}$

$\Rightarrow \mathbf{X} = \mathbf{Q}^{-1}\mathbf{P}^{-1}$

(ii) $\mathbf{P}^{-1} = \begin{pmatrix} -\dfrac{1}{9} & \dfrac{1}{6} & -\dfrac{4}{9} \\ \dfrac{2}{9} & \dfrac{1}{6} & -\dfrac{1}{9} \\ -\dfrac{1}{3} & \dfrac{1}{2} & -\dfrac{1}{3} \end{pmatrix}$

$$\mathbf{Q}^{-1} = \begin{pmatrix} \frac{3}{2} & -4 & \frac{1}{2} \\ 1 & -2 & 0 \\ -\frac{3}{2} & 5 & -\frac{1}{2} \end{pmatrix}$$

$$\left(\mathbf{PQ}\right)^{-1} = \mathbf{Q}^{-1}\mathbf{P}^{-1} = \begin{pmatrix} -\frac{11}{9} & -\frac{1}{6} & -\frac{7}{18} \\ -\frac{5}{9} & -\frac{1}{6} & -\frac{2}{9} \\ \frac{13}{9} & \frac{1}{3} & \frac{5}{18} \end{pmatrix}$$

10   (ii)   Multiplying only the first column by k equates to a stretch of scale factor k in one direction, so only multiplies the volume by k.

     (iii)   Multiplying any column by k multiplies the determinant by k.

11   (i) $10 \times 43 = 430$

     (ii) $4 \times 5 \times -7 \times 43 = -6020$

     (iii) $x \times 2 \times y \times 43 = 86xy$

     (iv) $x^4 \times \frac{1}{2x} \times 4y \times 43 = 86x^3 y$

Exercise 2.2 (Page 64)

1   (i)

$$\det \mathbf{M} = 20, \quad \mathbf{M}^{-1} = \begin{pmatrix} 0.2 & 0.4 & -0.6 \\ -0.25 & 0.25 & -0.25 \\ -0.15 & -0.05 & 0.45 \end{pmatrix}$$

2   (i)   Planes π_1 and π_3 are parallel and the second is not parallel to either, so will cross through both to form two parallel straight lines.

     (ii)   Planes π_1 and π_3 are parallel and π_2 is coincident to π_1.

     (iii)   All three planes are parallel.

3   (i) $\begin{pmatrix} 5 & 3 & -2 \\ 6 & 2 & 3 \\ 7 & 1 & 8 \end{pmatrix}\begin{pmatrix} x \\ y \\ z \end{pmatrix} = \begin{pmatrix} 6 \\ 11 \\ 12 \end{pmatrix}$

     (ii)   Eliminating y gives the equations $-8x - 13z = -13$ and $-8x - 13z = -15$ which are inconsistent so the planes form a prism.

4   (i)   Planes meet at the unique point $(3, -14, 8)$

     (ii)   Inconsistent, the planes form a prism

     (iii)   Consistent, then planes form a sheaf

     (iv)   Planes meet at the unique point $(-15, 24, -1)$

     (v)   Three coincident planes

5   (i)   The planes intersect in the unique point $(-0.8, 0.6, 1.5)$

     (ii)   Inconsistent, the planes form a prism

     (iii)   Consistent, the planes form a sheaf

6   (i) $\frac{1}{13k - 65}\begin{pmatrix} 13 & -26 & -13 \\ 7 & -2k-4 & -3k+8 \\ -4 & 3k-7 & -2k+14 \end{pmatrix}$;

     $k = 5$

     (ii)   Unique point of intersection at

$$\left(\frac{52 - 13p}{13}, \frac{20 - 7p}{13}, \frac{4p - 17}{13} \right)$$

     (iii)   Form a sheaf of planes when $p = 4$

Chapter 3

Discussion point (Page 68)

One is positive and one is negative, and they have the same numerator, so terms will cancel out.

Discussion point (Page 72)

The numerator of one fraction is the negative of the numerator of the other, so that fractions cancel out in pairs.

Exercise 3.1 (Page 73)

1   $2n^2 + n$

2   $n(n+1)^2$

3   $\frac{1}{2}n(n+1)(n^2 + n + 1)$

4   (i) $\frac{1}{r} - \frac{1}{r+2}$

     (ii) $\frac{3}{2} - \frac{1}{n+1} - \frac{1}{n+2}$

5   (i) $\frac{1}{4}n(n+1)(n+2)(n+3)$ (ii) $26\,527\,650$

6   (ii) $\frac{1}{3}(n+1)(n+2)(n+3) - 2$

     (iii) $\frac{1}{3}n(n^2 + 6n + 11)$

7   (i) $6 - \frac{5}{n+1} - \frac{2}{n+2}$ (ii)   6

8   (i) $\frac{5}{3} - \frac{1}{2n+1} - \frac{2}{2n+3}$ (iii) $\frac{5}{3}$

9   (i) $\frac{1}{3}S = \frac{1}{3} - 2\left(\frac{1}{3}\right)^2 + 3\left(\frac{1}{3}\right)^3 - 4\left(\frac{1}{3}\right)^4 + \dots$

     (ii) $\frac{4}{3}S = 1 - \frac{1}{3} + \left(\frac{1}{3}\right)^2 - \left(\frac{1}{3}\right)^3 + \left(\frac{1}{3}\right)^4 + \dots$

     This is a geometric series with common ratio $-\frac{1}{3}$. Sum to infinity is $\frac{3}{4}$ and so $S = \frac{9}{16}$.

Exercise 3.2 (Page 77)

1   (ii) $\mathbf{A}^k = \begin{pmatrix} 1 - 3(k+1) & 9(k+1) \\ -(k+1) & 1 + 3(k+1) \end{pmatrix}$

     (iii) $\mathbf{A}^k = \begin{pmatrix} -2 - 3k & 9k + 9 \\ -k - 1 & 3k + 4 \end{pmatrix}$

2 (ii) $\frac{1}{2}(k+1)(3(k+1)+1)$

(iii) $\frac{1}{2}(k+1)(3k+4)$

3 (i) $u_{k+1} = 2^{k+2} - 1$

(ii) $u_{k+1} = 2(2^{k+1} - 1) + 1$

14 (i) $\mathbf{M}^2 = \begin{pmatrix} 9 & 6 & -3 \\ 0 & 9 & 0 \\ 0 & 18 & 0 \end{pmatrix}$

$\mathbf{M}^3 = \begin{pmatrix} 27 & 18 & -9 \\ 0 & 27 & 0 \\ 0 & 54 & 0 \end{pmatrix}$

$\mathbf{M}^4 = \begin{pmatrix} 81 & 54 & -27 \\ 0 & 81 & 0 \\ 0 & 162 & 0 \end{pmatrix}$

(ii) $\mathbf{M}^n = \begin{pmatrix} 3^n & 2 \times 3^{n-1} & -3^{n-1} \\ 0 & 3^n & 0 \\ 0 & 2 \times 3^n & 0 \end{pmatrix}$

Chapter 4

Discussion point (Page 80)

The curve may be discontinuous, or it may approach the x-axis but never actually reach it.

Activity 4.1 (Page 83)

The graph of $y = \dfrac{1}{(x-2)^2}$ is above the x-axis for all values of x, so the value of the integral cannot be negative.

Karen has integrated from 1 to 3, but there is a discontinuity at $x = 2$.

Exercise 4.1 (Page 84)

1 (i)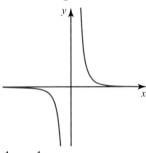

(ii) $\dfrac{1}{8} - \dfrac{1}{2a^2}$

(iii) $\dfrac{1}{8}$

2 (i) $3(b-1)^{\frac{1}{3}} + 3$, $3\sqrt[3]{2} - 3(c-1)^{\frac{1}{3}}$

(ii) $3\sqrt[3]{2} + 3$

(iii)

3 (i)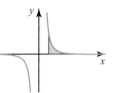

(ii) $1 - e^{-d}$

(iii) 1

4 Convergent, 0.5

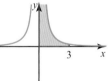

5 Divergent

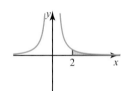

6 Divergent

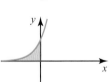

7 Convergent, 0.5

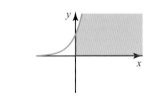

8 Convergent, 1

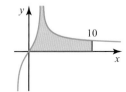

9 Divergent

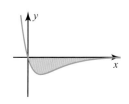

10 Convergent, 9.25 to s.f. **11** Convergent, -0.5

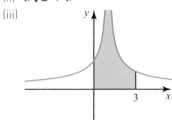

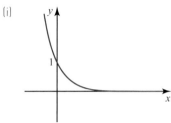

12 1

13 $-\ln 2$

14 1.5

Discussion point (Page 87)

If you replace x with $x\sqrt{3}$ you are doing integration by substitution, so you need to change the variable, e.g. let $u = x\sqrt{3}$ and then use $du = \sqrt{3}\,dx$.

Exercise 4.2 (Page 88)

1 arcsin has domain $[-1, 1]$ and range $[-\frac{\pi}{2}, \frac{\pi}{2}]$
arccos has domain $[-1, 1]$ and range $[0, \pi]$
arctan has domain $\mathbb{R}$ and range $[-\frac{\pi}{2}, \frac{\pi}{2}]$

2 (i) $\arcsin\frac{x}{5} + c$ (ii) $\frac{1}{4}\arctan\frac{t}{4} + c$

3 (i) 0.615 (3 s.f.)   (ii) 0.464 (3 s.f.)

4 (i) $\dfrac{3}{\sqrt{1 - 9x^2}}$ (ii) $-\dfrac{1}{\sqrt{4 - x^2}}$

(iii) $\dfrac{5}{25x^2 + 1}$ (iv) $\dfrac{6x}{\sqrt{1 - 9x^4}}$

(v) $\dfrac{e^x}{e^{2x} + 1}$ (vi) $-\dfrac{6x}{x^4 - 2x^2 + 2}$

5 (i) $\frac{1}{2}\arctan\frac{x}{2} + c$ (ii) $\frac{1}{2}\arctan 2x + c$

(iii) $\arcsin\frac{x}{2} + c$ (iv) $\frac{1}{2}\arcsin 2x + c$

6 (i) $\frac{5}{6}\arctan\frac{x}{6} + c$ (ii) $\frac{2}{5}\arctan\frac{2x}{5} + c$

(iii) $\frac{1}{2}\arcsin\frac{2x}{3} + c$ (iv) $\frac{7}{\sqrt{3}}\arcsin\left(x\sqrt{\frac{3}{5}}\right) + c$

7 (i) $\dfrac{\pi}{12}$ (ii) $\dfrac{\pi}{4}$

(iii) $\dfrac{7\pi}{36}$ (iv) $\dfrac{\pi}{12}$

10 $\arcsin(x^2) + \dfrac{2x^2}{\sqrt{1 - x^4}}$

11 π

12 $\dfrac{\pi}{12\sqrt{10}}$

Exercise 4.3 (Page 92)

1 (ii) $2\ln|x + 1| + \ln|5x - 1| + c$

2 (ii) $-\ln|2x + 3| + \ln|x + 1| - \dfrac{1}{x + 1} + c$

3 (ii) $\ln|x - 3| + \frac{1}{2}\ln(x^2 + 5) + c$

4 (i) $\dfrac{1}{x + 7} + \dfrac{2 - x}{x^2 + 3}$

(ii) $\ln|x + 7| - \frac{1}{2}\ln\left(\frac{1}{3}x^2 + 1\right) + \dfrac{2}{\sqrt{3}}\arctan\dfrac{x}{\sqrt{3}} + c$

5 (i) $2\ln|x + 1| - \ln(x^2 + 1) + 2\arctan x + c$

(ii) $\ln|x - 2| - \frac{1}{2}\ln(x^2 + 9) - \frac{1}{3}\arctan\frac{x}{3} + c$

(iii) $\frac{1}{8}\ln(4x^2 + 1) + \ln|x + 2| + \frac{1}{2}\arctan 2x + c$

6 (i) $\frac{1}{2}(\pi + \ln 2)$

(ii) 2.23

(iii) $-\dfrac{\pi}{3\sqrt{3}} - \ln\dfrac{3}{2}$

7 $\ln\dfrac{\sqrt{13}}{3}$

8 $\dfrac{1}{x + 1} - \dfrac{3}{x - 1} + \dfrac{2}{(x - 1)^2} + \dfrac{2x}{x^2 + 1}$

9 $1 - \ln 2$

10 (i) 0.75   (iii) 0.579   (iv) 9.3%

Activity 4.4 (Page 94)

The first one can be written as $\dfrac{1}{(x + 1)^2 - 3}$ and the second can be written as $\dfrac{1}{\sqrt{(x - 1)^2 + 2}}$. Neither of these are the correct form for using the standard integrals.

Exercise 4.4 (Page 96)

1 $\dfrac{1}{4\sqrt{15}}$

2 $\dfrac{9\pi}{4}$

3 $\dfrac{1}{4\sqrt{2}}$

4 $\frac{1}{2}\arctan\dfrac{3x + 1}{2} + c$

5 $\arcsin\dfrac{x - 1}{2} + c$

6 (i) $x\arcsin x + \sqrt{1 - x^2} + c$

(ii) (a) $x\arccos x - \sqrt{1 - x^2} + c$

(b) $x\arctan x - \frac{1}{2}\ln(x^2 + 1) + c$

(c) $x\,\text{arccot}\,x + \frac{1}{2}\ln(x^2 + 1) + c$

7 (i) $\dfrac{1}{16\sqrt{3}}$ (ii) $\dfrac{1}{\sqrt{2}}$ (iii) $\dfrac{\pi}{5}$

8 (i) $\frac{1}{2}b\sqrt{a^2 - b^2} + \frac{1}{2}a^2\arcsin\dfrac{b}{a}$

(ii)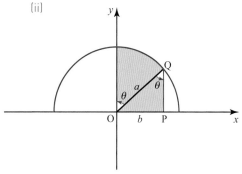

$\sin\theta = \dfrac{b}{a}$, so $\dfrac{1}{2}a^2 \arcsin\dfrac{b}{a}$ is the area of the sector OPQ and $\dfrac{1}{2}b\sqrt{a^2 - b^2}$ is the area of the triangle.

10 (i) $\dfrac{1}{2}\ln(x^2 + 1) + \arctan x + c$

(ii) $\dfrac{1}{2}\ln\left|x^2 + 2x + 3\right| + c$

(iii) $\dfrac{1}{\sqrt{2}}\arctan\dfrac{(x + 1)}{\sqrt{2}} + c$

11 (i) $\sqrt{4 - x^2} + 2\arcsin\dfrac{x}{2} + c$

(ii) $\arcsin\dfrac{x - 2}{2} + c$

(iii) $\sqrt{4x - x^2} + c$

12 $\dfrac{1}{|x|\sqrt{x^2 - 1}}$, $\operatorname{arcsec}\dfrac{x}{a} + c$

14 $\dfrac{1}{2}\left(\sqrt{2} + \ln(1 + \sqrt{2})\right)$

Practice questions 1 (Page 99)

1 $\begin{pmatrix} \frac{1}{2} & -\frac{\sqrt{3}}{2} \\ \frac{\sqrt{3}}{2} & \frac{1}{2} \end{pmatrix}$ or $\begin{pmatrix} \frac{1}{2} & \frac{\sqrt{3}}{2} \\ -\frac{\sqrt{3}}{2} & \frac{1}{2} \end{pmatrix}$ [4]

2 (i) $\dfrac{1}{(2r + 3)(2r + 5)} \equiv \dfrac{1}{2(2r + 3)} - \dfrac{1}{2(2r + 5)}$

[3]

(ii) $\displaystyle\sum_{r=1}^{n} \dfrac{1}{(2r + 3)(2r + 5)} = \dfrac{1}{2}\cdot\dfrac{1}{5} - \dfrac{1}{2}\cdot\dfrac{1}{2n + 5}$

[2]

$= \dfrac{n}{5(2n + 5)}$ [2]

(iii) $\dfrac{1}{10}$ [1]

3 Put $n = 1$. $7 + 2^2 + 1 = 12$ so the result is true for $n = 1$.   [1]

Assume true for $n = k$, so $7^k + 2^{2k} + 1 = 6\lambda$ for some $\lambda \in \mathbb{N}$.   [1]

Consider the expression with $n = k + 1$.

This becomes $7^{k+1} + 2^{2(k+1)} + 1 =$

$7(7^k + 2^{2k} + 1) - 7 \times 2^{2k} + 4 \times 2^{2k} - 6$

So $7^{k+1} + 2^{2(k+1)} + 1 = 42\lambda - 7 \times 2^{2k} + 4 \times 2^{2k} - 6 = 42\lambda - 3 \times 2^{2k} - 6$ by the assumption.

Now $42\lambda - 3 \times 2^{2k} - 6 = 6(7\lambda - 2^{2k-1} - 1)$ which is divisible by 6 since $k \geqslant 1$.

Hence if the result is true for $n = k$, it is true for $n = k + 1$.   [4]

Since the result is true for $n = 1$, by the principle of mathematical induction it is true for all $n \in \mathbb{N}$.   [1]

4 c_n is $\dfrac{1 + 3 + 5 + \cdots + (2n - 1)}{(2n + 1) + \cdots + (4n - 1)}$ in some form.   [1]

The sum of the n terms of the numerator is n^2.   [2]

The sum of the $2n$ combined terms of the numerator and denominator is $4n^2$.   [1]

Hence c_n is $\dfrac{n^2}{4n^2 - n^2} = \dfrac{n^2}{3n^2} = \dfrac{1}{3}$.   [2]

5 (i) $\Delta = \begin{vmatrix} 1 & 1 \\ k & 0 \end{vmatrix} + k\begin{vmatrix} k & 1 \\ 1 & k \end{vmatrix}$ [1]

$\Delta = -k + k(k^2 - 1) = k^3 - 2k$ [1]

No inverse if $\Delta = 0$.   [1]

No inverse for $k = 0$ or $k = \pm\sqrt{2}$.   [1]

Inverse $= \dfrac{1}{k^3 - 2k}\begin{pmatrix} k^2 & -k & -k \\ -k & k^2 - 1 & 1 \\ -k & 1 & k^2 - 1 \end{pmatrix}^{\mathrm{T}}$ [1]

Inverse $= \dfrac{1}{k^3 - 2k}\begin{pmatrix} k^2 & -k & -k \\ -k & k^2 - 1 & 1 \\ -k & 1 & k^2 - 1 \end{pmatrix}$ [1]

(ii) Two planes are parallel.   [1]

The other plane ($y + z = 5$) intersects each of them in a line.   [1]

(iii) If there is a point of intersection, coordinates are

$\dfrac{1}{k^3 - 2k}\begin{pmatrix} k^2 & -k & -k \\ -k & k^2 - 1 & 1 \\ -k & 1 & k^2 - 1 \end{pmatrix}\begin{pmatrix} k \\ 1 \\ 1 \end{pmatrix}$ [1]

$= \dfrac{1}{k^3 - 2k}\begin{pmatrix} k^3 - 2k \\ 0 \\ 0 \end{pmatrix}$ [1]

The point of intersection is $(1, 0, 0)$ and the coordinates are independent of k.   [1]

6 (i) L is parallel to $\begin{pmatrix} 1 \\ 3 \\ -2 \end{pmatrix}$.

M is parallel to $\begin{pmatrix} 5 \\ 1 \\ 4 \end{pmatrix}$.   [1]

$\begin{pmatrix} 1 \\ 3 \\ -2 \end{pmatrix}\cdot\begin{pmatrix} 5 \\ 1 \\ 4 \end{pmatrix} = 5 + 3 - 8 = 0$ so the lines are perpendicular.   [1]

Suppose $\dfrac{x-1}{1} = \dfrac{y+3}{3} = \dfrac{z-1}{-2} = s.$

A general point on L has coordinates

$(s + 1, 3s - 3, -2s + 1)$ [1]

If there is a point of intersection,

$\dfrac{x-3}{5} = \dfrac{y-2}{1} = \dfrac{z+1}{4}.$

So $\dfrac{s-2}{5} = \dfrac{3s-5}{1} = \dfrac{-2s+2}{4}$ [1]

$\dfrac{s-2}{5} = \dfrac{3s-5}{1}$

$s - 2 = 15s - 25$ [1]

$14s = 23$

$s = \dfrac{23}{14}$ [1]

$\dfrac{s-2}{5} = -\dfrac{1}{14}$

$\dfrac{-2s+2}{4} = -\dfrac{9}{28}$, which is not equal to

$-\dfrac{1}{14}$ so there is no point of intersection. [1]

(ii) Suppose A has coordinates
$(s + 1, 3s - 3, -2s + 1)$.
B has coordinates $(5t + 3, t + 2, 4t - 1)$. [1]

$\overrightarrow{AB} = \begin{pmatrix} 5t - s + 2 \\ t - 3s + 5 \\ 4t + 2s - 2 \end{pmatrix}$ [1]

For minimum distance, AB is perpendicular to L.

$\begin{pmatrix} 1 \\ 3 \\ -2 \end{pmatrix} \cdot \begin{pmatrix} 5t - s + 2 \\ t - 3s + 5 \\ 4t + 2s - 2 \end{pmatrix} =$

$= 5t - s + 2 + 3(t - 3s + 5) - 2(4t + 2s - 2)$

$= 0$ [1]

$-14s = -21$

$s = 1.5$

A has coordinates $(2.5, 1.5, -2)$. [1]
AB is perpendicular to M so

$\begin{pmatrix} 5 \\ 1 \\ 4 \end{pmatrix} \cdot \begin{pmatrix} 5t - s + 2 \\ t - 3s + 5 \\ 4t + 2s - 2 \end{pmatrix} = 0$

$42t = -7$ [1]

$t = -\dfrac{1}{6}$

B has coordinates $\left(2\dfrac{1}{6}, 1\dfrac{5}{6}, -1\dfrac{4}{6}\right)$. [1]

7   Require $\displaystyle\int_0^2 \left(\dfrac{2x(6 - x)}{(3x + 2)(x^2 + 4)}\right) dx.$ [1]

Using partial fractions gives

$\dfrac{2x(6 - x)}{(3x + 2)(x^2 + 4)} \equiv \dfrac{-2}{3x + 2} + \dfrac{4}{x^2 + 4}$ [3]

$\displaystyle\int_0^2 \left(\dfrac{-2}{3x + 2} + \dfrac{4}{x^2 + 4}\right) dx$

$= \left[-\dfrac{2}{3}\ln(3x + 2) + 2\tan^{-1}\dfrac{x}{2}\right]_0^2$ [4]

$= \left(-\dfrac{2}{3}\ln(8) + \dfrac{2\pi}{4}\right) - \left(-\dfrac{2}{3}\ln(2) + 0\right)$

$= -\dfrac{2}{3}\ln 4 + \dfrac{\pi}{2}.$ [2]

Chapter 5

Discussion point (Page 102)

There are an infinite number of pairs of polar co-ordinates (r, θ) to define a given point P.

Activity 5.1 (Page 102)

All the points are in the position shown below:

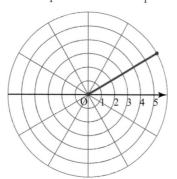

For example, $\left(6, \dfrac{-5}{\frac{\pi}{4}}\right), \left(6, \dfrac{11}{\frac{\pi}{4}}\right), \left(6, \dfrac{-13}{\frac{\pi}{4}}\right)$

Exercise 5.1 (Page 105)

1   (i) $(0, -8)$ (ii) $\left(-4\sqrt{2}, -4\sqrt{2}\right)$

     (iii) $\left(4, 4\sqrt{3}\right)$ (iv) $\left(-4\sqrt{3}, 4\right)$

2   (i) $(13, -1.18)$ (ii) $(5, \pi)$

     (iii) $\left(2, -\dfrac{5\pi}{6}\right)$ (iv) $(5, 0.927)$

3

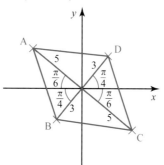

By symmetry the quadrilateral has two pairs of parallel sides.

$\frac{\pi}{6} + \frac{\pi}{4} = \frac{5\pi}{12}$ so the diagonals do not meet at right angles and therefore the shape is not a rhombus.

So ABCD is a parallelogram.

4

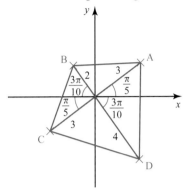

By symmetry AB and BC are the same length; similarly AD and CD are the same length.

The diagonals of the shape meet at right angles. ABCD is a kite.

5 (i) 4

(ii) $16 < r < 170 \quad \theta = -27°$

(iii) (a) $99 < r < 107 \quad 153° < \theta < 171°$

(b) $16 < r < 99 \quad -81° < \theta < -63°$ and
$107 < r < 162 \quad -81° < \theta < -63°$

(c) $162 < r < 170 \quad 45° < \theta < 63°$

6 (i) $B\left(4, -\frac{5\pi}{12}\right) \quad C\left(4, \frac{11\pi}{12}\right)$

(ii) $B(0,0)$ and C either $\left(4, -\frac{\pi}{6}\right)$ or $\left(4, \frac{7\pi}{12}\right)$

(iii) The points could be arranged as shown:

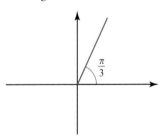

If O is the midpoint of AB then B has coordinates $\left(4, -\frac{3\pi}{4}\right)$ and C has coordinates either $\left(4\sqrt{3}, -\frac{\pi}{4}\right)$ or $\left(4\sqrt{3}, \frac{3\pi}{4}\right)$.

Alternatively the points could be arranged as shown below:

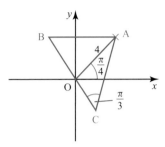

If O is the midpoint of BC then B has coordinates $\left(\frac{4\sqrt{3}}{3}, \frac{3\pi}{4}\right)$ and C has coordinates $\left(\frac{4\sqrt{3}}{3}, -\frac{\pi}{4}\right)$.

7 (ii) A(5.39, 0.38)   B(8.71, 1.01)
C(8.71, 1.64)   D(5.39, 2.26)

(iii) A(5.00, 2.00)   B(4.63, 7.38)
C(−0.52, 8.69)   D(−3.43, 4.16)

Activity 5.3 (Page 108)

$r = 7$ is a circle centre $(0, 0)$, radius 7:

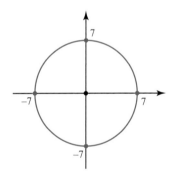

$\theta = \frac{\pi}{3}$ is a half line starting at the origin making an angle $\frac{\pi}{3}$ with the initial line:

Activity 5.4 (Page 109)

2 is the 'radius' of the curve; 3 is the number of petals on the curve.

Starting at 2 on the initial line, a point would move as shown by the arrows below.

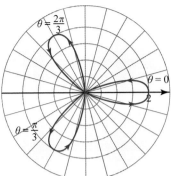

The diagrams below show the curve $r = k\sin n\theta$ for various values of k and n.

$k = 1, n = 1$

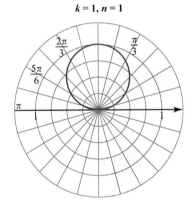

Circle, diameter 1

$k = 1, n = 2$

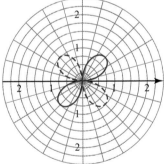

'Radius' of rhodonea is 1; four petals

$k = 1, n = 3$

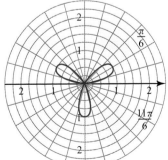

'Radius' of rhodonea is 1; three petals

$k = 5, n = 1$

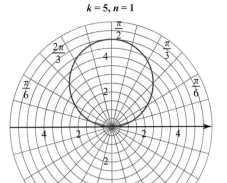

Circle, diameter 5

$k = 5, n = 2$

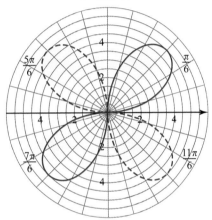

'Radius' of rhodonea is 5; four petals

$k = 1, n = 4$

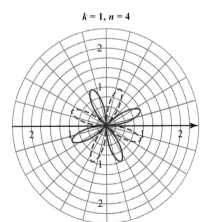

'Radius' of rhodonea is 1; eight petals

Generally, $r = k\sin n\theta$ has 'radius' of k (for $k > 0$). For $n > 1$, the curve has n petals if n is odd and $2n$ petals when n is even. For $k = 1, n = 1$ the curve is a circle of diameter 1 passing through the origin and the point $(0, 1)$, symmetrical about the y-axis.

Exercise 5.2 (Page 109)

1 (i) Circle centre O, radius 5

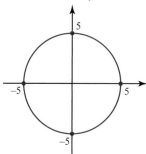

   (ii) Half line from the origin making an angle
        $-\dfrac{3\pi}{4}$ with the initial line

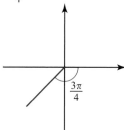

   (iii) Circle symmetrical about the x-axis, passing
        through the origin and the point $(3, 0)$

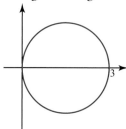

   (iv) Circle symmetrical about y-axis, passing
        through the origin and the point $(0, 2)$

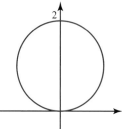

   (v) Spiral starting from the origin

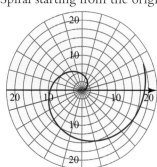

2

θ	0	$\frac{\pi}{12}$	$\frac{\pi}{6}$	$\frac{\pi}{4}$	$\frac{\pi}{3}$	$\frac{5\pi}{12}$	$\frac{\pi}{2}$	$\frac{7\pi}{12}$	$\frac{2\pi}{3}$	$\frac{3\pi}{4}$	$\frac{5\pi}{6}$	$\frac{11\pi}{12}$	π
r	0	2.07	4	5.66	6.93	7.73	8	7.73	6.93	5.66	4	2.07	0

The same values would be repeated for $\pi \leqslant \theta < 2\pi$;
the values of r would be negative and so would form
a separate loop around the circle.

The circle has centre $(0, 4)$, radius 4 so the cartesian
equation is $x^2 + (y - 4)^2 = 16$.

3

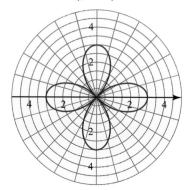

$r = 3\cos 2\theta$

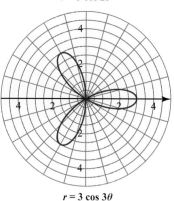

$r = 3\cos 3\theta$

$r = 3\cos 2\theta$ has four petals and $r = 3\cos 3\theta$ has
three petals.
Substituting in values at intervals of $\dfrac{\pi}{6}$ (say)
shows that for $r = 3\cos 3\theta$ the entire curve
is generated twice, once every π radians; for
$r = 3\cos 2\theta$ the entire curve is generated once
in the interval $0 \leqslant \theta \leqslant 2\pi$.

4

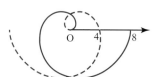

5 (i) The curve has a heart shape, hence the
       name cardiod.

(ii) The curve is a reflection of the curve in part (i) in the 'vertical axis'

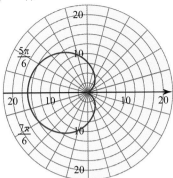

6 $r = a\sec\theta \Rightarrow r = \dfrac{a}{\cos\theta} \Rightarrow r\cos\theta = a \Rightarrow x = a$

$r = b\,\mathrm{cosec}\,\theta \Rightarrow r = \dfrac{b}{\sin\theta} \Rightarrow r\sin\theta = b \Rightarrow y = b$

7 (i)

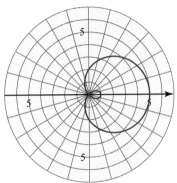

$r = 2 + 3\cos\theta$

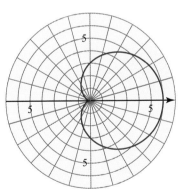

$r = 3 + 3\cos\theta$

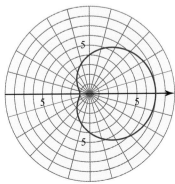

$r = 4 + 3\cos\theta$

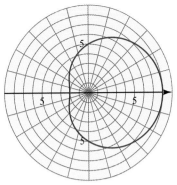

$r = 5 + 3\cos\theta$

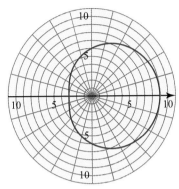

$r = 6 + 3\cos\theta$

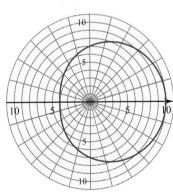

$r = 7 + 3\cos\theta$

(ii) (a) $a = b$ Cardioid
(b) $a < b$ Limacon has a 'loop' inside
(c) $a > b$ Limacon has a 'dimple'
(d) $a \geqslant 2b$ Convex limacon

(iii) The shape is the same but the curves are now symmetrical about the y-axis rather than the x-axis.

8 Values of r are only defined in the intervals $0 \leqslant \theta \leqslant \dfrac{\pi}{4},\ \dfrac{3\pi}{4} \leqslant \theta \leqslant \dfrac{5\pi}{4}$ and $\dfrac{7\pi}{4} \leqslant \theta \leqslant 2\pi$ (the value of r^2 is negative in the other intervals). Where the curve is defined, one half of each loop is created for each interval of $\dfrac{\pi}{4}$.

Taking negative square roots instead of positive square roots produces the same curve. For example, when $\theta = 0$, $r = -a$ when taking the negative square root; this is equivalent to the point (π, a) obtained when taking the positive square root.

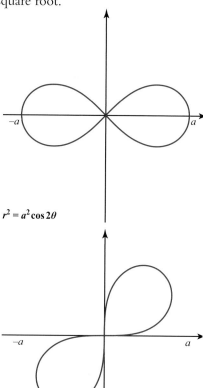

$r^2 = a^2 \cos 2\theta$

$r^2 = a^2 \sin 2\theta$

9 (i)

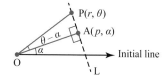

Let (r, θ) be a point on the line L that is perpendicualr to OA. In triangle OPA,
$$\cos(\theta - \alpha) = \frac{p}{r} \Rightarrow r\cos(\theta - \alpha) = p$$

(ii) $r\cos(\theta - \alpha) = p$
$\Rightarrow r\cos\theta\cos\alpha + r\sin\theta\sin\alpha = p$
$\Rightarrow x\cos\alpha + y\sin\alpha = p$

Exercise 5.3 (Page 112)

1 (i) Area is 25π
(ii) Area is 50π (i.e. two loops of the circle)

2 (i)

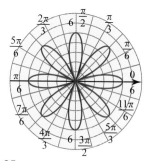

(ii) $\dfrac{25\pi}{16}$

3 (i)

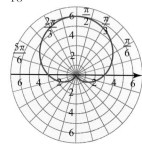

(ii) $\dfrac{27\pi}{2}$

4 $\dfrac{64\pi}{3}$

5 (i) $2\pi + \dfrac{3\sqrt{3}}{2}$ (ii) $\pi + 3\sqrt{3}$

6

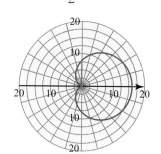

Larger part = $24\pi + 64$ Smaller part = $24\pi - 64$

7

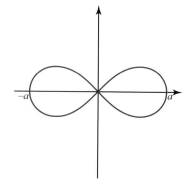

One loop has area $\dfrac{1}{2}a^2$.

8   $A = \dfrac{a^2}{4k}\left(e^{\frac{k\pi}{2}} - 1\right), B = \dfrac{a^2}{4k}e^{4k\pi}\left(e^{\frac{k\pi}{2}} - 1\right),$

     $C = \dfrac{a^2}{4k}e^{8k\pi}\left(e^{\frac{k\pi}{2}} - 1\right)$

     Common ratio $e^{4k\pi}$

9   1.15

Chapter 6

Activity 6.1 (Page 115)

2   $p(x) = 1 + x + \dfrac{1}{2}x^2 + \dfrac{1}{6}x^3 + \dfrac{1}{24}x^4 + \dfrac{1}{120}x^5$

Activity 6.2 (Page 116)

$\left(1 + x + \dfrac{1}{2}x^2 + \dfrac{1}{6}x^3\right)\left(1 - x + \dfrac{1}{2}x^2 - \dfrac{1}{6}x^3\right)$

$= 1 - \dfrac{1}{12}x^4 - \dfrac{1}{36}x^6$

If more terms were used, the terms in x^4 and x^6 and higher terms would cancel out, leaving 1.

Discussion point (Page 116)

It is often the case that if the terms of a series converge, the sum of the series also converges. However, it is not always the case: for example, the terms of the series $u_n = \dfrac{1}{n}$ converge to zero but the sum of the terms does not converge.

Activity 6.3 (Page 117)

Order = 7

Maclaurin approximation up to term in x^{10} is 7.388994709, percentage error = −0.00083%.

Exercise 6.1 (Page 121)

1   (ii)   0.995004   (iii) $1.4 \times 10^{-7}\%$

2   (i)   0.60653, 7 terms   (ii)   0.00024%

3   (i) $\sin x = x - \dfrac{x^3}{3!} + \ldots$

     (ii) $x^3 + 6x^2 - 6x = 0$

4   (i) $\ln x$ is not defined at $x = 0$

5   $f'(0) = 0, f''(0) = -3, f^{(3)}(0) = 15$

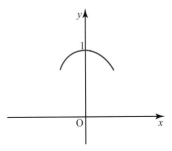

6   $\tan x = x + \dfrac{1}{3}x^3 + \ldots$

7   (iii)   0.24%

8   (i) $a_0 = 2$

     (iii) $2 + x - x^2 - \dfrac{x^3}{3} + \dfrac{x^4}{4} + \dfrac{x^5}{15} - \dfrac{x^6}{24}$

     (iv)

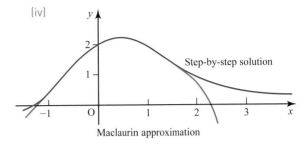

Maclaurin approximation

9   (i)   (a) $1 - \dfrac{\theta^2}{2!} + \dfrac{\theta^4}{4!} - \dfrac{\theta^6}{6!} + \ldots$

        (b) $\theta - \dfrac{\theta^3}{3!} + \dfrac{\theta^5}{5!} - \dfrac{\theta^7}{7!} + \ldots$

    (c) $1 + i\theta - \dfrac{\theta^2}{2!} - \dfrac{i\theta^3}{3!} + \dfrac{\theta^4}{4!} + \dfrac{i\theta^5}{5!} - \dfrac{\theta^6}{6!} - \dfrac{i\theta^3}{7!} + \ldots$

     (ii)

      $1 + i\theta - \dfrac{\theta^2}{2!} - \dfrac{i\theta^3}{3!} + \dfrac{\theta^4}{4!} + \dfrac{i\theta^5}{5!} - \dfrac{\theta^6}{6!} - \dfrac{i\theta^3}{7!} + \ldots$

10   (ii) $a_1 = 1, a_2 = 0$

     (iv) $x + \dfrac{x^3}{3!} + \dfrac{9x^5}{5!} + \ldots$

        $+ \dfrac{(2r-1)^2(2r-3)^2 \ldots 5^2 \times 3^2}{(2r+1)!}x^{2r+1} + \ldots$

Activity 6.5 (Page 123)

(i)   In each case you get the series for the derivative of the original function.

(ii) $\sin x$ has only odd powers of x. The graph of $y = \sin x$ has rotational symmetry about the origin, as do all graphs of the form $y = x^n$ for odd n.

     $\cos x$ has only even powers of x. The graph of $y = \cos x$ is symmetrical about the y-axis, as are all graphs of the form $y = x^n$ for even n.

Exercise 6.2 (Page 124)

1   (i) $1 - \dfrac{u^2}{2!} + \dfrac{u^4}{4!} - \dfrac{u^6}{6!} + \ldots$

     (ii) $1 - 2x^2 + \dfrac{2x^4}{3} - \dfrac{4x^6}{45} + \ldots$

2   (i) $3x - \dfrac{9}{2}x^3$

     (ii) $2x - 2x^2 + \dfrac{8}{3}x^3 - 4x^4$

     (iii) $1 + \dfrac{1}{2}x + \dfrac{1}{8}x^2 + \dfrac{1}{48}x^3 + \dfrac{1}{384}x^4$

3 (i) $x^2 - \frac{1}{3}x^4$

    (ii) $x - \frac{1}{2}x^2 + \frac{1}{6}x^3 - \frac{1}{12}x^4$

    (iii) $x - x^2 + \frac{1}{3}x^3$

    (iv) $1 + x + \frac{1}{2}x^2 - \frac{1}{8}x^4$

4 (i) $\frac{1}{2}\arcsin 2x + c$

    (ii) $2x + \frac{4}{3}x^3 + \frac{12}{5}x^5 + \frac{40}{7}x^7 + \dots$

    (iii) Using $x = 0.25$ gives $0.523525856\dots$,
       percentage error $= -0.0139\%$

5 (i) 0.6456

    (ii) 0.6911

    (iii) $2x + \frac{2x^3}{3} + \frac{2x^5}{5}$, $x = \frac{1}{3}$ gives $\ln 2 = 0.6930$

6 (iv) 3.14159

7 (i) $1 + x + x^2 + x^3 + \dots$ (ii) $\dfrac{1}{(1-x)^2}$

8 (i) $x + x^2 + \frac{1}{3}x^3 - \frac{1}{30}x^5 - \frac{1}{90}x^6$

    (ii) $x + x^2 + \frac{1}{3}x^3 - \frac{1}{12}x^5 - \frac{1}{36}x^6$, agrees with
       (i) as far as the term in x^4

    (iii) $1 + x - \frac{1}{3}x^3$; the x^4 term of the product is
       not correct

9 (ii) $f^{(3)}(x) = 4f''(x) - 13f'(x)$
       $f^{(4)}(x) = 4f^{(3)}(x) - 13f''(x)$

    (iii) $f(x) = 3x + 6x^2 + \frac{3}{2}x^3 - 5x^4 + \dots$

11 Given $0 < r < 1$, the area of the square

    $= \left(\dfrac{1}{1-r}\right)^2 = 1 + 2r + 3r^2 + 4r^3 + 5r^4 + \dots$

    Also the side of the square

    $= (1-r)(1 + 2r + 3r^2 + 4r^3 + 5r^4 + \dots)$

    $= (1-r)\left(\dfrac{1}{1-r}\right)^2 = \dfrac{1}{1-r}$

Review: Complex numbers

Review Exercise R.1 (Page 130)

1 (i) (a) -1 (b) $-i$ (c) 1 (d) i (e) -1

    (ii) The powers of i form a cycle:

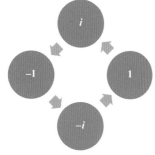

All numbers of the form i^{4n} are equal to 1.
All numbers of the form i^{4n+1} are equal to i.
All numbers of the form i^{4n+2} are equal to -1.
All numbers of the form i^{4n+3} are equal to $-i$.

2 (i) $9 + 12i$ (ii) $-21 - 20i$

    (iii) $53 - 89i$

3 (i) $-1 + 2i$ (ii) $\dfrac{9}{5} + \dfrac{7}{5}i$

    (iii) $\dfrac{9}{5} - \dfrac{7}{5}i$

4 $\dfrac{2}{29} - \dfrac{179}{29}i$

5 $a = 1, -2$ and $b = 2, 3$
    Possible complex numbers are
    $1 + 15i$, $1 + 10i$, $4 + 15i$ and $4 + 10i$

6 $\dfrac{4}{29}$, $\dfrac{4}{29}$

7 $z + z^* = 2x$ which is real
    $zz^* = x^2 - y^2$ which is real

8 $a = \dfrac{5}{3}$, $b = -\dfrac{5}{3}$

Review Exercise R.2 (Page 133)

1 (i)

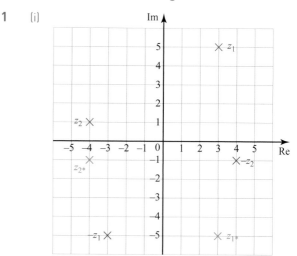

    (ii) $-z$ represents a rotation of z about the
       origin, through an angle of $180°$.

    (iii) z^* represents a reflection of z in the x-axis.

2

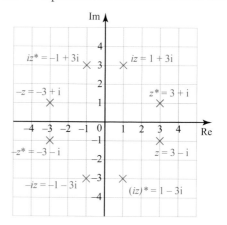

3

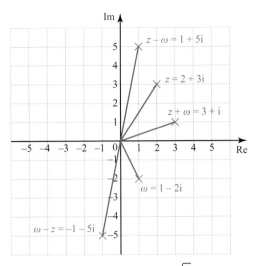

4 (i) (a) $z^0 = 1$, $z^1 = -\frac{1}{2} - \frac{\sqrt{3}}{2}i$,

$$z^2 = -\frac{1}{2} + \frac{\sqrt{3}}{2}i$$

(b)

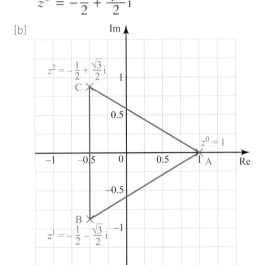

(ii) Using coordinate geometry:

Length AC $= \sqrt{\left(\frac{3}{2}\right)^2 + \left(\frac{\sqrt{3}}{2}\right)^2} = \sqrt{3}$

Length BC $= \sqrt{\left(\frac{3}{2}\right)^2 + \left(\frac{\sqrt{3}}{2}\right)^2} = \sqrt{3}$

Length AC $= 2 \times \frac{\sqrt{3}}{2} = \sqrt{3}$

So the triangle is equilateral.

5 (i) $z_1 = -\frac{7}{29} + \frac{3}{29}i$ $z_2 = -\frac{3}{29} - \frac{7}{29}i$

(ii)

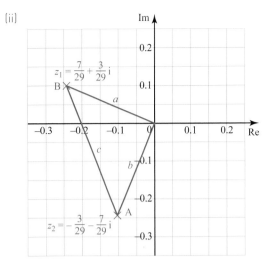

(iii) z_2 is a rotation of z_1 through an angle of $90°$ anticlockwise.

(iv) Length OA $= \sqrt{\left(\frac{7}{29}\right)^2 + \left(\frac{3}{29}\right)^2} = \frac{\sqrt{58}}{29}$

Length OB $= \sqrt{\left(\frac{3}{29}\right)^2 + \left(\frac{7}{29}\right)^2} = \frac{\sqrt{58}}{29}$

Length AB $= \sqrt{\left(\frac{4}{29}\right)^2 + \left(\frac{10}{29}\right)^2} = \frac{2\sqrt{29}}{29}$

So the triangle is isosceles.

Chapter 7

Discussion point (Page 136)

$\cosh t$ is positive for all values of t, so only the branch for which x is positive is given.

Exercise 7.1 (Page 139)

2 $\cosh x = \sqrt{5}$, $\tanh x = \frac{2}{\sqrt{5}}$

3 (ii) $3e^{2x} + 2e^x - 1 = 0$

4 (i) $4\cosh 4x$ (ii) $2x\sinh x^2$

  (iii) $2\cosh x \sinh x$ (iv) $\cosh^2 x + \sinh^2 x$

5 (i) $\frac{1}{3}\cosh 3x + c$ (ii) $x\cosh x - \sinh x + c$

  (iii) $\frac{1}{2}\sinh\left(1 + x^2\right) + c$

6 (i) $\ln\frac{3}{4}$, $\ln 2$ (ii) $-\ln 2$ (ii) no real root

7 $x = \ln 3$, $y = \ln 2$

8 $1.62\,\mathrm{m}$, $22.3°$

9 (iii) $\frac{1}{4}\sinh 2x + \frac{1}{2}x + c$, $\frac{1}{4}\sinh 2x - \frac{1}{2}x + c$

10 $\left(-\frac{1}{2}\ln 3, -\frac{1}{3\sqrt{3}}\right)$

11 $10 \ln 2 - 6$

12 $\cosh x = 1 + \dfrac{x^2}{2!} + \dfrac{x^4}{4!} + \dots + \dfrac{x^{2r}}{(2r)!} + \dots$

14   (i) $a + b, a - b, c$ all have the same sign and $b^2 + c^2 > a^2$

(ii) $a + b, a - b, c$ all have the same sign and $b^2 + c^2 = a^2$ or $a + b$ and $a - b$ have opposite signs and $b^2 + c^2 > a^2$

(iii) $b^2 + c^2 < a^2$ or $a + b$ and $a - b$ have the same signs and c has the opposite sign

15   (i) $x = 0$

(ii) $\left(\dfrac{1}{2} \ln(2 + \sqrt{3}), \dfrac{4}{3\sqrt{3}} \right),$

$\left(\dfrac{1}{2} \ln(2 - \sqrt{3}), -\dfrac{4}{3\sqrt{3}} \right)$

(iv)

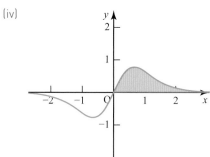

Activity 7.1 (Page 145)

$\operatorname{arcosh} 2 = \ln\left(2 + \sqrt{3} \right).$ This is one of the roots of the equation $\cosh x = 2,$ but there is a second root which is the negative of this.

Exercise 7.2 (Page 145)

1   (i) $\ln(3 + 2\sqrt{2})$ (ii) $\ln(1 + \sqrt{2})$ (iii) $\dfrac{1}{2} \ln 3$

(iv) $\ln(-2 + \sqrt{5})$ (iii) $\ln 2$ (vi) $-\dfrac{1}{2} \ln 5$

2   (i) $x = \ln(3 + \sqrt{10})$ (ii) $\pm \ln(3 + 2\sqrt{2})$

(iii) $\dfrac{1}{2} \ln \dfrac{3}{2}$

3   (i) $\dfrac{3}{\sqrt{1 + 9x^2}}$ (ii) $\dfrac{4}{\sqrt{16x^2 - 1}}$

(iii) $\dfrac{4x}{\sqrt{1 + 4x^4}}$ (iv) $\dfrac{1}{\sqrt{x^2 + x}}$

4   (i) $\dfrac{1 + x(x^2 - 1)^{-\frac{1}{2}}}{x + \sqrt{x^2 - 1}}$

6   (ii) $\operatorname{artanh} x = \dfrac{1}{2} \ln\left(\dfrac{1 + x}{1 - x} \right)$

7   (i) $x \operatorname{arcosh} x - \sqrt{x^2 - 1} + c$

(ii) $x \operatorname{arsinh} x - \sqrt{x^2 + 1} + c$

(iii) $x \operatorname{artanh} x - \dfrac{1}{2} \ln\left(1 - x^2 \right) + c$

8   (i) $\dfrac{1}{2} \cosh\left(x^2 \right) + c$

(ii) $\dfrac{1}{2} x^2 \cosh\left(x^2 \right) - \dfrac{1}{2} \sinh\left(x^2 \right) + c$

9 $1 - \dfrac{1}{2} \sqrt{3}$

Exercise 7.3 (Page 147)

1   (i) $\operatorname{arcosh} \dfrac{x}{2} + c$ (ii) $\operatorname{arsinh} \dfrac{x}{2} + c$

2   (i) $\ln\left(2 + \sqrt{3} \right)$ (ii) $\ln\left(\dfrac{2 + \sqrt{5}}{1 + \sqrt{2}} \right)$

3   (i) $\dfrac{1}{3} \operatorname{arcosh} 3x + c$ (ii) $\dfrac{1}{3} \operatorname{arsinh} 3x + c$

(iii) $\dfrac{1}{2} \operatorname{arsinh} \dfrac{2x}{3} + c$ (iv) $\dfrac{1}{2} \operatorname{arcosh} \dfrac{2x}{3} + c$

4   (i) $\dfrac{1}{5} \ln\left(\dfrac{1}{4}\left(5 + \sqrt{21} \right) \right)$ (ii) $\dfrac{1}{3} \ln\left(3 + \sqrt{10} \right)$

5   (ii) $\dfrac{1}{2} x \sqrt{x^2 + 4} + 2 \operatorname{arsinh} \dfrac{x}{2} + c$

6   (ii) $\dfrac{1}{2} x \sqrt{x^2 - 9} - \dfrac{9}{2} \operatorname{arcosh} \dfrac{x}{3} + c$

9   (i) $4\left(x + \dfrac{3}{2} \right)^2 - 49$ (ii) 0.322

Chapter 8

Discussion point (Page 150)

Find the volume of a cone with radius 2 and height 2, and subtract the volume of a cone with radius 1 and height 1.

Volumes of some other solids of revolution can be found using formulae, but many cannot.

Exercise 8.1 (Page 155)

1   For example; ball, tin of soup, bottle of wine, roll of sticky tape, dinner plate

2   (i)   Cone, radius 18 and height 6

(ii) 648π cubic units

(iii)   Cone, radius 2, height 6

(iv) 8π cubic units

3 $\dfrac{1}{2}\pi$

4   (i)

(ii) $\dfrac{56}{3}\pi$ cubic units

5 (i)

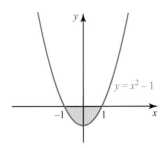

(ii) $\dfrac{16}{15}\pi$ cubic units

6 (i)

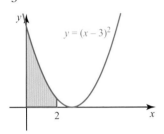

(ii) $\dfrac{32}{5}\pi$ cubic units

7 (i)

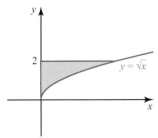

(ii) $\dfrac{26}{3}$ square units   (iii) $\dfrac{242}{5}\pi$ cubic units

8 (i)

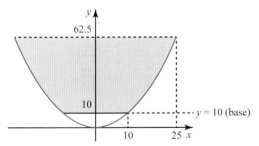

(ii) 45.9 litres

9 (i)

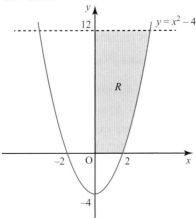

(ii) $\displaystyle\int_0^{12}\pi(y+4)\,dy$ (iii) 3 litres

10 (i)   A(-3, 4), B(3, 4)   (ii) 36π cubic units

11 (i) $\pi\left(1-\dfrac{1}{a}\right)$ cubic units   (ii) π cubic units

Activity 8.1 (Page 158)

(i)   Area $=\dfrac{26}{3}$ square units,
mean value $=\dfrac{13}{3}=4.33...$

(ii) $\dfrac{14}{3}=4.66...$ (iii) 4.5

Discussion point (Page 158)

Finding the mean of a function deals with a continuous variable, whereas finding the mean value of a set of numbers deals with a discrete set of data. Finding the mean of a function involves finding the limit of a sum of values.

Activity 8.2 (Page 159)

(i) $\dfrac{a}{\sqrt{2}}$

Exercise 8.2 (Page 159)

1   1.49 (3 s.f.)

2   0.25

3   (i)   0

(ii)   The graph has rotational symmetry about $x=1$, so the mean value is 0

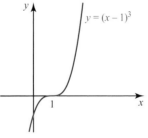

4   (i) $\dfrac{8}{3}$

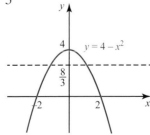

(ii) 1.26 (3 s.f.)

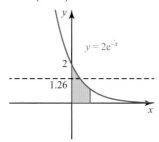

(iii) 0.549 (3 s.f.)

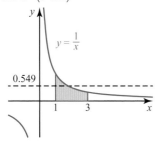

(iv) 0.637 (3 s.f.)

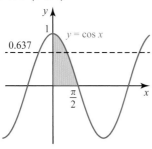

5 (i) 0.2 (ii) 0.2

(iii) The graph is symmetrical about $x = 1$ so the mean value is the same for the intervals $[0, 1]$ and $[1, 2]$ and therefore also for the interval $[0, 2]$.

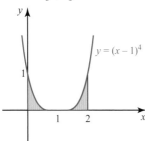

6 (i) 3 (ii) 1.5

(iii) 2.75, 1.375 (iv) 2.6875, 1.34375

(v) Area = $\frac{8}{3}$, mean value = $\frac{4}{3}$

The total area of the rectangles is approaching the area under the graph, as more rectangles are used.

The mean height of the rectangles is approaching the mean value of the function, as more rectangles are used.

7 (i) $\frac{1}{2}x - \frac{1}{4}\sin 2x + c$ (iii) $\frac{1}{2}$

(iv) $\frac{1}{\sqrt{2}}$ (v) 120 V

Exercise 8.3 (Page 163)

1 (i) $\arcsin \frac{x}{3} + c$ (ii) $\operatorname{arcosh} \frac{x}{3} + c$

(iii) $\operatorname{arsinh} \frac{x}{3} + c$ (iv) $\frac{1}{3}\arctan \frac{x}{3} + c$

(v) $\frac{1}{6}\ln\left|\frac{3+x}{3-x}\right| + c$

2 (i) $\frac{\pi}{20}$ (ii) $\frac{1}{5}\ln 3$

(iii) $\ln(1 + \sqrt{2})$ (iv) $\ln(2 + \sqrt{3})$

(v) $\frac{\pi}{2}$

3 (i) $\frac{1}{3}\arcsin \frac{3x}{2} + c$ (ii) $\frac{1}{3}\operatorname{arcsinh} \frac{3x}{2} + c$

(iii) $\frac{1}{3}\operatorname{arccosh} \frac{3x}{2} + c$

4 (i) $\frac{\pi}{3\sqrt{3}}$ (ii) $\frac{1}{\sqrt{3}}\log\frac{3 + \sqrt{5}}{2}$

(iii) $\frac{\pi}{3\sqrt{3}}$

6 $9\sqrt{3} - 12 = 3.59$ (3 s.f.)

7 $\frac{3}{320}$

8 $\frac{3}{\sqrt{2}}\arctan \frac{x}{\sqrt{2}} - \frac{1}{x-1} + c$

9 $\frac{3}{2}\ln 2 - \frac{1}{8}\pi\sqrt{2}$

10 (i) $\frac{\pi}{4} + \ln\frac{32\sqrt{2}}{9} \approx 2.40$

(ii) There are discontinuities at $x = -1$ and $x = -2$

(iii) Only (d), area $\frac{\pi}{2} + 2\ln 2$

11 (i) $\frac{4}{3}$ (ii) $\frac{2}{3}$ (iii) $\frac{128}{105}\pi$ (iv) 0.781 (3 s.f.)

12 $2\ln\left(\frac{4 + \sqrt{17}}{3 + \sqrt{10}}\right)$

Practice questions 2 (Page 166)

1 (i) Eventually y will be zero (or negative) and that is not realistic. [1]

(ii) $\frac{1}{30}\int_0^{30}(-0.884x^4 + 53x^3 - 840x^2 + 1332x + 403991)\,dx$ [1]

$11\,595\,390 \div 30 = 386\,513\,km$ [2]

(iii) The values are quite close. A suitable comment would be, the distances in January 2016 must be fairly representative, or the model is quite good for January 2016. [1]

2 $1 + i$ is a root so $(1 + i)^2 = 2i$ $(1 + i)^3 = -2 + 2i$ [1]

$(-2 + 2i)(m + 2) + 2i(m^2 - 8)$

$+ (1 + i)(m + 3) - 2 = 0$ [1]

Real parts: $-2(m + 2) + (m + 3) - 2 = 0$

so $m = -3$ [2]

Equation is $-x^3 + x^2 - 2 = 0$

Second root is $1 - i$ [1]

Sum of roots is 1 so third root is -1.   [1]

3   (a)   (i) $A = \displaystyle\int_1^a \frac{1}{x}\, dx$ [1]

$A = [\ln x]_1^a = \ln a$ [1]

(ii)   As $a \to \infty$, $\ln a \to \infty$ [1]

(b) $V = \pi\displaystyle\int_1^\infty \frac{1}{x^2}\, dx = \pi\left[-\frac{1}{x}\right]_1^\infty$ [2]

$V = \pi\displaystyle\lim_{N \to \infty}\left(-\frac{1}{N} + 1\right) = \pi$ [1]

4   (i) $2\sinh^2 x + 1 = 2\left(\dfrac{e^x - e^{-x}}{2}\right)^2 + 1$ [1]

$= 2\left(\dfrac{e^{2x} + e^{-2x} - 2}{4}\right) + 1$ [1]

$= \dfrac{e^{2x} + e^{-2x}}{2} = \cosh 2x$ [1]

(ii) $2\sinh^2 x - 3\cosh x = 0$

$2(\cosh^2 x - 1) - 3\cosh x = 0$ [1]

$2\cosh^2 x - 3\cosh x - 2 = 0$

$(2\cosh x + 1)(\cosh x - 2) = 0$

$\cosh x = 2$ [1]

$x = \pm\ln(2 + \sqrt{3})$ [1]

(iii) $\displaystyle\int_{-\ln(2+\sqrt{3})}^{\ln(2+\sqrt{3})} (2\sinh^2 x - 3\cosh x)\, dx$ [1]

$= \displaystyle\int_{-\ln(2+\sqrt{3})}^{\ln(2+\sqrt{3})} (\cosh 2x - 1 - 3\cosh x)\, dx$ [1]

$= \left[\dfrac{1}{2}\sinh 2x - x - 3\sinh x\right]_{-\ln(2+\sqrt{3})}^{\ln(2+\sqrt{3})}$ [1]

$= \sinh\left(2\ln(2 + \sqrt{3})\right) - 2\ln(2 + \sqrt{3})$

$- 6\sinh\left(\ln(2 + \sqrt{3})\right)$ [1]

$= \left(\dfrac{e^{\ln(2+\sqrt{3})^2} - e^{-\ln(2+\sqrt{3})^2}}{2}\right) - 2\ln(2 + \sqrt{3})$

$- 6\left(\dfrac{e^{\ln(2+\sqrt{3})} - e^{-\ln(2+\sqrt{3})}}{2}\right)$ [1]

$= \dfrac{1}{2}\left((2 + \sqrt{3})^2 - \dfrac{1}{(2 + \sqrt{3})^2}\right) - 2\ln(2 + \sqrt{3})$

$- 3\left((2 + \sqrt{3}) - \dfrac{1}{(2 + \sqrt{3})}\right)$ [1]

$= -2\sqrt{3} - 2\ln(2 + \sqrt{3})$ (simplification of surds by calculator)   [1]

Area positive, so area is

$2\sqrt{3} + 2\ln(2 + \sqrt{3})$.   [1]

5   (i) $y_1 = -\tan x$; $y_2 = -\sec^2 x$;

$y_3 = -2\sec^2 x \tan x$ [2]

so $y_3 + 2y_2 y_1 = 0$.   [1]

(ii) $y(0) = 0$; $y_1(0) = 0$; $y_2(0) = -1$; $y_3(0) = 0$.   [2]

Find y_4 and obtain $y_4(0) = -2$.   [1]

Establish $\ln(\cos x) \approx -\dfrac{x^2}{2} - \dfrac{x^4}{12} + \dots$ [2]

(iii) $\ln\left(\cos\dfrac{\pi}{3}\right) = \ln\dfrac{1}{2} = -\ln(2)$ [1]

so $\ln(2) \approx \dfrac{\left(\dfrac{\pi}{3}\right)^2}{2} + \dfrac{\left(\dfrac{\pi}{3}\right)^4}{12} + \dots$

$= \dfrac{\pi^2}{18}\left(1 + \dfrac{\pi^2}{54}\right)$ [2]

(iv) $\ln\left(\cos\dfrac{\pi}{4}\right) = \ln\dfrac{1}{\sqrt{2}} = -\dfrac{1}{2}\ln(2)$,

so $\ln(2) \approx \dfrac{2\left(\dfrac{\pi}{4}\right)^2}{2} + \dfrac{2\left(\dfrac{\pi}{4}\right)^4}{12} + \dots$

$= \dfrac{\pi^2}{16}\left(1 + \dfrac{\pi^2}{96}\right)$ [2]

$\ln(2) \approx 0.69315\dots$; using $\dfrac{\pi}{4}$ in the expansion gives $0.68027\dots$ relative error $\approx 2\%$; using $\dfrac{\pi}{3}$ in the expansion gives $0.64853\dots$ relative error $\approx 6\%$ [1]

Using $\dfrac{\pi}{4}$ is better as it is smaller.   [1]

6   (i)   Angle CQA is 90° since QA is tangent to the circle. The result follows from angle sum of triangles CQA and OPA as angle OAP is common.   [1]

(ii) $OA = CA - \dfrac{a}{2}$ so $r\sec\theta = a\sec\phi - \dfrac{a}{2}$.

Since $\theta = \phi$ this gives $r = a\left(1 - \dfrac{1}{2}\cos\theta\right)$.   [4]

(iii)   For $\dfrac{\pi}{2} \leqslant \theta \leqslant \pi$, we have $OA = CA + \dfrac{a}{2}$ but as $-1 \leqslant \cos\theta \leqslant 0$ for $\dfrac{\pi}{2} \leqslant \theta \leqslant \pi$, the same equation is required.   [2]

For $-\pi \le \theta \le 0$, the same equation is required as the curve is a reflection in the line through COA and $\cos(-\theta) = \cos\theta$. [2]

(iv) Anything like an ellipse with one end flattened and with correct common points with the circle. [2]

(v) Since $r = a\left(1 - \frac{1}{2}\cos\theta\right)$, we have

$r^2 = ar\left(1 - \frac{1}{2}\cos\theta\right) = ar - \frac{1}{2}ar\cos\theta$. [1]

The cartesian equation is

$x^2 + y^2 = \sqrt{x^2 + y^2} \times a - \frac{1}{2}ax$. [1]

Review: Roots of polynomials

Exercise R.1 (Page 171)

1 (i) $\frac{3}{2}$ (ii) $-\frac{5}{2}$

2 (i) $-\frac{1}{3}$ (ii) $\frac{4}{3}$ (iii) $\frac{7}{3}$

3 (i) 2   (ii) -5 (iii) 0   (iv) -3

4 $z^3 - 3z^2 - 18z + 40 = 0$

5 (i) $3w^2 + 8w + 9 = 0$

   (ii) $3w^2 + 4w + 16 = 0$

   (iii) $3w^2 - 14w + 31 = 0$

6 (i) $w^3 - 9w^2 + 23w - 17 = 0$

   (ii) $8w^3 - 12w^2 - 2w + 1 = 0$

   (iii) $w^3 + 6w^2 - 23w - 10 = 0$

7 (i) Roots are $3, 3, -2$ and $k = 18$

     or roots are $-\frac{1}{3}, -\frac{1}{3}, \frac{14}{3}$ and $k = -\frac{14}{27}$

   (ii) $k = 18$

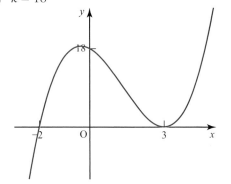

$k = -\frac{14}{27}$

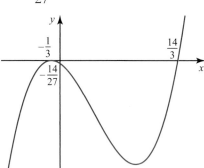

9 $z = 3, 6$ or $-2, k = 36$

10 $p = -3, q = -6$, roots are $1, 2, \pm\sqrt{2}$
   or $p = 3, q = 6$, roots are $-1, -2, \pm\sqrt{2}$

Exercise R.2 (Page 174)

1 (i) $2 + 5i$ (ii) $p = -4, q = 29$

2 (ii) $3 + 4i, 3 - 4i$

   (iii)

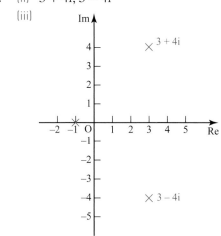

3 $z^4 - 13z^3 + 65z^2 - 151z + 130 = 0$

4 $p = 0$, roots are $z = -2, 1 + i$ and $1 - i$

5 $z = \frac{2}{3}, 2 + 3i, 2 - 3i$

6 $z^4 - 2z^3 + 2z^2 - 10z + 25 = 0$

7 $z = -2, -1 + 3i, -1 - 3i$

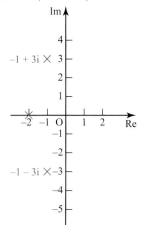

8 (i) $a = -9, b = -5$ (ii) $z = 2 - i, 2 + i, \dfrac{1}{2}$

9 $z = 1 + i, 1 - i, \dfrac{1}{2}i, -\dfrac{1}{2}i$

10 $p = 3, q = -10, z = -1 + 2i, -1 - 2i, 3 + i, 3 - i$

Chapter 9

Discussion point (Page 176)

The first is the aeroplane landing and the second is the aeroplane taking off. When landing, $\dfrac{dh}{dt}$ is decreasing, reaching almost zero at the point of landing. When taking off, $\dfrac{dh}{dt}$ increases rapidly.

Exercise 9.1 (Page 182)

1 $\dfrac{dm}{dt} = -km$

2 $\dfrac{dP}{dt} = \dfrac{P}{34}$

3 $\dfrac{dV}{dt} = -2\sqrt{h}$

5 (iii) (i) is the coffee, (ii) is the juice

6 (ii) 69.3 hours   (iii) 20 mg

(iv) $m = 50e^{\frac{-t}{100}}$

7 $\dfrac{dr}{dt} = -\dfrac{1}{100\pi}$

8 $\dfrac{dp}{dh} = \begin{cases} 9800(1 + 0.001h) & 0 \leqslant h \leqslant 100 \\ 10780 & h > 100 \end{cases}$

9 $\dfrac{dV}{dt} = -\sqrt{20h},\ \dfrac{dh}{dt} = -\dfrac{\sqrt{5h}}{2}$

11 $y = x^3 - \dfrac{1}{2}x^2 + x + c,\ y = x^3 - \dfrac{1}{2}x^2 + x + 2.5$

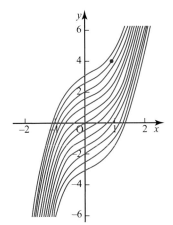

12 (i) and (v)

13 (ii) $\alpha + A, \alpha$ (iii) $\alpha = 25, A = 65$

(iv)

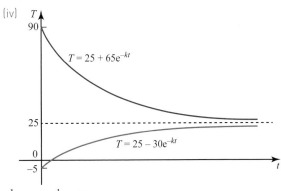

14 $\dfrac{dy}{dx} = \dfrac{b - y}{a + vt - x}$

15 $\dfrac{dh}{dt} = \dfrac{4.8}{\pi h^2}$

Exercise 9.2 (Page 190)

1 (i) $y = Ae^x$ (ii) $y = Ae^{\frac{1}{2}x^2}$

(iii) $y = Ae^{\frac{1}{4}x^4}$ (iv) $y = \ln(x^3 + c)$

(v) $\ln|1 + y| = \dfrac{x^2}{2} + c$ or $y = A^{\frac{1}{2}x^2} - 1$

(vi) Not possible   (vii) $e^x + e^{-y} = c$

(viii) $y = \dfrac{A(x - 1)}{x}$ (ix) Not possible

(x) $y^3 = K - 3\cos x$

(xi) $y = A(x - 2) - 2$

(xii) $\ln\left|\dfrac{x - 8}{x}\right| = 8t + c$ or $x = \dfrac{8}{1 - Ae^{8t}}$

2 (i) $y = \dfrac{10}{1 - 10\ln x}$ (ii) $y = \dfrac{2}{1 - 2\ln x}$

(iii) $y^2 = \dfrac{2x^3 + 298}{3}$ (iv) $p = \ln\left(\dfrac{3 - \cos 2s}{2}\right)$

(v) $y = \ln\left(e^{10} + \dfrac{x^3}{3}\right)$

(vi) $e^{-y}(1 + y) = 3e^{-2} - t$

3 (i) $m = Ae^{-5t}$ (ii) $m = 10e^{-5t}$

4 (i) $P = Ae^{0.7t}$ (ii) $P = 100e^{0.7t}$

(iii) 0.99 minutes

5 (i) $v = \dfrac{20}{1 + 2t}$ (ii) 4.5 seconds

6 (i) $h = \dfrac{4}{\pi t + k}$ (ii) $t = \dfrac{4}{9\pi H}$

7 (i) $h = \left(2 - \dfrac{\sqrt{5}}{4}t\right)^2$

(ii) $t = \dfrac{8}{\sqrt{5}} \approx 3.58 \text{ minutes}$

8 (i) $\left|10 - 0.2v\right| = Ae^{-0.2t}$

   (ii) (a) $v = 50(1 - e^{-0.2t}), 50$

       (b) $v = 50 + 30e^{-0.2t}, 50$

9 (i) $T = 20 + 80e^{-0.5t}$ (ii) $t = 1.96$ minutes

10 $v = \dfrac{40e^{-0.2t}}{41 - 40e^{-0.2t}}$

11 (i) $i = 0.2 - Ae^{-2500t}$ (ii) $i = 0.2(1 - e^{-2500t})$

   (iii) $i = \dfrac{V}{R} - Ae^{\frac{-Rt}{L}}$

12 6 minutes before midnight

13 (i) In running out at 8 litres per minute, salt is
        removed at a rate of $8C$ kg per minute and as
        the salt inflow is zero $\Rightarrow \dfrac{dM}{dt} = -8C$.

        C kg of salt per litre $\Rightarrow 2000C$ kg of salt in
        the tank $\Rightarrow M = 2000C$

   (ii) $\dfrac{dC}{dt} = -0.004C$ (iii) 101 minutes

Activity 9.1 (Page 193)

(i) (a)   (ii) (a), (b), (c)

Exercise 9.3 (Page 198)

1 (i) $e^{\frac{1}{3}x^3}$ (ii) $e^{-\cos x}$ (iii) $x^{-\frac{1}{4}}$

   (iv) x (v) e^{7x} (vi) $\sec x$

2 (i) $\dfrac{dy}{dx} + \dfrac{1}{x}y = \dfrac{1}{x^2}$ (ii) x

   (iii) $x\dfrac{dy}{dx} + y = \dfrac{1}{x}$ (iv) $\dfrac{d}{dx}(xy) = \dfrac{1}{x}$

   (v) $xy = \ln x + c$ (vi) $y = \dfrac{1}{x}\ln x + \dfrac{c}{x}$

   (vi) $c = 0$ (vii) $y = \dfrac{1}{x}\ln x$

3 (i) $y = \dfrac{x^2}{4} - \dfrac{1}{4x^2}$ (ii) $y = 4 - 2e^{-\frac{1}{2}x^2}$

   (iii) $y = 3e^{-3(x^2-1)}$ (iv) $y = \dfrac{1}{2}(3e^{x^2} - 1)$

   (v) $y = x^3 - x^2$ (vi) $y = \dfrac{1}{3}(1 - 4e^{-x^3})$

4 (i) $v = 25 + Ae^{-0.4t}$

   (ii)-(iii) $v = 25(1 - e^{-0.4t})$

   (iv) The method of separation of variables is
         usually preferred as it involves less work.

5 (i) $k = \dfrac{1}{3}$ (ii) $\dfrac{dv}{dt} = 10 - \dfrac{1}{3}v$

   (iii) $v = 30 + Ae^{-\frac{1}{3}t}$ (iv) $v = 30\left(1 + e^{-\frac{1}{3}t}\right)$

6 (i) $y = \sin x + \dfrac{\cos x}{x} + \dfrac{A}{x}$

   (ii) $y = \sin x + \dfrac{1 + \cos x}{x}$ (iii) $y = \sin x$

   (iv) As $x \to 0$, $y \to \sin x$

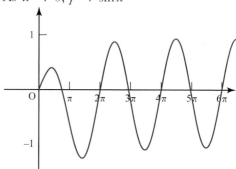

7 (i) $e^{k_2 t}y = Ae^{-k_2 t} + \dfrac{k_1 a}{k_2 - k_1}e^{-k_1 t}$

   (ii) $y = \dfrac{k_1 a}{k_2 - k_1}\left(e^{-k_1 t} - e^{-k_2 t}\right)$

   (iii) $\dfrac{dx}{dt} = -k_1 x$ (iv) $x = ae^{-k_1 t}$

        The amount y of Th-234 is affected by both
        its own decay (the $k_2 y$ part of the differential
        equation) and the rate at which U-238
        decays into it – given by $k_1 x$, i.e. $ak_1 e^{-k_1 t}$.

   (v) $y = kate^{-kt}$

   (vi) $\dfrac{dz}{dt} = -ky$, $z = a(kt + 1)e^{-kt} - a$

8 (i) $y = \dfrac{1 - x}{1 + x}\left(A + x + \dfrac{x^2}{2}\right)$

   (ii) (a) $y = \dfrac{1 - x}{1 + x}\left(x + \dfrac{x^2}{2}\right)$

        As $x \to -1$, $y \to -\infty$

       (b) $y = \dfrac{1 - x}{1 + x}\left(1 + x + \dfrac{x^2}{2}\right)$

        As $x \to -1$, $y \to \infty$

   (iii) $y = \dfrac{1}{2}(1 - x^2)$

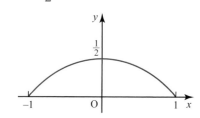

9 (i) $x^2y = A + \dfrac{1}{3}(1 + x^2)^{\frac{3}{2}}$

(ii) $x^2y = \left(1 - \dfrac{2\sqrt{2}}{3}\right) + \dfrac{1}{3}(1 + x^2)^{\frac{3}{2}}$

$y \to \infty$ as $x \to 0$

(iii) $1 + \dfrac{3x^2}{2} + \dfrac{3x^4}{8}$

(iv) $y \approx \dfrac{A}{x^2} + \dfrac{1}{3x^2}\left(1 + \dfrac{3x^2}{2} + \dfrac{3x^4}{8} + \ldots\right)$

Take $A = -\dfrac{1}{3} \Rightarrow y \approx -\dfrac{1}{3x^2} + \dfrac{1}{3x^2}(1 + x^2)^{\frac{3}{2}}$

Chapter 10

Discussion point (Page 203)

Similarity – both use r and θ

Difference – Modulus argument form is used to represent a complex number on the Argand diagram, polar coordinates are used to represent a real number on the coordinate plane

Activity 10.1 (Page 206)

1 (i)

$i = \cos\dfrac{\pi}{2} + i\sin\dfrac{\pi}{2}, \ -2 = 2\big(\cos(-\pi) + i\sin(-\pi)\big)$

(ii) (a) Anticlockwise rotation of the vector z through $\dfrac{\pi}{2}$ radians

(b) Anticlockwise rotation of the vector z through π radians (which is equivalent to two successive rotations through $\dfrac{\pi}{2}$ radians) and enlargement by scale factor 2

Review Exercise (Page 210)

1 (i) $3\sqrt{2} - 3\sqrt{2}i$ (ii) $\dfrac{3\sqrt{3}}{2} - \dfrac{1}{2}i$

(iii) $-\sqrt{2} + \sqrt{2}i$ (iv) $-\dfrac{7\sqrt{3}}{2} - \dfrac{7}{2}i$

2 (i) $4 \quad \dfrac{\pi}{6} \quad 4\left(\cos\dfrac{\pi}{6} + i\sin\dfrac{\pi}{6}\right)$

(ii) $4 \quad \dfrac{5\pi}{6} \quad 4\left(\cos\left(\dfrac{5\pi}{6}\right) + i\sin\left(\dfrac{5\pi}{6}\right)\right)$

(iii) $4 \quad -\dfrac{\pi}{6} \quad 4\left(\cos\left(-\dfrac{\pi}{6}\right) + i\sin\left(-\dfrac{\pi}{6}\right)\right)$

(iv) $4 \quad -\dfrac{5\pi}{6} \quad 4\left(\cos\left(-\dfrac{5\pi}{6}\right) + i\sin\left(-\dfrac{5\pi}{6}\right)\right)$

3 (i) $\dfrac{3\sqrt{2}}{2} + \dfrac{3\sqrt{2}}{2}i$

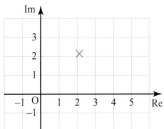

(ii) $5i$

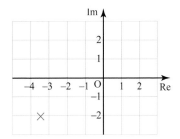

(iii) $-2\sqrt{3} - 2i$

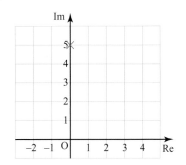

(iv) $3\sqrt{2} - 3\sqrt{2}i$

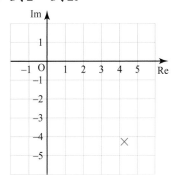

4 (i) $15\left(\cos\dfrac{\pi}{2} + i\sin\dfrac{\pi}{2}\right)$

(ii) $\dfrac{3}{5}\left(\cos\left(-\dfrac{5\pi}{6}\right) + i\sin\left(-\dfrac{5\pi}{6}\right)\right)$

(iii) $\dfrac{5}{3}\left(\cos\dfrac{5\pi}{6} + i\sin\dfrac{5\pi}{6}\right)$

(iv) $\dfrac{1}{5}\left(\cos\dfrac{\pi}{3} + i\sin\dfrac{\pi}{3}\right)$

5 (i) Enlargement scale factor $5\sqrt{2}$ and a clockwise rotation through $\frac{\pi}{4}$ radians

(ii) Enlargement scale factor $\frac{1}{2}\sqrt{2}$ and a clockwise rotation through $\frac{2\pi}{3}$ radians

6 (i) $1 + \sqrt{3}i$

(ii) $-\frac{1}{6} - \frac{1}{6}\sqrt{3}i$

7 (i)

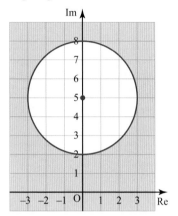

(ii)

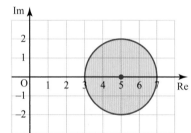

(iii)

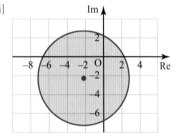

8 (i)

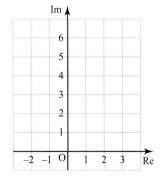

(ii)

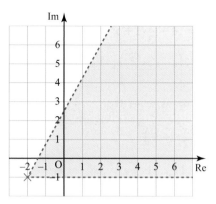

(iii)

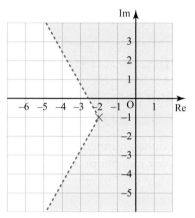

9 (i)

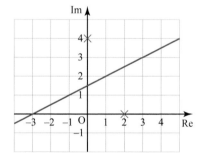

(ii)

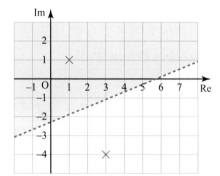

(iii)

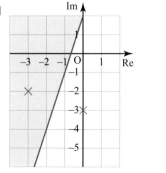

10 (i) $\left|z - (1 + 2i)\right| < 3$

(ii) $\left|z - 3i\right| \geqslant \left|z - (6 + i)\right|$

(iii) $-\dfrac{\pi}{4} < \arg\left(z - (-2 + 2i)\right) < 0$

11

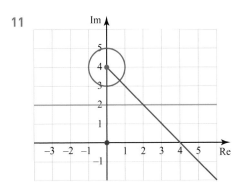

12 $a = \pm 2, b = \pm 3$ or $a = \pm 3, b = \pm 2$

13 (i) (a) Enlarge z scale factor 3, centre O

(b) Enlarge z scale factor 2, centre O, and rotate through $\dfrac{\pi}{2}$ radians anticlockwise

(c) Enlarge z scale factor $\sqrt{13}$, centre O, and rotate through 0.84 radians anticlockwise

(ii) For example, if $z = 1 + 2i$

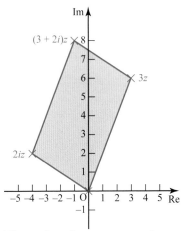

The points form a rectangle.

Activity 10.2 (Page 212)

$z^2 = 4(\cos 0.2 + i\sin 0.2)$

$z^3 = 8(\cos 0.3 + i\sin 0.3)$

$z^4 = 16(\cos 0.4 + i\sin 0.4)$

$z^5 = 32(\cos 0.5 + i\sin 0.5)$

$z^n = 2^n(\cos 0.1n + i\sin 0.1n)$

$z^n = r^n(\cos n\theta + i\sin n\theta)$

Discussion point (Page 213)

The lines will continue all the way round. What happens then depends on whether or not $\dfrac{\theta}{\pi}$ is a rational number. If it is rational, a limited number of lines get repeated over and over again as $n \to \infty$. If it is not rational the number of lines $\to \infty$ but no line ever gets repeated.

Activity 10.3 (Page 215)

Using the two results given.

$\cos(-\phi) + i\sin(-\phi) = \cos\phi - i\sin\phi$

$\Rightarrow \left[\cos(-\phi) + i\sin(-\phi)\right]^n = \left[\cos\phi - i\sin\phi\right]^n$

$\Rightarrow \cos(-n\phi) + i\sin(-n\phi) = \left[\cos\phi - i\sin\phi\right]^n$

Using de Moivre's theorem on the left-hand side.

$\Rightarrow \cos(n\phi) - i\sin(n\phi) = \left[\cos\phi - i\sin\phi\right]^n$

Using the two results given in the question.

Exercise 10.1 (Page 215)

1 (a) (i) $\cos\dfrac{2\pi}{3} + i\sin\dfrac{2\pi}{3}$

(ii) $\cos\left(-\dfrac{2\pi}{3}\right) + i\sin\left(-\dfrac{2\pi}{3}\right)$

(iii) $\cos\left(-\dfrac{5\pi}{6}\right) + i\sin\left(-\dfrac{5\pi}{6}\right)$

(b) (i) $-\dfrac{1}{2} + \dfrac{\sqrt{3}}{2}i$ (ii) $-\dfrac{1}{2} - \dfrac{\sqrt{3}}{2}i$

(iii) $-\dfrac{\sqrt{3}}{2} - \dfrac{1}{2}i$

2 (i) $z_1 = w^3$ (ii) $z_2 = w^2$ (iii) $z_3 = w^4$

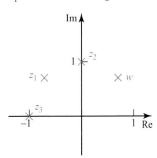

3 (i) $\cos(-3\theta) + i\sin(-3\theta)$

(ii) $\cos\left(\dfrac{5\pi}{12}\right) + i\sin\left(\dfrac{5\pi}{12}\right)$ (iii) -1

4 (i) $z_1 = w^{-1}$

   (ii) $z_2 = w^{-3}$

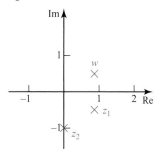

5 (i) $\left[2\left(\cos\left(-\dfrac{\pi}{3}\right) + i\sin\left(-\dfrac{\pi}{3}\right)\right)\right]^4$

$$= 16\left(-\dfrac{1}{2} + \dfrac{\sqrt{3}}{2}i\right) = -8 + 8\sqrt{3}i$$

   (ii) $\left[2\sqrt{2}\left(\cos\left(\dfrac{3\pi}{4}\right) + i\sin\left(\dfrac{3\pi}{4}\right)\right)\right]^7$

$$= 1024\sqrt{2}\left(-\dfrac{\sqrt{2}}{2} - \dfrac{\sqrt{2}}{2}i\right) = -1024 - 1024i$$

   (iii) $\left[6\left(\cos\dfrac{\pi}{6} + i\sin\dfrac{\pi}{6}\right)\right]^6$

$$= 46656(-1 + 0i) = -46656$$

6 $64\left(\sqrt{3} + i\right)$

7 (i) $81\left(\cos 8\theta + i\sin 8\theta\right)$

   (ii) $i\cos 15\theta - \sin 15\theta$

   (iii) $\dfrac{1}{8}\left(i\cos(-21\theta) - \sin(-21\theta)\right)$

8 $k = 41472$

9 (i) $z_1 = \cos 0 + i\sin 0,\ z_2 = \cos\dfrac{2\pi}{3} + i\sin\dfrac{2\pi}{3},$

$$z_3 = \cos\left(-\dfrac{2\pi}{3}\right) + i\sin\left(-\dfrac{2\pi}{3}\right)$$

   (iii)

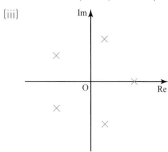

$$\cos 0 + i\sin 0,\ \cos\dfrac{2\pi}{5} + i\sin\dfrac{2\pi}{5},\ \cos\dfrac{4\pi}{5}$$

$$+ i\sin\dfrac{4\pi}{5},\ \cos\left(-\dfrac{2\pi}{5}\right) + i\sin\left(-\dfrac{2\pi}{5}\right),$$

$$\cos\left(-\dfrac{4\pi}{5}\right) + i\sin\left(-\dfrac{4\pi}{5}\right),$$

Activity 10.4 (Page 219)

$n = 2$ $1 + (-1) = 0$

$n = 3$ $1 + \dfrac{-1 + \sqrt{3}i}{2} + \dfrac{-1 - \sqrt{3}i}{2} = 0$

$n = 4$ $1 + (-1) + i + (-i) = 0$

Exercise 10.2 (Page 223)

1 $\omega^0 = 1,\ \ \omega^1 = 0.309 + 0.951i,$

   $\omega^2 = -0.809 + 0.588i,\ \ \omega^3 = -0.809 - 0.588i,$

   $\omega^4 = 0.309 - 0.951i$

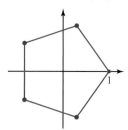

The points form a regular pentagon with one vertex at the point 1.

2 $\omega^0 = 1,\ \ \omega = \dfrac{\sqrt{2}}{2} + \dfrac{\sqrt{2}}{2}i,\ \ \omega^2 = i,$

   $\omega^3 = -\dfrac{\sqrt{2}}{2} + \dfrac{\sqrt{2}}{2}i,\ \ \omega^4 = -1,$

   $\omega^5 = -\dfrac{\sqrt{2}}{2} - \dfrac{\sqrt{2}}{2}i\ \ \omega^6 = -i,\ \omega^7 = \dfrac{\sqrt{2}}{2} - \dfrac{\sqrt{2}}{2}i$

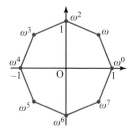

The points form a regular octagon with one vertex at the point 1.

3 $0.90 + 2.79i,\ -0.90 - 2.79i$

4 $1 + i,\ 1 - i,\ -1 + i,\ -1 - i$

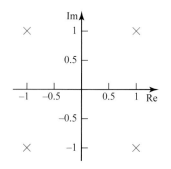

5 (i) $\left(2^{\frac{1}{6}}, -\dfrac{\pi}{12}\right)\left(2^{\frac{1}{6}}, \dfrac{7\pi}{12}\right)\left(2^{\frac{1}{6}}, -\dfrac{3\pi}{4}\right)$

(ii) $(1.38, 0.25)$ $(1.38, 1.82)$

$(1.38, -2.90)$ $(1.38, -1.36)$

(iii) $(1.38, 0.44)$ $(1.38, 1.70)$ $(1.38, 2.96)$

$(1.38, -2.07)$ $(1.38, -0.81)$

6 The fifth roots give alternate tenth roots and their negatives (given by a half turn about the origin) fill the gaps.

7 $w = -119 - 120\mathrm{i}$

$-3 + 2\mathrm{i}, -2 - 3\mathrm{i}, 3 - 2\mathrm{i}$

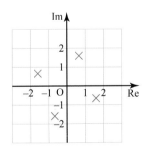

8 $2\left(\cos\dfrac{\pi}{18} + \mathrm{i}\sin\dfrac{\pi}{18}\right), 2\left(\cos\dfrac{13\pi}{18} + \mathrm{i}\sin\dfrac{13\pi}{18}\right),$

$2\left(\cos\left(-\dfrac{11\pi}{18}\right) + \mathrm{i}\sin\left(-\dfrac{11\pi}{18}\right)\right)$

9 $(z + 1 - 3\mathrm{i})^7 = 2187$

10 (i) $\sqrt{3}\left(\cos\dfrac{3\pi}{8} + \mathrm{i}\sin\dfrac{3\pi}{8}\right), \sqrt{3}\left(\cos\dfrac{7\pi}{8} + \mathrm{i}\sin\dfrac{7\pi}{8}\right),$

$\sqrt{3}\left(\cos\dfrac{11\pi}{8} + \mathrm{i}\sin\dfrac{11\pi}{8}\right), \sqrt{3}\left(\cos\dfrac{15\pi}{8} + \mathrm{i}\sin\dfrac{15\pi}{8}\right)$

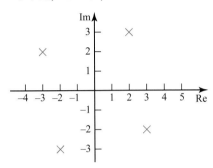

(ii) Midpoint of PS has argument $\dfrac{\pi}{8}$ and

modulus $\sqrt{\dfrac{3}{2}}$.

So $\arg(w) = 4 \times \dfrac{\pi}{8} = \dfrac{\pi}{2}$ and

$|w| = \left(\sqrt{\dfrac{3}{2}}\right)^4 = \dfrac{9}{4}$

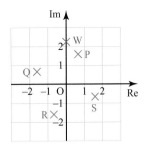

Activity 10.5 (Page 224)

$\sin 5\theta = 5c^4s - 10c^2s^3 + s^5$

$= 5s\left(1 - s^2\right)^2 - 10s^3\left(1 - s^2\right) + s^5$

$= 5s - 10s^3 + 5s^5 - 10s^3 + 10s^5 + s^5$

$\sin 5\theta = 16\sin^5\theta - 20\sin^3\theta + 5\sin\theta$

Activity 10.6 (Page 225)

$2\mathrm{i}\sin\theta = z - z^{-1}$

$\Rightarrow \quad (2\mathrm{i}\sin\theta)^5 = \left(z - z^{-1}\right)^5$

$\Rightarrow \quad 32\mathrm{i}\sin^5\theta = z^5 - 5z^3 + 10z - 10z^{-1}$
$+ 5z^{-3} - z^{-5}$

$= \left(z^5 - z^{-5}\right) - 5\left(z^3 - z^{-3}\right)$
$+ 10\left(z - z^{-1}\right)$

$= 2\sin 5\theta - 10\sin 3\theta + 20\sin\theta$

$\Rightarrow \qquad \sin^5\theta = \dfrac{\sin 5\theta - 5\sin 3\theta + 10\sin\theta}{16}$

Exercise 10.3 (Page 225)

1 (ii) $\tan 3\theta = \dfrac{3\tan\theta - \tan^3\theta}{1 - 3\tan^2\theta}$

2 (i) $z^3 = \cos 3\theta + \mathrm{i}\sin 3\theta,$

$z^{-3} = \cos 3\theta - \mathrm{i}\sin 3\theta$

3 (i) $z^{-1} = \cos\theta - \mathrm{i}\sin\theta$

(ii) (b) $\dfrac{\cos 4\theta + 4\cos 2\theta + 3}{8}$

(iii) (b) $\dfrac{\sin 5\theta - 5\sin 3\theta + 10\sin\theta}{16}$

4 $\cos 6\theta = 32\cos^6\theta - 48\cos^4\theta + 18\cos^2\theta - 1$

$\dfrac{\sin 6\theta}{\sin\theta} = 32\cos^5\theta - 32\cos^3\theta + 6\cos\theta$

5 $\sin^6\theta = \dfrac{-\cos 6\theta + 6\cos 4\theta - 15\cos 2\theta + 10}{32}$

$\displaystyle\int \sin^6\theta \, d\theta = -\dfrac{1}{192}\sin 6\theta + \dfrac{3}{64}\sin 4\theta$

$- \dfrac{15}{64}\sin 2\theta + \dfrac{5}{16}\theta + c$

6 $\cos^4\theta\sin^3\theta = \dfrac{-\sin 7\theta - \sin 5\theta + 3\sin 3\theta + 3\sin\theta}{64}$

$\displaystyle\int_0^\pi \cos^4\theta\sin^3\theta\,\mathrm{d}\theta = \dfrac{4}{35}$

7 $\dfrac{203}{480}$

8 (ii) $\cos^2\theta = \dfrac{5 \pm \sqrt{5}}{8}$

(iii) $\theta = 18° \Rightarrow \cos 5\theta = 0$

$\cos\theta = \pm\left(\dfrac{5 + \sqrt{5}}{8}\right)^{\frac{1}{2}}$ or $\pm\left(\dfrac{5 - \sqrt{5}}{8}\right)^{\frac{1}{2}}$

$\cos 18°$ is close to 1 so $\cos 18° = \left(\dfrac{5 + \sqrt{5}}{8}\right)^{\frac{1}{2}}$

and using $\cos^2\theta + \sin^2\theta \equiv 1$ gives

$\sin 18° = \left(\dfrac{3 - \sqrt{5}}{8}\right)^{\frac{1}{2}}$

9 $a = \dfrac{1}{2}z^{-(2n-1)}$, $r = z^2$, $2n$ terms

10 (i) $z^n = \cos n\theta + \mathrm{i}\sin n\theta$

$\dfrac{1}{z^n} = \cos n\theta - \mathrm{i}\sin n\theta$

$z^n + \dfrac{1}{z^n} = 2\cos n\theta$

$z^n - \dfrac{1}{z^n} = 2\mathrm{i}\sin n\theta$

(ii) $p = \dfrac{1}{16}$, $q = -\dfrac{1}{32}$, $r = -\dfrac{1}{16}$, $s = \dfrac{1}{32}$

Discussion point (Page 227)

$\cos(-\theta) = \cos\theta$ and $\sin(-\theta) = -\sin\theta$

so

$r(\cos\theta - \mathrm{i}\sin\theta) = r(\cos(-\theta) + \mathrm{i}\sin(-\theta))$

Therefore

$r(\cos\theta - \mathrm{i}\sin\theta) = re^{-\mathrm{i}\theta}$

Activity 10.7 (Page 228)

(i) $e^{x+y\mathrm{i}} = e^x e^{y\mathrm{i}} = e^x(\cos y + \mathrm{i}\sin y)$

(ii) $e^{z+2\pi n\mathrm{i}} = e^z e^{2\pi n\mathrm{i}} = e^z(\cos 2\pi n + \mathrm{i}\sin 2\pi n)$

$= e^z \times 1 = e^z$

(iii) Using (i) with $x = 0$, $y = \pi$ gives

$e^0(\cos\pi + \mathrm{i}\sin\pi) = -1$

Discussion point (Page 228)

The earlier results $\cos\theta = \dfrac{z + z^{-1}}{2}$ and

$\sin\theta = \dfrac{z - z^{-1}}{2\mathrm{i}}$, where $z = \cos\theta + \mathrm{i}\sin\theta$, can

be rewritten with $z = e^{\mathrm{i}\theta}$, so $z^{-1} = e^{-\mathrm{i}\theta}$, giving

$\cos\theta = \dfrac{e^{\mathrm{i}\theta} + e^{-\mathrm{i}\theta}}{2}$ and $\sin\theta = \dfrac{e^{\mathrm{i}\theta} - e^{-\mathrm{i}\theta}}{2\mathrm{i}}$.

Activity 10.8 (Page 230)

$S = 2^n\cos^n\dfrac{\theta}{2}\sin\dfrac{n\theta}{2}$

Exercise 10.4 (Page 230)

1 (i) $4e^{\frac{\pi}{3}\mathrm{i}}$ (ii) $\sqrt{3}e^{-\frac{5\pi}{6}\mathrm{i}}$ (iii) $5e^{-\frac{\pi}{2}\mathrm{i}}$

(iv) $3\sqrt{2}e^{-\frac{3\pi}{4}\mathrm{i}}$ (v) $2e^{-\frac{\pi}{6}\mathrm{i}}$

2 (i) -5 (ii) $-1 + \mathrm{i}$

(iii) $-1 - \mathrm{i}$ (iv) $\dfrac{5\sqrt{2}}{2} - \dfrac{5\sqrt{2}}{2}\mathrm{i}$

3 $zw = 6e^{-\frac{1}{12}\pi\mathrm{i}}$

$\dfrac{z}{w} = \dfrac{2}{3}e^{\frac{5\pi}{12}\mathrm{i}}$

4 (i) $w = 32e^{\frac{\pi}{2}\mathrm{i}}$

(ii) $2e^{\frac{\pi}{10}\mathrm{i}}$, $2e^{\frac{\pi}{2}\mathrm{i}}$, $2e^{\frac{9\pi}{10}\mathrm{i}}$, $2e^{-\frac{3\pi}{10}\mathrm{i}}$, $2e^{-\frac{\pi}{10}\mathrm{i}}$

5 (iii) $\dfrac{\sin\theta + \sin(n-1)\theta - \sin n\theta}{2 - 2\cos\theta}$

6 (ii) $S = 2^n\cos^n\theta\sin n\theta$

7 (i) $a = 16$, $b = -20$, $c = 5$

8 $z^* = a - b\mathrm{i}$ so

$e^{z^*} = e^{a-b\mathrm{i}} = e^a\big(\cos(-b) + \mathrm{i}\sin(-b)\big)$

$= e^a(\cos b - \mathrm{i}\sin b)$

$e^z = e^{a+b\mathrm{i}} = e^a(\cos b + \mathrm{i}\sin b)$ so

$\big(e^z\big)^* = e^a(\cos b - \mathrm{i}\sin b)$

9 (i) $C = \dfrac{e^{3x}(3\cos 2x + 2\sin 2x)}{13} + c_1$

$S = \dfrac{e^{3x}(-2\cos 2x + 3\sin 2x)}{13} + c_2$

10 $C = \dfrac{2\cos\theta + 1}{5 + 4\cos\theta}$ $S = \dfrac{2\sin\theta}{5 + 4\cos\theta}$

Chapter 11

Discussion point (Page 237)

Use the scalar product to check that the vector $\mathbf{a} \times \mathbf{b}$ is perpendicular to both $\mathbf{a}$ and $\mathbf{b}$:

$(\mathbf{a} \times \mathbf{b}).\mathbf{a} = \begin{pmatrix} 24 \\ -1 \\ -14 \end{pmatrix} . \begin{pmatrix} 3 \\ 2 \\ 5 \end{pmatrix}$

$= (24 \times 3) + (-1 \times 2) + (-14 \times 5) = 0$

$$(\mathbf{a} \times \mathbf{b}).\mathbf{b} = \begin{pmatrix} 24 \\ -1 \\ -14 \end{pmatrix}.\begin{pmatrix} 1 \\ -4 \\ 2 \end{pmatrix}$$

$$= (24 \times 1) + (-1 \times -4) + (-14 \times 2) = 0$$

Activity 11.1 (Page 238)

$\mathbf{j} \times \mathbf{i} = -\mathbf{k}$ $\mathbf{j} \times \mathbf{j} = 0$ $\mathbf{j} \times \mathbf{k} = \mathbf{i}$

$\mathbf{k} \times \mathbf{i} = \mathbf{j}$ $\mathbf{k} \times \mathbf{k} = 0$ $\mathbf{k} \times \mathbf{j} = -\mathbf{i}$

Exercise 11.1 (Page 239)

1   (i) $\begin{pmatrix} -23 \\ 13 \\ 2 \end{pmatrix}$ (ii) $\begin{pmatrix} 37 \\ 41 \\ 19 \end{pmatrix}$

    (iii) $\begin{pmatrix} -8 \\ 34 \\ 27 \end{pmatrix}$ (iv) $\begin{pmatrix} 21 \\ -29 \\ 9 \end{pmatrix}$

2   (i) $\begin{pmatrix} 5 \\ 19 \\ -2 \end{pmatrix}$ (ii) $\begin{pmatrix} 14 \\ -62 \\ -9 \end{pmatrix}$

    (iii) $\begin{pmatrix} -3 \\ -2 \\ 3 \end{pmatrix}$ (iv) $\begin{pmatrix} -18 \\ 57 \\ 47 \end{pmatrix}$

3   (i) $\overrightarrow{AB} = \begin{pmatrix} 1 \\ -4 \\ 3 \end{pmatrix}, \overrightarrow{AC} = \begin{pmatrix} 4 \\ -1 \\ 0 \end{pmatrix}$

    (ii) $\begin{pmatrix} 3 \\ 12 \\ 15 \end{pmatrix}$ or $\begin{pmatrix} 1 \\ 4 \\ 5 \end{pmatrix}$

    (iii) $x + 4y + 5z = 7$

4   $\dfrac{1}{\sqrt{635}} \begin{pmatrix} 19 \\ 15 \\ -7 \end{pmatrix}$

5   $\sqrt{74}$

6   (i) $5x + 4y - 8z = 5$
    (ii) $24x + y - 29z = 1$
    (iii) $19x + 40y + 3z = 188$
    (iv) $30x - 29y - 24z = 86$

7   (i) $-8\mathbf{j}$ (ii) $6\mathbf{j} - 4\mathbf{k}$
    (iii) $2\mathbf{i} - 12\mathbf{j}$ (iv) $6\mathbf{i} + 14\mathbf{j} - 2\mathbf{k}$

9   (i) $\frac{1}{2}\sqrt{717}$ (ii) $\sqrt{717}$

Activity 11.2 (Page 243)

(i) $A = \left(0, -\dfrac{c}{b}\right),$ $A' = \left(0, -\dfrac{c}{b}, 0\right)$

(ii) $y = -\dfrac{a}{b}x - \dfrac{c}{b},$ $d = b,$ $e = -a,$ $f = 0$

Activity 11.3 (Page 246)

Because it is perpendicular to both l_1 and l_2.

Exercise 11.2 (Page 248)

1   (i) $\sqrt{29}$ (ii)   7   (iii) $2\sqrt{26}$

2   (i)   13   (ii) $3\sqrt{5}$ (iii)   1

3   (i)   5   (ii)   5   (iii) $5\sqrt{2}$

4   (i) $\mathbf{r} = \begin{pmatrix} 3 \\ 1 \\ 0 \end{pmatrix} + \lambda \begin{pmatrix} 1 \\ -2 \\ -1 \end{pmatrix}$ (ii) $\sqrt{\dfrac{7}{3}}$

5   (ii) $\dfrac{1}{2}\sqrt{38}$

6   (i)   4, the lines are skew
    (ii)   0.4, the lines are skew
    (iii)   0, the lines intersect
    (iv) $\sqrt{\dfrac{77}{6}}$ the lines are parallel

7   (i) $2\sqrt{69}$ (ii)   M$(-3, 2, 6)$

8   (i) $\dfrac{2\sqrt{5}}{15}$

    (ii) $\mathbf{r} = \begin{pmatrix} 2 \\ 0 \\ -5 \end{pmatrix} + \lambda \begin{pmatrix} 4 \\ -5 \\ 2 \end{pmatrix}$

    (iii)   M$\left(\dfrac{82}{45}, \dfrac{10}{45}, -\dfrac{229}{45}\right)$

9   (i)   AB: $\mathbf{r} = \begin{pmatrix} 2 \\ -3 \\ 4 \end{pmatrix} + \lambda \begin{pmatrix} -1 \\ 0 \\ 1 \end{pmatrix}$ and

       CD: $\mathbf{r} = \begin{pmatrix} 0 \\ 3 \\ -2 \end{pmatrix} + \lambda \begin{pmatrix} 2 \\ 0 \\ 7 \end{pmatrix}$ or

       equivalent; 6
    (ii)   3.08 m – No, the cable is not long enough

10   (ii) $k = \dfrac{-21}{2}$

    (iv) $\mathbf{r} = \begin{pmatrix} 4 \\ 12 \\ 5 \end{pmatrix} + \alpha \begin{pmatrix} 2 \\ 10 \\ 11 \end{pmatrix}$ or equivalent

11   (i) $\begin{pmatrix} 4k - 4 \\ 2 - 2k \\ 4k - 4 \end{pmatrix}$

(ii) (a) $k = 1$ (b) $\dfrac{45}{\sqrt{26}}$

(c) $8x - y + 4z + 5 = 0$

(iii) $\dfrac{1}{3}|12 - 2k|$

(iv) $(-6, 13, 14), k = 6$

12 (i) (b) $\begin{pmatrix} 2 \\ 2 \\ -1 \end{pmatrix}$

(c) $\mathbf{r} = \begin{pmatrix} 0 \\ -2 \\ 0 \end{pmatrix} + \lambda \begin{pmatrix} 2 \\ 2 \\ -1 \end{pmatrix}$

(ii) $\dfrac{46}{5}$ (iii) 15

(iv) $k = 50; (16, 8, 4)$

Chapter 12

Activity 12.1 (Page 256)

(i) (a) and (d) are linear, (a), (b), (d) and (e) have constant coefficients

(ii) (a) Can be solved by any of the three methods, $y = Ae^{17t}$

(b) Separation of variables, $y = \dfrac{1}{c - x}$

(c) Separation of variables or integrating factor, $y = Ae^{-\frac{1}{2}x^2}$

(d) Can be solved by any of the three methods, $y = Ae^{3t}$

(e) After division by y, any of the three methods can be used, $y = Ae^t$

Exercise 12.1 (Page 263)

1 (i) $y = Ae^{3x}$ (ii) $y = Ae^{-7x}$
(iii) $x = Ae^{-t}$ (iv) $p = Ae^{0.02t}$
(v) $z = Ae^{0.2t}$

2 (i) $y = 3e^{-2x}$ (ii) $y = e^{\frac{5}{2}x}$
(iii) $x = 2e^{\frac{1}{3}(1-t)}$ (iv) $P = P_0e^{kt}$
(v) $m = m_0e^{-kt}$

3 (i) $y = Ae^{4x} + Be^{-4x}$
(ii) $y = Ae^{1.79x} + Be^{-2.79x}$
(iii) $x = Ae^{-1.71t} + Be^{-0.29t}$
(iv) $y = A + Be^{-\frac{2}{7}x}$

(v) $y = e^{\frac{2}{3}x}(Ax + B)$

4 (i) $\lambda^2 - 3\lambda + 2 = 0$
(ii) $\lambda = 1, 2$ (iii) $y = Ae^x + Be^{2x}$
(iv) $\dfrac{dy}{dx} = Ae^x + 2Be^{2x}$
(v) $y = 2e^x$

5 (i) $\lambda^2 - 8\lambda + 16 = 0, \lambda = 4$
(ii) $y = A + Be^{4x}$ (iii) $y = 1 - \dfrac{1}{4}e^{4x}$

6 (i) $x = \dfrac{4}{5}\left(1 - e^{-5t}\right)$
Initially moves quickly towards its limiting position $x = 0.8$
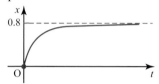

(ii) $x = \left(1 + t\left(\sqrt{e} - 1\right)\right)e^{-\frac{1}{2}t}$ Initially increases briefly then decays to zero
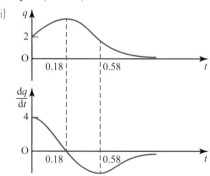

7 (i) $q = (2 + 9t)e^{-2.5t}$
(ii)
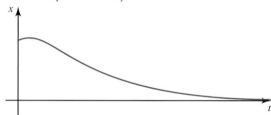

(iii) $q \to 0$ and $\dfrac{dq}{dt} \to 0$ as $t \to \infty$

8 (i) $T = 37.3 + 12.7e^{-0.5t}$
(ii) $T = 42.0\,°C$
(iii)
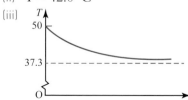

(iv) $37.3\,°C$

9 (i) $T = Ae^{2x} + Be^{-2x}$

    (ii) $T = -52.3e^{2x} + 152.3e^{-2x}$

    (iii) 87.9 °C

10 (ii) $x = 2 - 2e^{-50t}$

    (iii) 49.9 seconds, 50 km h^{-1}

    (iv) The car takes an infinite amount of time to stop.

Exercise 12.2 (Page 273)

1 (i) $x = A\sin 3t + B\cos 3t$

    (ii) $x = \dfrac{1}{3}\sin 3t$

    (iii) Period $= \dfrac{2\pi}{3}$, amplitude $= \dfrac{1}{3}$

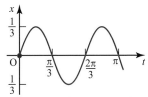

2 (i) $x = A\sin\dfrac{1}{2}t + B\cos\dfrac{1}{2}t$

    (ii) $x = 4\cos\dfrac{1}{2}t$

    (iii) Period $= 4\pi$, amplitude $= 4$

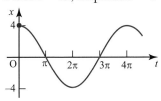

3 (i) $y = e^{2x}(A\sin x + B\cos x)$

    (ii) $y = e^{x}(A\sin 2x + B\cos 2x)$

    (iii) $x = e^{-t}\left(A\sin\sqrt{3}t + B\cos\sqrt{3}t\right)$

    (iv) $x = e^{-0.5t}(A\sin t + B\cos t)$

4 (i) $x = A\sin\dfrac{2}{3}t + B\cos\dfrac{2}{3}t$

    (ii) $x = 5\sin\left(\dfrac{2}{3}t + 0.927\right)$

    (iii)

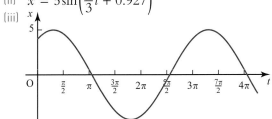

5 (i) $x = \dfrac{\sqrt{2}}{4}\sin 2\sqrt{2}t + \cos 2\sqrt{2}t$

    $= \dfrac{3\sqrt{2}}{4}\sin\left(2\sqrt{2}t + 1.23\right)$

Oscillations with constant amplitude

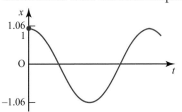

(ii) $x = e^{\frac{t}{2}}\left(\cos\dfrac{\sqrt{3}}{2}t - \dfrac{1}{\sqrt{3}}\sin\dfrac{\sqrt{3}}{2}t\right)$

    $= \dfrac{2}{\sqrt{3}}e^{\frac{t}{2}}\cos\left(\dfrac{\sqrt{3}}{2}t + 0.52\right)$

Exponentially increasing oscillations

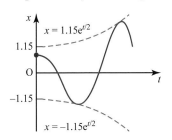

(iii) $x = 2e^{-t}\sin t$

Decaying oscillations

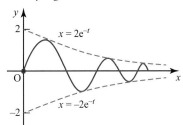

(iv) $x = e^{t}\left(2\cos\dfrac{1}{2}t - 4\sin\dfrac{1}{2}t\right)$

    $= \sqrt{20}e^{t}\cos\left(\dfrac{1}{2}t + 1.11\right)$

Increasing oscillations

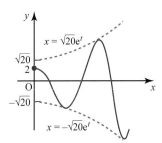

(v) $x = 3e^{-t+\frac{\pi}{4}} \sin 2t = 6.58 e^{-t} \sin 2t$

Decaying oscillations

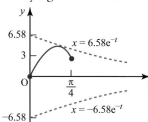

6 (i) $x = A \sin 8t + B \cos 8t$

(ii) $x = 0.1 \cos 8t$

(iii) Period $= \dfrac{\pi}{4}$, amplitude $= 0.1$

(iv)

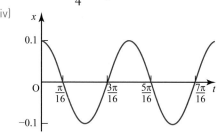

7 (i) $\theta = e^{-2t}(A \sin t + B \cos t)$ or
$\theta = e^{-2t} A \cos(t + \varepsilon)$

(ii) $\theta = \dfrac{\pi}{4} e^{-2t}(2 \sin t + \cos t)$ or

$\theta = 1.76 e^{-2t} \cos(t - 1.11)$

(iii) The motion effectively ends at about
$t = 2.5$. Although θ subsequently does
become negative, the amplitude of the
oscillation is very small and decreases rapidly.

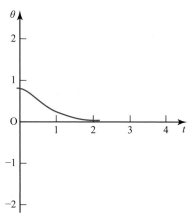

(iv) $\theta \to 0$ as $t \to \infty$

8 (i) $k = \sqrt{20}$

(ii) (a) Underdamped   (b) Overdamped

10 $y = Ae^{2t} + Be^{-2t} + C \sin 2t + D \cos 2t$

Exercise 12.3 (Page 280)

1 (i) $y = -2x - 5$

(ii) $x = \dfrac{1}{4}t + \dfrac{1}{2}$

(iii) $y = -0.08 \cos 3x + 0.06 \sin 3x$

(iv) $x = \dfrac{3}{4}e^{-2t}$

2 (i) $y = A \cos 2x + B \sin 2x + \dfrac{1}{8}e^{2x}$

(ii) $x = Ae^{3t} + Be^{-t} + e^{-2t}$

(iii) $y = e^{-x}(A \cos 2x + B \sin 2x)$
$+ 0.1 \text{six}(x) + 0.2 \cos(x)$

(iv) $y = e^{-x}(A \cos 2x + B \sin 2x) + 0.6x + 0.16$

3 (i) $x = A \cos t + B \sin t - \dfrac{1}{2}t \cos t$

(ii) $y = Ae^{-4x} + Be^{x} + \dfrac{1}{5}xe^{x}$

(iii) $x = Ae^{t} + Be^{3t} + \dfrac{1}{2}te^{3t}$

(iv) $x = (At + B)e^{3t} + 2t^2 e^{3t}$

4 (i) $y = -6e^{3x} + e^{2x} + 6x + 5$

(ii) $x = -2 \cos 3t + \sin 3t + 2e^{-t}$

(iii) $y = e^{-x}(-\cos 2x + 1)$

(iv) $y = -e^{-2t} + 4e^{t} - \cos 2t - 3 \sin 2t$

(v) $x = \dfrac{1}{2}e^{2t} - e^{t} + \dfrac{1}{2} + te^{t}$

(vi) $y = 2 \sin 2x - 3x \cos 2x$

(vii) $y = e^{-2x}(2 \cos x + 3 \sin x) - \cos x + \sin x$

(viii) $y = \dfrac{1}{2}e^{-2x} + \dfrac{1}{2}e^{2x} + x - 1$

5 (i) $P = Ae^{-t}$

(ii) $P = 100 - 25(\cos t - \sin t)$

(iii) $P = 100 - 25(\cos t - \sin t) - 55e^{-t}$

(iv) 100

(v) $25\sqrt{2}$

6 (i) $x = e^{-1.5t}\left(A \cos \dfrac{\sqrt{31}}{2}t + B \sin \dfrac{\sqrt{31}}{2}t\right) + 0.05$

(ii)
$x = e^{-1.5t}\left(0.05 \cos \dfrac{\sqrt{31}}{2}t + \dfrac{0.15}{\sqrt{31}} \sin \dfrac{\sqrt{31}}{2}t\right) + 0.05$

$x \to 0.05$ as $t \to \infty$

(iii) 1.16

7 (i) $e^{-50t}\left(A \sin 50\sqrt{3}t + B \cos 50\sqrt{3}t\right)$

(ii) Underdamped

(iii) $-0.1 \cos 100t$

$q = e^{-50t}\left(A \sin 50\sqrt{3}t + B \cos 50\sqrt{3}t\right)$
$-0.1 \cos 100t$

(iv) $q = e^{-50t}\left(\dfrac{1}{10\sqrt{3}} \sin 50\sqrt{3}t + \dfrac{1}{10} \cos 50\sqrt{3}t\right)$
$- 0.1 \cos 100t$

(v)
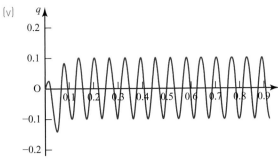

The oscillations quickly settle to almost constant amplitude.

8 (i) (a) $x = 4 - \dfrac{20}{3}e^{-t} + \dfrac{8}{3}e^{-2.5t}$

(b) $x = 4.9 - (7.01t + 4.9)e^{-1.43t}$

(c) $x = 8 - e^{-0.875t}(10.0\sin 0.70t + 8\cos 0.70t)$

(ii) (a) (b)

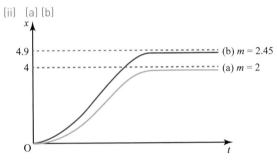

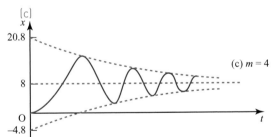

9 (i) (a) (b) $y = Ae^{kt} + \dfrac{1}{p - k}e^{pt}$

(ii) (a) (b) $y = Ae^{kt} + te^{kt}$

10 (i) $x = 2\cos 3t$

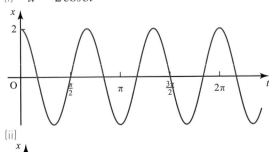

(ii)

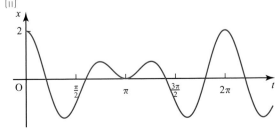

(iii) $x = 2\cos 3t + \dfrac{5}{6}t\sin 3t$

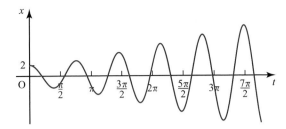

The machine will shake itself to bits (this situation is described as **resonance**).

Exercise 12.4 (Page 287)

1 (i) $x = Ae^{2t} + Be^{t},\ y = Ae^{2t} + 2Be^{t}$

(ii) $x = e^{t},\ y = 2e^{t}$

(iii) $(x, y) \to \infty$ as $t \to \infty$ along the line $y = 2x$

2 (i) $x = Ae^{5t} + Be^{-t},\ y = Ae^{5t} - Be^{-t}$

(ii) $x = \dfrac{1}{2}\left(3e^{5t} - e^{-t}\right),\ y = \dfrac{1}{2}\left(3e^{5t} + e^{-t}\right)$

(iii) $(x, y) \to \infty$ as $t \to \infty$ along the line $y = x$

3 (i) $x = e^{-t}(A\sin t + B\cos t),$

$y = \dfrac{e^{-t}}{5}((A - 2B)\cos t - (2A + B)\sin t)$

(ii) $x = e^{-t}(12\sin t + \cos t),$

$y = e^{-t}(2\cos t - 5\sin t)$

(iii) $(x, y) \to (0, 0)$ as $t \to \infty$

4 (i) $x = e^{2t}(A + Bt),\ y = -Be^{2t} + 3$

(ii) $x = 2 + (t - 1)e^{2t},\ y = 3 - e^{2t}$

(iii) $x \to \infty,\ y \to -\infty$ as $t \to \infty$

5 (i) $x = \dfrac{\sqrt{2}}{4}\left(e^{\sqrt{2}t} - e^{-\sqrt{2}t}\right),$

$y = \dfrac{1}{4}\left(2 - \sqrt{2}\right)e^{\sqrt{2}t} + \dfrac{1}{4}\left(2 + \sqrt{2}\right)e^{-\sqrt{2}t}$

(ii) $(x, y) \to \infty$ as $t \to \infty$ along the line $y = \left(\sqrt{2} - 1\right)x$

6 (i) $x = 1 - e^{t}\cos\sqrt{6}t,\ y = \dfrac{\sqrt{6}e^{t}}{2}\sin\sqrt{6}t + 1$

(ii) Spirals away from $(0, 1)$ to ∞

7 (i) $0, -20$

(ii) $\dfrac{d^2x}{dt^2} - 3\dfrac{dx}{dt} - 4x = 0$

(iii) $x = 3e^{4t} + 12e^{-t}$

(iv) $y = 12e^{-t} - 2e^{4t}$

(v) The second species becomes extinct after nearly 35.8 years.

8 (i) $p = \dfrac{28}{5}, q = 7$

(iii) $p = Ae^{-10t} + Be^{-3t} + \dfrac{28}{5}$

$p = \dfrac{28}{5} - \dfrac{3}{5}e^{-10t} - 5e^{-3t}$

$q = 7 + 3e^{-10t} - 10e^{-3t}$

(iv)

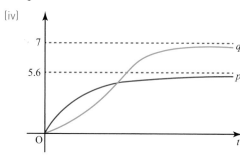

$p \to \dfrac{28}{5} = 5.6, q \to 7$

9 (ii) $x = e^{0.01t}(49\,900t + 10\,000)$

(iii) $y = e^{0.01t}(49\,900t + 5\,000\,000)$

(iv) $x = 273\,000, y = 5.52$ million

10 (i) $x = 8e^{-4t}, y = 16e^{-2t} - 16e^{-4t}$,
$z = 8e^{-4t} - 16e^{-2t} + 8$

(ii) 4

(iii) $z \to 8$ as $t \to \infty$

11 (iii) $a > b + c$, limit is $\dfrac{a - b - c}{a - c}$

Practice questions 3 (Page 292)

1 (i) $x + y = -1$

(ii) Between 1.5 and 1.8 (the analytic solution gives 1.678)

2 $\text{Re}(z) \geqslant -2, \quad \text{Im}(z) \geqslant 1, \quad \arg(z) \geqslant \dfrac{\pi}{4},$
$|z| \leqslant 4.$

3 $y = \dfrac{1}{2}e^{x^2}(e^{2x} + 1)$

4 (i) $-\dfrac{1}{2} + i\dfrac{\sqrt{3}}{2}.$

(ii) $-\dfrac{5}{2} + i\dfrac{\sqrt{3}}{2}, \quad \dfrac{1}{2} - i\dfrac{3\sqrt{3}}{2}.$

(iii) $\left(-\dfrac{1}{4} + i\dfrac{3\sqrt{3}}{4}\right)^3 = \dfrac{5}{4} - i\dfrac{9\sqrt{3}}{8}$

5 (ii) $\dfrac{e^{i\theta}\left(e^{2ni\theta} - 1\right)}{e^{2i\theta} - 1}$

(iii) $\dfrac{\sin^2 n\theta}{\sin\theta}$

6 (i) $x = e^{-3t}(A\cos 5t + B\sin 5t) - \dfrac{2}{3}\cos 4t + \dfrac{1}{2}\sin 4t$

(ii) $x = -\dfrac{2}{3}\cos 4t + \dfrac{1}{2}\sin 4t$ as the initial conditions affect only A and B, and these are in terms tending to zero because of the factor e^{-3t}.

(iii) The auxiliary equation has roots $3 \pm 5i$, and so the exponential factor is e^{3t} which tends to infinity. This leads to the model breaking down (infinite current is not possible) unless the initial conditions give $A = B = 0$.

7 (i) $\begin{pmatrix} 10 \\ 9 \\ 1 \end{pmatrix} \times \begin{pmatrix} 4 \\ -8 \\ 2 \end{pmatrix} = \begin{pmatrix} 26 \\ -16 \\ -116 \end{pmatrix}$

(ii) Q $(5, 7.5, 1.5)$
Distance is $49.80\,\text{m}$

(iii) New distance is $47.14\,\text{m}$
Charlie has walked $16.08\,\text{m}$

Answers

Index

A

absolute reference 117
acceleration due to gravity (g) 183, 254
accuracy, ever increasing 114, 116
adjugate/adjoint matrix 57, 59, 66
aeroplane taking off 176, 240–3
altitude, rate of change 177–8
amplitude 268, 269
angles, finding 2–3, 7–8, 17–18, 23–4, 27
answers 294–333
anti-commutative property 237
approximating e^x 115–16, 117
arc length, builders' rule 121
area scale factor 48–9, 57, 66
arccos 18, 86
arccosh 143, 144, 145, 149, 161
arcsin 85, 86, 94, 98, 149, 161
arcsinh 141, 142, 146–7, 161, 165
arctan 86, 88, 95, 98, 161, 165
arctanh 149
area under curves 81–4, 151, 165
area of limaçons 111
area enclosed by polar curves 110–11
area scale factor 48–9, 57, 66
area of sectors of circles 111, 113
area of triangles 4
Argand diagram 131, 173, 202–3, 206–7, 217–18
    and complex numbers 219
    loci in 207–9
    polygons on 219–23
argument 203, 204–5, 207, 232
    of zero 203, 208
associative 29, 30, 44
asymptote 189
atmosphere, Earth's 178, 185
auxiliary equations 255–60, 261–2, 266–73, 290

B

Bernoulli, Johann 189
boundary conditions 258, 260
boundary value problem 258
bridges 134, 253

C

cable, hanging 140
calculus, further 80–8, 89–91, 93–6, 98
cardioids 109
cartesian equations 27, 106, 107
    of a line 12, 13–14
    of a plane 5–7, 61
cartesian/polar coordinates, converting 103–4, 113
catenary 134–9, 140
chain rule 87, 146–7, 179
change, rates of 177–82, 184
circles 107, 135, 207–9, 218, 233

circular functions 135, 136, 137, 139
circular substitutions 160–2
cofactors 56, 57, 58, 66
commutative 3, 29, 30, 44
complementary function 257–9, 261–2, 269, 276–7, 279, 290–1
complex conjugate 128, 133
complex exponentials 266
complex numbers (C) 128–9, 131–2, 133, 202–9, 226–8, 232–3
    in Argand diagrams 219
    argument (θ) 203, 204–5, 207
    de Moivre's theorem 212–15, 224–5, 229–30
    in exponential form 227, 232, 233
    imaginary part (Im(z)) 128, 129, 133
    in modulus-argument form 202–9, 212, 221
    nth roots of 216–22, 233
    polar form, multiplying 206–7
    real part (Re(z)) 128, 129, 133
    representing geometrically 130–2, 133
    simplifying powers 213–15
complex plane
    see Argand diagram
complex roots 172–3, 175, 266–73
composition of transformations 38, 46
compound angle formulae 205, 268
cone, volume of 151
conformable 29, 30
conjecture 74
conjugate pairs 129, 172
constant, arbitrary 258, 276, 277, 278, 291
constant coefficients 254, 255, 256, 257
constant of integration 179, 181, 196
constant, unknown 255
convergent integrals 81–2, 98
cooling 188–9
cos 2, 86, 96
cosh 136, 142–3, 144, 145, 149
    inverse 143, 144, 145, 149, 161
critical damping 273, 291
cubic equations 169, 171, 173, 175
cyclical interchange 59

D

damping 271–3, 291
de Moivre's theorem 212–15, 218, 219, 220, 224–5, 233
    to sum series 229–30
density of air 178
derivatives 115, 117–18, 194–5
determinants 47–52, 55, 56–8, 66
    zero 49, 51, 67
differential equations 177, 200, 201
    complementary function of 257–9, 261–2, 269, 276–7, 279, 290–1
    with constant coefficients 254

general solution 179–81, 186–8, 194–5, 196, 201
    homogeneous and non- homogeneous 254–8, 275–80, 290–1
    linear 193–5, 201, 254, 284–6
    first order 176–82, 185, 193, 195–6, 201, 255, 276–7
    second order 177, 179, 181, 253–4, 256–8, 275–8, 290
    third order 177, 181, 254
    systems of 283–6, 291
    verification of solutions 181–2
differentiation 85–6, 93, 177, 178
direction vectors 10–11, 13, 18, 27
distance finding
    between parallel lines 245–6, 251
    between skew lines 246–7, 252
    from a point to a line 240–3, 251
    from a point to a plane 243–5, 251
divergent integrals 82, 84, 98
divisibility proofs 75–6
double angle formulae 95, 106, 159
double-loop curves 108–9

E

e^x, approximating 115–16, 117
electricity 159, 160
elements of matrices 28
enlargement 33, 35, 45, 57, 207, 232
    then rotation (spiral dilation) 207
equal matrices 29
equations 10–12, 62, 170–1
    linear 193–5, 201
    parametric 136, 285
    of a plane 5–7, 61, 235, 238–9
    polar 106, 107–9
    through three points 239
    vector, of lines 10–18, 27
equiangular spiral 101, 112
Euler, Leonhard 189
Euler's formula 125
ever increasing accuracy 114, 116
expansion of the determinant 55, 56, 57–8
exponential decay 271–3
exponential functions 115, 255
exponentials, complex 266

F

factor theorem 172
family of solution curves 180, 181
fluxions, method of 189
function, mean value 158–9, 165

G

'Gabriel's Horn' 157
Gauss, Carl Friedrich 216
geometrical interpretation in two dimensions 51–2

Girard, Albert 216
graphs and properties 136–7, 141
Gregory's series 125

H

half lines 208, 209, 233
hyperbola, rectangular 136, 148
hyperbolic functions 134–9
    inverse 141–7, 149
hyperbolic identities 138, 161
hyperbolic substitutions 160–2

I

i 128, 133
identity/unit matrices 28, 30, 44
if...then 74, 79
images 33, 45
imaginary axis (Im) 131
imaginary numbers 131
infinity 81, 98
    sum to 120
    tends to 81, 82, 84, 98, 285
initial conditions 258
initial line 101, 102
initial value problem 258
integers (Z) 128
integrals 80–4, 87, 88, 98, 161–2
    particular 276, 277, 278–80, 291
    standard 147, 161
integrating factors 193–7
integration 86–8, 93–6, 98
    applications of 150–4, 160–2, 165
    constant of 179, 181, 196
    general 160–2, 165
    using inverse hyperbolic functions
    146–7, 149
    partial fractions in 89–91, 98
intersection of lines 14–17, 22–3
intersection of planes 22–3, 60–3, 67
invariance 41–3, 46

L

lead build-up 283
Leibniz, Gottfried 189
Leibniz's series 125
lemniscate 110
limaçons 108–9, 110, 111
line of invariant points 42
linear equations 193–5, 201
lines, coincident 52
lines, intersection of 14–17, 22–3
lines and plane s 22–4
lines in three dimensions 12–14, 15–17
lines, vector equation of 10–18, 27
loci in Argand diagrams 207–9
logarithmic form 147
logarithmic function 82

M

Machin's formula 125
Maclaurin series 114–20, 123–4, 126,
127, 140

and polynomial approximations 115–20
    validity of 120, 123–4, 127
mapping 32, 35, 37, 41, 44
mathematical induction 74–7, 79
mathematicians 189, 216
matrices 28–30, 32–6, 37, 44, 50
    adjugate/adjoint 57, 59, 66
    determinant of 48–9
    identity/unit 28, 30, 44, 55–9, 66
    multiplication 29–30, 44, 46
    singular 49, 62, 63, 67
    square 28, 30, 32, 44, 47–52, 55–9, 66
    representing transformations 32–8,
    41–3, 45–6
    zero 28, 44, 49
mean value of a function 158–9, 165
method of fluxions 189
minor of an element 56, 57, 66
missile path 126, 184
modelling rates of change 177–82, 184
modulus of complex numbers 202–3,
204–5, 207
modulus-argument form 202–9, 218, 232
Moon, distance from Earth 166
multiple angle identities 224–5

N

n̂ 236, 237, 238, 242, 244, 251
natural logarithm function 82
natural numbers (N) 128
nautilus shell 101
Newton, Isaac 189
Newton's law of cooling 184, 188
Newton's second law 254
notation 170, 175, 177–8, 254
n^{th} derivative 118
n^{th} roots of a complex number 216–22,
233
n^{th} roots of unity 218–20, 233
n^{th} term of a sequence 77
numbers, rational (Q) 128
numbers, real (R) 128, 131

O

orthogonal projection 23, 27
oscillations 69, 266–73, 276
    amplitude 268, 269
    increasing 273, 282
    period of 159, 268, 269
overdamping 273, 291

P

parachutist 253–4, 276
parallel lines 17, 51, 245–6, 251
parallel planes 6
parallelograms 47–8
parametric equations 136, 285
partial fractions 79, 89–91
    in integration 89–91, 98
    summing series 71–2
particular integrals 276, 277, 278–80, 291
Pascal's triangle 68

pendulum 267
perfect derivative 194, 195
perpendicular bisector 208, 209, 233
perpendicular to a plane 5, 6, 239
phase shift (ε) 267–8
pi(π), use of 153
planes 6, 36
    angle between 7–8, 23–4, 27
    arrangement of 61, 62, 63, 67
    equation of 5–7, 61, 235, 238–9
    intersection of 60–3, 67
    intersection with lines 22–3
plotting points 107
polar coordinates 101–4, 107–9, 110–11,
113
polar curves, area enclosed by 110–11
polar equations 106, 107–9
polar form (r, θ) 203–5, 232
polar/cartesian coordinates, converting
103–4, 113
pole (fixed point) 101
polygons on Argand diagram 219–23
polynomial approximations and Maclaurin
series 115–20
polynomial equations, real coefficients
172–3
    complex roots 169–71, 172–3, 175
polynomial functions 115
population increase 182
position vectors 10, 14–15, 27, 35, 44
principal argument (arg) 203, 232
product rule 194, 261
projectile path 126, 184
proof by induction 74–7, 79, 213
Pythagoras' theorem 202

Q

quadratic equations 129, 170
    roots and coefficients 169, 170, 175
quadratic expressions 90–1
quadratic formula 129, 141–2, 217
quartic equations, roots and coefficients
169, 175
questions, practice 99–100, 166–8, 292–3

R

r (length) 101, 107, 111
radians 111, 120, 203
radioactive decay 182, 199
raindrop, volume 178–9
rates of change 177–82, 184
real axis (Re) 131
rectangular hyperbola 136, 148
reflection 32, 33, 35, 36–8, 46
relative reference 117
resonance 282
revolution, volume of 150–4, 157, 165
rhodonea 109
right-handed set of vectors 235, 238, 251
root mean square value 159
roots and coefficients 169–71, 173, 175
roots, imaginary parts 269

roots, real 142, 143, 258–60, 269, 290
roots, three and four 217–18
rotation 34, 35, 37–8, 45, 46, 232
   round the x-axis 150–3, 165
   round the y-axs 153–4, 165
rth derivative 116

S

sag 140
scalar product 2–4, 17, 27, 235
scale factor 33, 34, 232
second derivative 115
sector of a circle, area 111, 113
separable variables 195
separation of variables 185–9, 193, 201
sequences 69, 77
series 69, 74–5, 79, 120, 125, 226–7
   Maclaurin 114–20, 123–4, 126, 127, 140
   summing 69–72, 79
sheaf of planes 61, 62, 63, 67
shear 35–6, 45
sigma (sum of (Σ)) 170, 175
simple harmonic motion (SHM) 266–73, 290–1
simultaneous equations 15, 41, 50, 51, 67, 284–6
sin 86, 96, 98
singular matrix 49, 62, 63
sinh 135, 137, 142, 149
   inverse 141, 142
skew lines 15, 16–17, 18, 246–7, 252
solid of revolution 150–1
   volume of 151–4, 157, 165
solution curves 180, 181, 285
spheres, volume and surface area 178

spiral of Archimedes 109
spiral dilation 207
spiral, equiangular 101, 112
spreadsheet 117
spring displacement 266
square matrices 28, 30, 32, 44, 49
standard form 87
standard formulae (for sum) 69–70, 71
standard integrals 147, 161
stretch 34, 35, 45
sum to infinity 120
summing series 69–72, 74–5, 79, 116
suspension bridge 134

T

tan 96, 98
tangent field 180, 181
tanh 137, 149
   inverse 142
target expression 74–7, 79
telescoping sum 79
'Torricelli's Trumpet' 157
torus, volume of 157
transformations 32–8, 41–3, 45–6
translation 222
trial function 276–7, 278–9, 291
triangles, right-angled 4
triangular prism of planes 61, 62, 63, 67
trigonometric functions 135, 136, 137, 139
   inverse 85–8, 93
trigonometric identities 161, 224
trigonometric substitutions 94–6

U

underdamping 273, 291
unit matrices 28, 30, 44

unit modulus 218
unit squares 33–4, 35, 47
unit vectors 33, 45, 235, 236, 237, 238
unknown constant 255

V

variables, separation of 185–9, 193, 201
vector equation of lines 10–18, 27
vector equation of planes 5–7
vector length/modulus 2
vector product 234–9, 241, 242, 251
vectors 1–8, 10–18, 22–4, 27, 234–52
   direction 10–11, 13, 18, 27
   multiplying two 235–9
   perpendicular 4, 239
   position 10, 14–15, 27, 35, 44
   right-handed set 235, 238, 251
   unit 33, 45, 235, 236, 237, 238
volume scale factor 57, 66
volumes of revolution 150–4, 157, 165

W

warming 188–9

Y

$y = mx + c$ 12

Z

$z = re^{i\theta}$ 226–30
$z = re10$ 227, 232, 233
zero, argument of 203, 208
zero determinant 49, 51, 67
zero matrices 28, 44, 49
zero, tends to 81, 83, 152, 285, 286